C. Cattaneo (Ed.)

Relativistic Fluid Dynamics

Lectures given at a Summer School of the
Centro Internazionale Matematico Estivo (C.I.M.E.),
held in Bressanone (Bolzano), Italy,
June 7-16, 1970

Springer

FONDAZIONE
CIME
ROBERTO CONTI

C.I.M.E. Foundation
c/o Dipartimento di Matematica "U. Dini"
Viale Morgagni n. 67/a
50134 Firenze
Italy
cime@math.unifi.it

ISBN 978-3-642-11097-9 e-ISBN: 978-3-642-11099-3
DOI:10.1007/978-3-642-11099-3
Springer Heidelberg Dordrecht London New York

Reprint of the 1st ed. C.I.M.E., Ed. Cremonese, Roma 1971
With kind permission of C.I.M.E.

Printed on acid-free paper

Springer.com

CENTRO INTERNAZIONALE MATEMATICO ESTIVO

(C. I. M. E)

I Ciclo - Bressanone - dal 7 al 16 Giugno 1970

"RELATIVISTIC FLUID DYNAMICS"

Coordinatore: Prof. C CATTANEO

CENTRO INTERNAZIONALE MATEMATICO ESTIVO

(C.I.M.E.)

PROBLEMS MATHEMATIQUES
EN
HYDRODYNAMIQUE RELATIVISTE

PHAM MAU QUAN

Corso tenuto a Bressanone dal 7 al 16 Giugno 1970

Chapitre 1

LES SCHEMAS FLUIDES EN HYDRODYNAMIQUE RELATIVISTE

§1. GENERALITES SUR LA DYNAMIQUE RELATIVISTE DES FLUIDES

---

1. Le cadre géométrique.

La mécanique relativiste des fluides a pour cadre géométrique l'espace-temps qui est une variété differentiable V de dimension 4, de classe C^{∞}, sur laquelle est donnée une structure pseudo-riemannienne g de signature $+---$. La géométrie de l'espace-temps (V, g) est celle de la connexion riemannienne canoniquement associée à g.

La métrique définie par g est dite de type hyperbolique normal. Elle induit sur l'espace vectoriel tangent $T_x(V)$ en chaque point x de V une structure d'espace-temps plat de Minkowski. En coordonnées locales (x^{α}) on a

$$g = g_{\alpha\beta}\, dx^{\alpha} \otimes dx^{\beta} \qquad (\alpha, \beta = 0, 1, 2, 3). \tag{1.1}$$

Le tenseur $g_{\alpha\beta}$ dit tenseur fondamental de gravitation est assujetti à vérifier un système d'équations aux dérivées partielles du second ordre qui généralise les équations de Laplace.-Poisson et qui donne naissance aux conditions de conservation. Ces équations sont les dix équations d'Einstein

$$S_{\alpha\beta} = \chi\, T_{\alpha\beta} \tag{1.2}$$

où $S_{\alpha\beta}$ ne dépend que de la structure riemannienne g de l'espace-temps, $T_{\alpha\beta}$ est de signification purement mécanique et χ un facteur constant.

Le tenseur $T_{\alpha\beta}$ dit tenseur d'impulsion-énergie du fluide doit décrire au mieux la distribution énergétique dans l'espace-temps. Le tenseur $S_{\alpha\beta}$ est astreint aux deux conditions suivantes:

1. $S_{\alpha\beta}$ ne dépendent que de $g_{\alpha\beta}$, de leurs dérivées des deux premiers ordres, sont linéaires par rapport aux dérivées du second ordre
2. $S_{\alpha\beta}$ est conservatif, c'est-à-dire tel que

$$\nabla_\alpha S^\alpha_\beta = 0 \tag{1.3}$$

On démontre[1] qu'on a nécessairement

$$S_{\alpha\beta} = h(R_{\alpha\beta} - \frac{1}{2}(R+k)g_{\alpha\beta})$$

où $R_{\alpha\beta}$ est la courbure de Ricci, R la courbure scalaire de (V,g), h et k deux constantes arbitraires. k est la constante cosmologique, ne joue pas de rôle dans la description des fluides. On peut supposer $k = 0$. En supprimant d'autre part le facteur surabondant h, on prendra pour premier membre des équations d'Einstein

$$S_{\alpha\beta} = R_{\alpha\beta} - \frac{1}{2} R g_{\alpha\beta} \tag{1.4}$$

$S_{\alpha\beta}$ sera dit tenseur d'Einstein.

Le tenseur d'Einstein $S_{\alpha\beta}$ étant conservatif, il en est de même du tenseur d'impulsion-énergie $T_{\alpha\beta}$. Les équations

$$\nabla_\alpha T^{\alpha\beta} = 0 \tag{1.5}$$

---

(1) E. CARTAN. - J. Math. pures et appliquées, 1. p. 141-203. (1922)

expriment alors la conservation de l'impulsion-énergie et définiront l'évolution du fluide.

2. Le tenseur d'impulsion-énergie.

Dans toute théorie relativiste des fluides, le premier pas consiste à choisir l'expression du tenseur d'impulsion-énergie $T_{\alpha\beta}$. Chaque expression de $T_{\alpha\beta}$ définit un schéma de fluide. $T_{\alpha\beta}$ doit être symétrique si l'on veut satisfaire aux équations d'Einstein. Mais pour que $T_{\alpha\beta}$ puisse décrire un fluide physique, il faut qu'il existe un champ de vecteurs unitaires u^α orientés dans le temps

(2.1) $$g_{\alpha\beta}\, u^\alpha u^\beta = +1$$

pour lequel le scalaire $T_{\alpha\beta}\, u^\alpha u^\beta$ soit positif. u^α est dit vecteur vitesse unitaire du fluide et ses trajectoires définissent les lignes de courant.

En fait les fluides réels sont doués de propriétés diverses. Les forces de liaisons internes qui jouent un rôle fondamental dans l'étude dynamique se traduisent par le tenseur des pressions propres. Les phénomènes calorifiques introduisent un scalaire θ dit champ de température propre. Les propriétés électromagnétiques sont susceptibles d'être représentées par deux champs de tenseurs antisymétriques $H_{\alpha\beta}$, $G_{\alpha\beta}$ comme on le verra. Il convient d'autre part d'étudier l'évolution thermodynamique du fluide. Ces diverses propriétés peuvent être envisagées dans une décomposition géométrique du tenseur d'impulsion-énergie.

On est aussi conduit à mettre $T_{\alpha\beta}$ sous la forme

(2.2) $$T_{\alpha\beta} = \rho\, u_\alpha u_\beta - \pi_{\alpha\beta} - Q_{\alpha\beta} + \tau_{\alpha\beta}$$

où ρ est un scalaire positif représentant la densité propre de matière-énergie ponderable, $\pi_{\alpha\beta}$ les pressions propres, $Q_{\alpha\beta}$ les échanges thermiques par conduction et $\tau_{\alpha\beta}$ le tenseur d'énergie électromagnétique. Si l'on néglige certaines propriétés, les termes correspondants ne figurent pas dans la décomposition. De même on peut introduire de nouveaux termes pour étudier de nouvelles propriétés.

A chaque expression de $T_{\alpha\beta}$, correspond alors un schéma fluide. Dans chaque cas, l'évolution du fluide sera défini par les équations de conservation (1.5) qui, tenant compte du caractère unitaire de u^{α}, conduisent aux équations suivantes

(2.3) $$u_{\beta} \nabla_{\alpha} T^{\alpha\beta} = 0$$

(2.4) $$(g^{\beta}_{\rho} - u^{\beta} u_{\rho}) \nabla_{\alpha} T^{\alpha\rho} = 0$$

(2.3) est dite l'équation de continuité et (2.4) constitue le système différentiel aux lignes de courant.

A ces équations on adjoindra éventuellement d'autres équations telles que les équations thermodynamiques, les équations du champ électromagnétique. On obtiendra de cette manière le système fondamental des équations du schéma considéré. Ainsi le schéma fluide pur a fait l'objet de nombreuses études divenues classiques, en particulier celles de L. P. Eisenhart et de A. Lichnerowicz. Le schéma fluide thermodynamique a été étudiée par C. Eckart et par l'auteur dans sa thèse en 1954. Le schéma fluide champ électromagnétique a fait l'objet des travaux de A. Lichnerowicz, de ceux de l'auteur datant de 1955 qui ont suscité depuis de nombreux travaux, notamment ceux de G. Pichon. Il a conduit dans un cas particulier à la magnétohydrodynamique re lativiste dont l'étude a fait l'objet de très beaux travaux de Y. Choquet-Bruhat, A. Lichnerowicz.

C'est l'étude mathématique de quelques uns de ces schémas qui constitue le sujet de ces conférences.

3. Repère propre.

On appelle repère propre en un point x de l'espace-temps (V_4, g) un repère orthonormé $(V_{\lambda'})$ dont le premier vecteur $V_{0'}$ coincide avec le vecteur vitesse unitaire u et dont les trois autres vecteurs $V_{i'}$ définissent l'espace associé à la direction de temps u.

On peut rapporter l'espace-temps dans le voisinage de tout point à un champ de repères propres qu'on supposera différentiable (mais non nécessairement intégrable). La métrique d'univers prend alors la forme canonique

$$g = \eta_{\lambda'\mu'} \omega^{\lambda'} \otimes \omega^{\mu'} = \omega^{0'} \otimes \omega^{0'} - \omega^{1'} \otimes \omega^{1'} - \omega^{2'} \otimes \omega^{2'} - \omega^{3'} \otimes \omega^{3'} \tag{3.1}$$

où les $\omega^{\lambda'}$ sont les 1-formes duales des champs de vecteurs $V_{\lambda'}$ i.e. telles que $\langle \omega^{\lambda'}, V_{\mu'} \rangle = \delta^{\lambda'}_{\mu'}$, $\delta^{\lambda'}_{\mu'}$, étant le symbole de Kronecker égal à 1 si $\lambda' = \mu'$ et 0 si $\lambda' \neq \mu'$. Les $\omega^{\lambda'}$ constituent donc quatre formes de Pfaff linéairement indépendantes.

La considération du repère propre est fort utile. En effet l'espace vectoriel tangent $T_x(V)$ a une structure d'espace-temps de Minkowski le repère propre $V_{\lambda'}$ doit être identifié à un repère galiléen local où le fluide a une vitesse nulle. Si on connait les composantes d'un tenseur t relativement au repère propre, ses composantes dans un repère quelconque (e_α) se déduisent des premières par des formules de transformation connues.

En effet si $(A^{\lambda'}_{\alpha})$ est la matrice de passage du repère (e_α) au repère propre $(V_{\lambda'})$ et $(A^{\alpha}_{\lambda'})$ la matrice inverse, on a

(3.2) $$A^{\alpha}_{0'} = u^{\alpha} \qquad A^{\alpha}_{i'} = V_{(i')}{}^{\alpha}$$

(3.2') $$A^{0'}_{\alpha} = u_{\alpha} \qquad A^{i'}_{\alpha} = - V^{(i')}{}_{\alpha}$$

Si t est un tenseur d'ordre 2, ses composantes $t_{\alpha\beta}$ dans le repère (e_{α}) se déduisent de ses composantes $t_{\lambda'\mu'}$ dans le repère propre par les formules

(3.3) $$t_{\alpha\beta} = A^{\lambda'}_{\alpha} A^{\mu'}_{\beta} t_{\lambda'\mu'}$$

On a en particulier pour le tenseur métrique

(3.4) $$g_{\alpha\beta} = u_{\alpha} u_{\beta} - V_{(1')\alpha} V_{(1')\beta} - V_{(2')\alpha} V_{(2')\beta} - V_{(3')\alpha} V_{(3')\beta}$$

Ainsi pour déterminer l'expression du tenseur d'impulsion-énergie d'un fluide pur, on le rapporte d'abord au repère propre. Le fluide y est caractérisé par sa densité propre de matière-énergie ρ, son tenseur des pressions partielles $\pi_{i'j'}$. Son tenseur d'impulsion-énergie a pour composantes dans le repère propre

$$T_{0'0'} = \rho \qquad T_{i'j'} = - \pi_{i'j'}$$

Rapportons maintenant l'espace-temps à des coordonnées locales x^{α}, on a $\omega^{\lambda'} = A^{\lambda'}_{\alpha} dx^{\alpha}$ et l'application des formules (3.3) donne

(3.5) $$T_{\alpha\beta} = \rho u_\alpha u_\beta - \pi_{\alpha\beta}$$

où $\pi_{\alpha\beta} = \sum_{i',j'} \pi_{i'j'} A^{i'}_\alpha A^{j'}_\beta$ satisfait aux identités

(3.6) $$\pi_{\alpha\beta} u^\alpha = 0$$

On voit que dans le cas d'un fluide pur, le tenseur d'impulsion-énergie se décompose relativement à u^α une composante temporelle $\rho u_\alpha u_\beta$ et en une composante spatiale $\pi_{\alpha\beta}$.

Définition - On dit que le fluide est parfait si la quadrique des pressions dans le repère propre est une sphère i. e. si $\pi_{i'j'} = p\,\delta_{i'j'}$, p est dite pression scalaire du fluide.

Pour un fluide parfait, on a $\pi_{\alpha\beta} = p \sum_{i'} A^{i'}_\alpha A^{i'}_\beta$, soit en tenant compte de (3.2) et (3.4) $\pi_{\alpha\beta} = p(g_{\alpha\beta} - u_\alpha u_\beta)$. Ainsi le tenseur d'impulsion-énergie d'un fluide parfait est donné par

(3.7) $$T_{\alpha\beta} = (\rho + p) u_\alpha u_\beta - p g_{\alpha\beta}.$$

Appelons repère principal en x le repère orthonormé (W_λ) dont chaque vecteurs W_λ est vecteur propre de la matrice $(R_{\alpha\beta})$ par rapport à la matrice $(g_{\alpha\beta})$. Les directions définies par W_λ ne sont autres que les directions principales de Ricci. Or en vertu des équations d'Einstein, W_λ sont aussi vecteurs propres de la matrice $(T_{\alpha\beta})$ relativement à la matrice $(g_{\alpha\beta})$.

On peut exprimer les composantes de $T_{\alpha\beta}$ à partir des valeurs propres et vecteur propres comme

(3.8) $$T_{\alpha\beta} = s_0 W_{(0)\alpha} W_{(0)\beta} - \sum_{i=1}^{3} s_i W_{(i)\alpha} W_{(i)\beta}$$

On voit qu'en général le repère propre d'un fluide thermodynamique chargé est different du repère principal, sauf dans le cas d'un fluide pur où

$$T_{\alpha\beta} = \rho\, u_\alpha u_\beta - \sum_{i',j'} \pi_{i'j'} V_{(i')\alpha} V_{(j')\beta}$$

pour lequel $\rho = s_0$ et on peut faire une rotation du 3-plan espace de manière à amener $V_{i'}$ sur W_i.. s_i sont alors les valeurs propres de la matrice $(\pi_{i'j'})$.

§2. LE FLUIDE THERMODYNAMIQUE

4. Le fluide parfait et les variables thermodynamiques.

Le tenseur d'impulsion-énergie d'un fluide parfait non conducteur de chaleur est

$$T_{\alpha\beta} = (\rho+p)u_\alpha u_\beta - p g_{\alpha\beta} \tag{4.1}$$

Il est clair que u^α est vecteur propre orienté dans le temps et ρ la valeur propre correspondante de $(T_{\alpha\beta})$. Tout repère propre de ce fluide coincide avec un repère principal qui est indéterminé à cause de la valeur propre triple $-p$.

En vue de l'étude énergétique, on décompose la densité propre ρ en la somme d'une densité de matière r et d'une densité d'énergie vitesse $r\varepsilon$ où ε est l'énergie interne spécifique

$$\rho = r(1+\varepsilon) \tag{4.2}$$

On est amené à introduire l'indice f du fluide défini par

$$f = 1 + \varepsilon + \frac{p}{r} \qquad (4.3)$$

Dans ces formules et dans la suite, les unités physiques ont été choisies de manière que la vitesse limite c soit égale à 1. Autrement, il faudra remplacer ε, p par εc^{-2}, pc^{-2}.

Le tenseur d'impulsion-énergie d'un fluide parfait non conducteur de la chaleur prend alors la forme

$$T_{\alpha\beta} = rfu_{\alpha}u_{\beta} - pg_{\alpha\beta} \; . \qquad (4.4)$$

Du point de vue thermodynamique, la température propre θ et l'entropie spécifique propre S peuvent être définis comme en hydrodinamique classique par la relation

$$\theta dS = d\varepsilon + pd\tau \qquad (4.5)$$

où $\tau = \frac{1}{r}$ est le volume spécifique. En tenant compte de (4.3), et en prenant f, s, p comme variables thermodynamiques (non indépendantes) on peut écrire (4.5) sous la forme équivalente

$$r\theta dS = rdf - dp \qquad (4.6)$$

Ces relations expriment qu'il existe parmi les variables r, θ, f, S, p, seulement deux variables indépendantes pour lesquelles on choisira souvent f et S ou S et p.

Si on prend f et S comme variables indépendantes et si l'on se donne p en fonction de f et S, la relation (4.6) entraine

$$r = \frac{\partial p}{\partial f} \qquad\qquad r\theta = -\frac{\partial p}{\partial S}$$

La première relation définit l'équation d'état du fluide sous la forme

$$r = r(f, S)$$

et la seconde relation définit la température.

Des conditions de conservation appliquées au tenseur d'impulsion-énergie (4.4) on tire l'équation de continuité et le système différentiel aux lignes de courant

(4.7) $$\nabla_\alpha (rfu^\alpha) - u^\alpha \partial_\alpha p = 0$$

(4.8) $$rfu^\alpha \nabla_\alpha u^\beta - (g^{\alpha\beta} - u^\alpha u^\beta)\partial_\alpha p = 0$$

Tenant compte de (4.6) on peut écrire l'équation de continuité sous la forme

(4.7') $$f\nabla_\alpha (ru^\alpha) + r\theta\, u^\alpha \partial_\alpha S = 0$$

d'où l'on déduit

Théorème - Pour un fluide parfait, il est équivalent de dire qu'il y a conservation de la matière ou que l'entropie est constante le long des lignes de courant.

$$u \nabla_\alpha (r u^\alpha) = 0 \Leftrightarrow u^\alpha \partial_\alpha S = 0$$

un fluide tel que $u^\alpha \partial_\alpha S = 0$ est dit adiabatique.
De même tenant compte de l'équation thermodynamique, le système différentiel aux lignes de courant s'écrit avec les variables f, S

$$f u^\alpha \nabla_\alpha u^\beta - (g^{\alpha\beta} - u^\alpha u^\beta)\partial_\alpha f + \theta g^{\alpha\beta}\partial_\alpha S = 0 \tag{4.8'}$$

On en déduit

Théorème - Si le mouvement du fluide est isentropique (S = const.) le système différentiel aux lignes de courant se réduit à

$$u^\alpha \nabla_\alpha u^\beta - (g^{\alpha\beta} - u^\alpha u^\beta) \frac{\partial_\alpha f}{f} = 0 \tag{4.9}$$

Nous montrerons que pour un tel fluide isentropique, il existe un principe extrémal pour les lignes de courant

5. Le fluide visqueux.

Pour caractériser la déformation locale du fluide, nous introduisons la derivée de Lie du tenseur métrique g suivant le vecteur vitesse $\underline{u}$ nitaire

$$(L_u g)_{\alpha\beta} = \nabla_\alpha u_\beta + \nabla_\beta u_\alpha$$

et nous posons

$$\varepsilon_{\alpha\beta} = \frac{1}{2} \gamma^\rho_\alpha \cdot \gamma^\mu_\beta \ (\nabla_\rho u_\mu + \nabla_\mu u_\rho) \tag{5.1}$$

où $\gamma_{\alpha\beta}$ est le projecteur d'espace.

Les lois contraintes-déformations sont supposées linéaires, si le milieu est isotrope, les phénomènes de viscosité sont décrits par le tenseur

(5.2) $$\sigma_{\alpha\beta} = c_{\alpha\beta}{}^{\rho\mu}\varepsilon_{\rho\mu}$$

où

(5.3) $$c_{\alpha\beta\rho\mu} = \lambda\gamma_{\alpha\beta}\gamma_{\rho\mu} + \mu(\gamma_{\alpha\rho}\gamma_{\beta\mu} + \gamma_{\alpha\mu}\gamma_{\beta\rho})$$

ou compte tenu de $g_{\alpha\beta}\, u^\alpha u^\beta = 1$

(5.3') $$c_{\alpha\beta\rho\mu} = \lambda\gamma_{\alpha\beta}\gamma_{\rho\mu} + \mu(g_{\alpha\rho}\, g_{\beta\mu} + g_{\alpha\mu}\, g_{\beta\rho})$$

Le tenseur d'impulsion-énergie du fluide visqueux homogène est alors donné par

(5.4) $$T_{\alpha\beta} = (\rho + p)u_\alpha\, u_\beta - p g_{\alpha\beta} + \lambda(\nabla_\sigma u^\sigma)\gamma_{\alpha\beta} + 2\mu\varepsilon_{\alpha\beta}$$

Cette expression est utilisée par C. Eckart, G. Pichon. A. Lichnerowicz proposait une autre expression de $\varepsilon_{\alpha\beta}$ où intervient dans la définition de la viscosité le vecteur $C_\alpha = fu_\alpha$. Le tenseur d'impulsion-énergie devient alors

(5.5) $$T_{\alpha\beta} = (\rho + p - \lambda\nabla_\rho C^\rho)\, u_\alpha u_\beta - (p - \lambda\nabla_\rho C^\rho)\, g_{\alpha\beta} + \mu\varepsilon_{\alpha\beta}$$

avec

$$2\bar{\varepsilon}_{\alpha\beta} = \bar{\nabla}_\alpha C_\beta + \bar{\nabla}_\beta C_\alpha - \bar{C}^\lambda (\bar{\nabla}_\lambda C_\alpha C_\beta + \bar{\nabla}_\lambda C_\beta C_\alpha)$$

$\bar{\nabla}$ désignant la dérivée covariante dans la métrique $\bar{g} = f^2 g$.

6. Le fluide conducteur de chaleur.

On tient compte des échanges thermiques par conduction. Celle-ci est définie par un vecteur q_α orthogonal au vecteur u^α. C'est l'expression de q_α, sa présence dans le tenseur d'impulsion-énergie qui caractérise les points de vue.

Eckart a choisi le tenseur:

$$T_{\alpha\beta} = (\rho + p) u_\alpha u_\beta - p g_{\alpha\beta} + \theta_{\alpha\beta} - (u_\alpha q_\beta + u_\beta q_\alpha) \tag{9.1}$$

Les équations qui régissent l'évolution du fluide sont données par les conditions de conservation du tenseur d'impulsion-énergie, la conservation du courant de matière, l'équation de définition de q_α :

$$\nabla_\alpha T^{\alpha\beta} = 0$$

$$\nabla_\alpha (r u^\alpha) = 0$$

$$q_\alpha = - \varkappa (g_\alpha^\beta - u_\alpha u^\beta)(\partial_\beta \theta - \theta u^\rho \nabla_\rho u_\beta) \tag{9.2}$$

et une équation thermodynamique.

L'auteur a proposé en 1954 le tenseur d'impulsion-énergie

$$T_{\alpha\beta} = (\rho + p)\, u_\alpha u_\beta - p g_{\alpha\beta} - (u_\alpha q_\beta + u_\beta q_\alpha) \tag{9.3}$$

où l'on a négligé la viscosité. q_α est défini par

$$q_\alpha = -\varkappa\, (g_\alpha^\beta - u_\alpha u^\beta)\, \partial_\alpha \theta \tag{9.4}$$

Les équations du mouvement sont constituées par les conditions de conservation $\nabla_\alpha T^{\alpha\beta} = 0$ et l'équation thermodynamique est remplacée par l'équation de conduction qui généralise celle de Fourier

$$\nabla_\alpha q^\alpha = C u^\alpha \partial_\alpha \theta - \frac{\ell}{\rho}\, u^\alpha \partial_\alpha \rho \tag{9.5}$$

C est la chaleur spécifique à volum constant et ℓ la chaleur de dilatation du fluide. Pichon a repris ce modèle en ajoutant le terme de viscosité $\theta_{\alpha\beta}$.

Pour Landau et Lifchitz, le tenseur d'impulsion-énergie d'un fluide conducteur de chaleur est identique à celui d'un fluide parfait, le vecteur courant de chaleur q_α apporte sa contribution à travers l'équation de conservation d'un certain vecteur P_α . Il pose

$$T_{\alpha\beta} = r f u_\alpha u_\beta - p g_{\alpha\beta} \tag{9.6}$$

$$P_\alpha = r u_\alpha - q_\alpha \tag{9.7}$$

Les équations du mouvement sont données par

$$\nabla_\alpha T^{\alpha\beta} = 0$$

$$\nabla_\alpha P^\alpha = 0$$

auxquelles ont ajoute l'équation de définition de q_α :

$$q_\alpha = -\varkappa\theta^2 (g_\alpha^\beta - u_\alpha u^\beta)\partial_\beta\left(\frac{1+G}{\theta}\right) \tag{9.8}$$

où G est la fonction de Gibbs définie par

$$G = \varepsilon + \frac{p}{r} - \theta S. \tag{9.10}$$

Ces modèles se justifient par des considérations physiques et cinétiques et ont tous le mérite de se réduire à la limite à la description classique non relativiste. L'étude du problème de Cauchy montre que les les systèmes d'équations auxquels ils donnent lieu, sont mixtes et comportent une partie parabolique provenant soit de la viscosité, soit de la définition du vecteur courant de chaleur q_α . Ce qui conduit à une vitesse de propagation infinie.

Pour lever la difficulté provenant de q_α , Cattaneo et Vernotte ont suggéré de modifier l'hypothèse de Fourier par un terme de Relaxation. Kranys a fait la traduction de cette hypothèse en relativité

$$q^\alpha + \nu u^\beta \nabla_\beta q^\alpha = -\varkappa (g^{\alpha\beta} - u^\alpha u^\beta)\partial_\beta\theta \tag{9.11}$$

Ce vecteur q_α n'est alors plus orthogonal à u_α . Adoptant le point de vue de Landau-Lifchitz et celui de Cattaneo-Vernotte-Kranys, Mahjoreb a proposé une nouvelle théorie cette année.

§3. LE CHAMP ELECTROMAGNETIQUE

10. Représentation du champ électromagnétique.

S'il existe un champ électromagnétique, le fluide est soumis à des inductions électromagnétiques qu'on peut décrire à l'aide de deux 2-formes: la 2-forme champ électrique-induction magnétique H et la 2-forme induction électrique-champ magnétique G. On note $*H$, $*G$ leurs formes duales au sens de l'élément de volume riemannien η de l'espace-champ. On a en composantes

(10.1) $$(*H)_{\alpha\beta} = \frac{1}{2}\eta_{\alpha\beta\lambda\mu} H^{\lambda\mu}$$

(10.2) $$(*G)_{\alpha\beta} = \frac{1}{2}\eta_{\alpha\beta\lambda\mu} H^{\lambda\mu}$$

On appelle vecteurs champs et inductions électriques et magnétiques les vecteurs définis par les 1-formes

(10.3) $$e = i_u H \qquad d = i_u G \qquad h = i_u(*G) \qquad b = i_u(*H)$$

où i_u est le produit intérieur par le vecteur vitesse unitaire u. En composantes, on a

(10.4) $$e = u^{\rho} H_{\rho\alpha} \qquad d_{\alpha} = u\, G \qquad h_{\alpha} = u^{\rho}(*G)_{\rho\alpha} \qquad b_{\alpha} = u^{\rho}(*H)_{\rho\alpha}$$

Ces vecteurs sont orthogonaux à u^{α}.

Inversement, H, G, $*H$, $*G$ s'expriment en fonction de e, d, h, b par les formules

(10.5) $$H = u\wedge e - *(u\wedge b) \qquad G = (u\wedge d) - *(u\wedge h)$$

(10.6) $$*H = u\wedge b + *(u\wedge e) \qquad *G = u\wedge h + *(u\wedge d)$$

Dans les deux dernières relations le signe + provient du fait que sur une variété riemannienne l'opérateur $*$ satisfait à la relation $*^2 = \varepsilon_g (-1)^{p\,(n-p)}$ où $n = \dim V$, p = degré de la forme et ε_g le signe du det. g. On en déduit les relations suivantes qui donnent lescom composantes

$$H_{\alpha\beta} = u_\alpha e_\beta - u_\beta e_\alpha - \eta_{\alpha\beta\lambda\mu} u^\lambda b^\mu$$

$$G_{\alpha\beta} = u_\alpha d_\beta - u_\beta d_\alpha - \eta_{\alpha\beta\lambda\mu} u^\lambda h^\mu$$

En théorie électromagnétique de Maxwell, les inductions dépendent linéairement des champs. Dans le cas isotrope où le fluide a une permitivité diélectrique λ et une perméabilité magnétique μ, on a

(10.7) $$d = \lambda e \qquad b = \mu h$$

Les deux relations (10.5) donnent alors

(10.8) $$G = \frac{1}{\mu} H + \frac{\lambda\mu - 1}{\mu} u\wedge i_u H$$

soit en composantes

(10.8')
$$G_{\alpha\beta} = \frac{1}{\mu} H_{\alpha\beta} + \frac{\lambda\mu - 1}{\mu} (u_\alpha u^\rho H_{\rho\beta} - u_\beta u^\rho H_{\rho\alpha})$$

qu'on peut mettre sous la forme

(10.9)
$$G_{\alpha\beta} = \frac{1}{\mu} \varepsilon_{\alpha\beta}^{\rho\sigma} H_{\rho\sigma}$$

où

$$\varepsilon_{\alpha\beta}^{\rho\sigma} = \bar{g}_\alpha^\rho \bar{g}^\sigma_\beta - \bar{g}_\alpha^\sigma \bar{g}_\beta^\rho \qquad \text{avec} \qquad \bar{g}^\lambda_\mu = g^\lambda_\mu - (1-\lambda\mu)\, u^\lambda u_\mu .$$

L'induction électromagnétique H, G satisfait aux équations de Maxwell

(10.10)
$$dH = 0$$

(10.11)
$$\delta G = J$$

où δ est la codifférentielle, J une 1-forme dont le vecteur associé définit le courant électrique. L'équation (10.10) signifie que la 2-forme H est localement exacte i.e. qu'il existe localement une 1-forme Φ telle que $H = d\Phi$. Φ s'appelle le potentiel vecteur électromagnétique, (10.11) donne en remarquant que $\delta^2 = 0$

(10.12)
$$\delta J = 0$$

équation qui exprime la conservation du courant électrique.

En composantes, les équations (10.10), (10.11), (10.12) s'écrivent

$$\frac{1}{2}\eta^{\alpha\beta\gamma\delta}\;\nabla_\alpha H_{\beta\gamma} = 0$$

$$\nabla_\alpha G^{\alpha\beta} = J^\beta$$

$$\nabla_\alpha J^\alpha = 0$$

On décompose le courant électrique J en un courant de convection colinéaire à u et un courant de conduction Γ orthogonal à u. Γ peut être défini par l'hypothèse d'Ohm $\Gamma = \sigma e$ où σ est la conductivité électrique du fluide. On a alors

(10.13) $$J^\alpha = \gamma u^\alpha + \sigma e^\alpha$$

γ s'appelle la densité de charge.

11. Le tenseur d'énergie électromagnétique.

A partir de $H_{\alpha\beta}$, $G_{\alpha\beta}$ on construit le tenseur d'impulsion-énergie électromagnétique $\tau_{\alpha\beta}$ dont la divergence donne la densité de force électromagnétique agissant sur le fluide. En généralisant un résultat connu dans le cas non inductif $\lambda = \mu = 1$, on obtient le tenseur donné par Minkowski

Pham

(11.1) $$\tau_{\alpha\beta} = \frac{1}{4} g_{\alpha\beta} (G_{\varrho\sigma} H^{\varrho\sigma}) - G_{\varrho\alpha} H^{\varrho}{}_{\beta}$$

Pour interpréter ce tenseur, on va l'exprimer à l'aide des vecteurs e, d, h, b:
On obtient

(11.2) $$\tau_{\alpha\beta} = (e_\varrho d^\varrho + h_\varrho b^\varrho)(u_\alpha u_\beta - \frac{1}{2} g_{\alpha\beta}) - (e_\alpha d_\beta + h_\alpha b_\beta) -$$

$$+ (P_\alpha u_\beta + u_\alpha Q_\beta)$$

où

(11.3) $$P_\alpha = \eta_{\alpha\lambda\mu\nu} e^\lambda h^\mu u^\nu \qquad Q_\alpha = \eta_{\alpha\lambda\mu\nu} d^\lambda b^\mu u^\nu$$

P_α est le vecteur de Poynting et $Q_\alpha = \lambda_\mu P_\alpha$. Sur (11.2) on voit la signification de chaque groupe de termes.

$\tau_{\alpha\beta}$ n'est pas symétrique. On peut prendre l'expression proposée par Abraham

(11.4) $$\tau_{\alpha\beta} = -(e_\varrho d^\varrho + h_\varrho b^\varrho)(u_\alpha u_\beta - \frac{1}{2} g_{\alpha\beta}) - (e_\alpha d_\beta - h_\alpha b_\beta) -$$

$$+ (P_\alpha u_\beta + u_\alpha P_\beta).$$

On peut penser à le symétriser [ ], mais les raisons physiques sont obscures.

Pham

Nous conservons l'expression (11.1) . En prenant la divergence de ce tenseur, nous avons

$$(11.5) \qquad \nabla_\alpha \tau^\alpha{}_\beta = \nabla_\alpha G^{\alpha\rho} H_{\rho\beta} + G^{\alpha\rho} \nabla_\alpha H_{\rho\beta} + \frac{1}{4} (G^{\rho\sigma} \nabla_\beta H_{\rho\sigma} + H_{\rho\sigma} \nabla_\beta G^{\rho\sigma})$$

Or les équations de Maxwell du 1er groupe s'écrivent encore

$$\nabla_\alpha H_{\rho\beta} + \nabla_\rho H_{\beta\alpha} + \nabla_\beta H_{\alpha\rho} = 0$$

Par multiplication contractée avec $G^{\alpha\rho}$, il vient

$$2 G^{\alpha\rho} \nabla_\alpha H_{\rho\beta} = - G^{\alpha\rho} \nabla_\beta H_{\alpha\rho}$$

En portant dans (11.4), on a alors en tenant compte de la définition de J

$$(11.5') \qquad \nabla_\alpha \tau^\alpha{}_\beta = J^\rho H_{\rho\beta} + \frac{1}{4} (G^{\rho\sigma} \nabla_\beta H_{\rho\sigma} - H_{\rho\sigma} \nabla_\beta G^{\rho\sigma})$$

On peut transformer la parenthèse en utilisant les équations de liaison ce qui donne finalement

$$(11.6) \qquad \nabla_\alpha \tau^\alpha{}_\beta = J^\rho H_{\rho\beta} + (\lambda\mu - 1) \nabla_\beta u^\rho P_\rho + \frac{1}{2} (e_\rho e^\rho \partial_\beta \lambda + h_\rho h^\rho \partial_\beta \mu)$$

où

$$J^\rho H_{\rho\beta} = \gamma e_\beta - \sigma (e_\rho e^\rho) u_\beta + \sigma\mu P_\beta$$

La signification du groupe $J^\rho H_{\rho\beta}$ est claire. Le tenseur supplémentaire $(\lambda\mu - 1)\nabla_\beta u^\rho P_\rho$ est nul si $\lambda\mu = 1$ ou $P_\alpha = 0$, c'est-à-dire si $\tau_{\alpha\beta}$ est symétrique. Il est encore nul si u^α est un champ à dérivée covariante nulle. Le terme supplémentaire $\frac{1}{2}(e_\rho e^\rho \partial_\beta \lambda + h_\rho h^\rho \partial_\beta \mu)$ correspond aux phénomènes de magnétostriction et d'électrostriction. En fait λ, μ dépendent des variables d'état.

12. Cas où $\tau_{\alpha\beta}$ défini par (11.1) est symétrique.

Le tenseur $\tau_{\alpha\beta}$ n'est pas symétrique en général. Son antisymétrisé est $(\lambda\mu - 1)(u_\alpha P_\beta + u_\beta P_\alpha)$. Comme u_α et P_α sont orthogonaux, l'antisymétrisé est nul i.e. que $\tau_{\alpha\beta}$ est symétrique si:

1.- $\lambda\mu = 1$, cas non inductif

2.- $P_\alpha = 0$, ce qui est vérifié soit que $e_\alpha = 0$ soit que $h_\alpha = 0$.

Dans le cas non inductif, on prend $\lambda = \mu = 1$, alors $H_{\alpha\beta} = G_{\alpha\beta} = F_{\alpha\beta}$. Le tenseur d'énergie électromagnétique s'écrit

(12.1) $$\tau_{\alpha\beta} = \frac{1}{4} g_{\alpha\beta}(F_{\rho\sigma} F^{\rho\sigma}) - F_{\rho\alpha} F^\rho{}_\beta$$

et

(12.2) $$\nabla_\alpha \tau^\alpha{}_\beta = J^\rho H_{\rho\beta}$$

Le cas $e_\alpha = 0$ correspond à celui de la magnétohydrodynamique ou fluides de conductivité $\sigma = \infty$. Comme le courant électrique doit être borné $\sigma e < \infty$, on a nécessairement $e = 0$. Le tenseur d'impulsion-énergie électromagnétique s'écrit alors

$$\tau_{\alpha\beta} = \mu\left\{|h|^2 (u_\alpha u_\beta - \frac{1}{2} g_{\alpha\beta}) - h_\alpha h_\beta\right\} \tag{12.3}$$

Le cas $h_\alpha = 0$ conduit à

$$\tau_{\alpha\beta} = \lambda\left\{|e|^2 (u_\alpha u_\beta - \frac{1}{2} g_{\alpha\beta}) - e_\alpha e_\beta\right\} \tag{12.4}$$

Il convient pour la description d'un électron considéré comme une boule continue.

Dans chacun de ces cas, le tenseur d'impulsion-énergie du fluide s'obtient à ajoutant $\tau_{\alpha\beta}$ à l'expression déjà connue. On obtient un tenseur d'impulsion-énergie du fluide qui est symétrique, par suite on peut écrir les équations d'Einstein.

13. Le fluide parfait chargé sans inductions.

On a dans ce cas le tenseur d'impulsion-énergie total

$$T_{\alpha\beta} = (\rho + p) u_\alpha u_\beta - p g_{\alpha\beta} + \tau_{\alpha\beta} \tag{13.1}$$

où

$$\tau_{\alpha\beta} = \frac{1}{4} g_{\alpha\beta} (F_{\rho\sigma} F^{\rho\sigma}) - F_{\rho\alpha} F^{\rho}{}_{\beta}$$

Les équations du mouvement sont données par les conditions de

Pham

conservation

(13.2) $$\nabla_\alpha T^{\alpha\beta} = 0$$

l'équation thermodynamique

(13.3) $$r\theta\, dS = rdf - dp$$

et les équations de Maxwell

(13.4) $$\frac{1}{2}\eta^{\alpha\beta\gamma\delta}\nabla_\alpha F_{\beta\gamma} = 0$$

(13.5) $$\nabla_\alpha F^{\alpha\beta} = J^\beta$$

Où supposons la conductivité $\sigma = 0$, de sorte que $J^\beta = \gamma u^\beta$ et

(13.6) $$\nabla_\alpha(\gamma u^\alpha) = 0$$

En prenant f, S comme variables thermodynamiques, les conditions de conservation (13.2) donnent l'équation de continuité et le système différentiel aux lignes de courant

(13.7) $$rfu^\alpha \nabla_\alpha u^\beta - (g^{\alpha\beta} - u^\alpha u^\beta)\partial_\alpha p - \gamma u^\alpha F_\alpha{}^\beta = 0$$

(13.8) $$f\nabla_\alpha(ru^\alpha) + r\theta u^\alpha \partial_\alpha S = 0$$

On en déduit que si le mouvement est adiabatique $(u^\alpha \partial_\alpha S = 0)$, il y a conservation de matière

(13.9) $$\nabla_\alpha (r u^\alpha) = 0$$

De (13.6) et (13.9) on tire

$$u^\alpha \frac{\partial_\alpha \gamma}{\gamma} + \nabla_\alpha u^\alpha = 0 \qquad\qquad u^\alpha \frac{\partial_\alpha r}{r} + \nabla_\alpha u^\alpha = 0$$

d'où par différence $u^\alpha \partial_\alpha \log \frac{\gamma}{r} = 0$. Le rapport $\frac{\gamma}{r}$ est donc constant le long des lignes de courant, on pose

(13.10) $$K = \frac{\gamma}{r}$$

Le système différentiel aux lignes de courant se met alors sous la forme

(13.11) $$f u^\alpha \nabla_\alpha u^\beta - (g^{\alpha\beta} - u^\alpha u^\beta) \partial_\alpha f + \theta g^{\alpha\beta} \partial_\alpha S = K u^\alpha F_\alpha{}^\beta$$

Chapitre 2

LE PROBLEME DE CAUCHY

Pham

Pour chaque modèle de fluide, on obtient un système fondamental d'équations aux dérivées partielles pour étudier l'évolution du fluide. Un problème essentiel est de voir dans quelles mesures ces équations déterminent les fonctions qui représentent les grandeurs physiques envisagées. Comme c'est un problème d'évolution, le problème mathématique posé est le problème de Cauchy. Les données initiales portées par une hypersurface Σ de l'espace-temps déterminent - elles les grandeurs dans le voisinage de Σ.

Le seul théorème général classiquement connu en réponse à cette question est le théorème d'existence et d'unicité de Cauchy-Kowaleski valable dans le cas analytique pour un système de N équations aux dérivées partielles à N fonctions inconnus dont le polynôme caractéristique n'est pas identiquement nul.

L'hypothèse d'analyticité restreint considérablement la portée de ce théorème en Physique. On peut maintenant se passer de l'hypothèse d'analyticité pour des systèmes quasi-linéaires hyperboliques stricts. Pour de tels systèmes, Leray a démontré un théorème d'existence et d'unicité pour le problème de Cauchy non analytique. Toute solution de ce problème possède un domaine d'influence, c'est-à-dire que la valeur en un point ne dépend que d'une partie des données initiales, celles se trouvant à l'intérieur d'un certain cônoide de sommet ce point. C'est cette notion d'hyperbolicité stricte et son critère que nous allons exposer en vue de l'appliquer aux différents systèmes d'équations trouvées dans le chapitre 1.

Pham

§I. THEOREME D'EXISTENCE ET D'UNICITÉ POUR LES SYSTEMES STRICTEMENT HYPERBOLIQUES

1. Systèmes strictement hyperboliques.

Soit V_n une variété différentiable de classe C^k (k suffisamment grand) et de dimension n.

Soit a(x, D) un opérateur différentiel d'ordre m agissant sur les fonctions. Localement a dépend des coordonnées locales x^α et des dérivées partielles ∂_α. Pour $\xi \in T^*_x(V_n)$, a (x, ξ) est un polynome réel en ξ de degré m. On désignera par h (x, ξ) la partie principale de a (x, ξ) c'est-à-dire la partie homogène de degré m de a (x, ξ). Soit $V_x(h)$ le cône projectif défini dans $T^*_x(V_n)$ par l'équation h (x, ξ) = 0.

Définition 1. - L'opérateur différentiel a (x, D) est dit strictement hyperbolique au point $x \in V_n$ si l'hypothèse suivante est vérifiée:

(H) Il existe dans $T^*_x(V_n)$ des points ξ tels que toute droite issue de ξ et ne passant pas par le sommet du cône $V_x(h)$ le coupe en m points réels distincts.

S'il en est ainsi, l'ensemble des points ξ forme l'intérieur de deux demi-cônes convexes apposés non vides $\Gamma^+_x(a)$ et $\Gamma^-_x(a)$ dont les bords appartiennent à $V_x(h)$.

Considérons maintenant un opérateur différentiel matriciel diagonal A(x, D) suffisamment différentiable en x

$$A(x,D) = \begin{pmatrix} a_1(x,D) & 0 & 0 \\ 0 & a_2(x,D) & 0 \\ \cdots & \cdots & \cdots \\ 0 & 0 & a_N(x,D) \end{pmatrix}$$

où les $a_i(x,D)$ sont des opérateurs différentiels d'ordre $m(i)$.

Définition 2. - On dit que l'opérateur différentiel diagonal $A(x,D)$ est strictement hyperbolique en un point x si

1) les $a_i(x,D)$ sont strictement hyperboliques en x

2) les deux demi-cônes convexes apposés

$$\Gamma_x^+(A) = \bigcap_i \Gamma_x^+(a_i) \qquad \Gamma_x^-(A) = \bigcap_i \Gamma_x^-(a_i)$$

ont un intérieur non vide.

Pour définir l'hyperbolicité stricte dans un domaine Ω (ouvert connexe) de V_n, introduisons le cône $C_x^+(A)$ dual du cône $\Gamma_x^+(A)$. $C_x^+(A)$ est fermé de l'ensemble des vecteurs $X \in T_x(V_n)$ tels que $\langle \xi, X \rangle \geqslant 0$ pour tout $\xi \in \Gamma_x^+(A)$. Le cône $C_x^-(A)$ dual de $\Gamma_x^-(A)$ est défini de manière analogue. Soit

(1.3) $$C_x(A) = C_x^+(A) \cup C_x^-(A)$$

Un chemin $\gamma : [0,1] \to V_n$ différentiable est dit orienté dans le temps relativement à A si la demi-tangente positive en chaque points de γ est dans $C_x^+(A)$. Une hypersurface différentiable Σ est dite orientée dans l'espace relativement à A si le sous expace vectoriel tangent $T_x(\Sigma)$ en chaque point x de Σ est extérieur à C(A).

Définition 3. - On dit que l'opérateur A(x, D) est strictement hyperbolique dans un domaine $\Omega \subset V_n$ si les deux conditions suivantes sont satisfaites:

1) A(x, D) est strictement hyperbolique en tout point $x \in \Omega$
2) l'ensemble des chemins temporels joignant deux points x_0, x_1 quelconques de Ω est compact ou vide pour la topologie de la convergence uniforme de l'ensemble $\{\gamma : [0, 1] \to V_n,\ \gamma(0) = x_0, \gamma(1) = x_1\}$.

Si A(x, D) est différentiable en x et si A(x, D) est strictement hyperbolique en un point x_0, on peut montrer qu'il existe un voisinage ouvert connexe Ω de x_0 homéomorphe à une boule de R^n dans lequel A(x, D) est strictement hyperbolique. Un tel ouvert sera dit simple.

2. Systèmes de Leray.

Considérons un système d'équations aux dérivées partielles de N équations à N inconnues (u^i) et n variables (x^α) qqe l'on écrit symboliquement

$$A(x, u, D)\, u + B(x, u) = 0 \tag{2.1}$$

où A(x, u, D) est une matrice diagonale d'éléments $a_i(x, u, D)$, $i = 1, \dots, N$ et B(x, u) une matrice colonne d'éléments $b_i(x, u)$. Les $a_i(x, u, D)$ sont des opérateurs différentiels d'ordre m(i).

Associons à chaque inconnu u^i un entier $s(i) \geqslant 1$ et à chaque équa tion de rang j un entier $t(j) \geqslant 1$ tels que:

$$m(i) = s(i) - t(i) + 1 \qquad (2.2)$$

les entiers $s(i)$, $t(j)$ ne sont définis qu'à une constante additive près.

Définition. - On dit que le système diagonal (2.1) est quasi-linéaire au sens de Leray si pour tout i , l'opérateur différentiel $a_i(x,u,D)$ est linéaire par rapport aux dérivées d'ordre $m(i)$, si les relations (2.2) sont vérifiées et si les a_i, b_i sont des fonctions suffisamment régulières de x^α, de u^j et des dérivées des u^j d'ordre $\leqslant s(i) - t(j)$, si $s(i) - t(j) < 0$ a_i et b_i sont indépendants de u^j.

Ceci étant, le problème de Cauchy pour le système (2.1) se pose de la manière suivante. Soit Ω un domaine simple de V_n. Les données de Cauchy sur une hypersurface Σ plongée dans Ω consistent en les valeurs des fonctions u^i, de leurs dérivées d'ordre $< m(i)$. Il existe (toujours) des fonctions w^i admettant des dérivées d'ordre $\leqslant s(i) + 1$ de carrés localement intégrables dont les traces sur Σ sont les données de Cauchy envisagérs.

Une solution du problème de Cauchy posé est alors une solution (u^i) de (2.1) dont les dérivées d'ordre $\leqslant s(i)$ sont de carrés localement intégrables et coincident su Σ avec celles de w^i. Pour ce problème, J. Leray a démontré un théorème d'existence et d'unicité que nous allons énoncer sans démonstration.

Théorème. - Si les données de Cauchy sur Σ sont définies par des fonctions w^i vérifiant les hypothèses:

1) l'opérateur différentiel $A(x, w, D)$ est hyperbolique strict dans Ω et l'hypersurface Σ orientée dans l'espace relativement à $A(x, w, D)$

2) les $a_i(x, w, D)w + b_i(x, w)$ s'annulent sur Σ ainsi que leurs dérivées jusqu'à l'ordre $t(i) - 1$.

Alors pour tout $x \in \Sigma$, le problème de Cauchy pour (2.1) admet au moins une solution dans un voisinage de x. Si $(\bar{u}^i)$ et (u^i) sont deux solutions et si $\bar{u}^i$ et u^i ont des dérivées d'ordre $\leq s(i)+1$ de carrés localement intégrables, alors elles coincident.

Définition. - Un système quasi-linéaire qui vérifie les hypothèses précedentes sera dit un système quasi-linéaire strictement hyperbolique, ou système de Leray.

Les systèmes quasi-linéaires qu'on rencontre en Physique ne sont pas toujours des systèmes diagonaux. Pour appliquer le théorème de Leray, il faudra les ramener à la forme diagonale. On sait qu'on le peut toujours. Mais le caractère strictement hyperbolique doit être démontré. Nous allons exposer la méthode dans un cas.

§2. APPLICATION AUX EQUATIONS DE L'HYDRODYNAMIQUE DES FLUIDES PARFAITS

3. Les coordonnées harmoniques.

Dans l'étude du problème de Cauchy relatif au système des équations d'Einstein, les coordonnées harmoniques sont un outil précieux.

Définition. - Un système de coordonnées locales (x^ρ) est dit harmonique si chaque fonction coordonnée x^ρ est une solution de l'é quation de Laplace

(3.1) $$\Delta f = -g^{\alpha\beta} (\partial_{\alpha\beta} f - \Gamma^{\gamma}_{\alpha\beta} \partial_{\gamma} f) = 0$$

$\Gamma^{\alpha}_{\beta\gamma}$ étant les coefficients de la connexion riemannienne.

On remarque que les variétés caractéristiques de (3.1) sont tangentes en chaque point au cône élémentaire C_x de l'espace-temps.

Si (x^{ρ}) est un système de coordonnées locales, on pose

(3.2) $$F^{\rho} = \Delta x^{\rho} = g^{\alpha\beta} \Gamma^{\rho}_{\alpha\beta}$$

F^{ρ} dépendent de $g_{\alpha\beta}$ et de leurs dérivées premières. Si $F^{\rho} = 0$, le système de coordonnées locales est harmonique.

On associe à F^{ρ} les quantités $L_{\alpha\beta}$ définies par

(3.3) $$L_{\alpha\beta} = \frac{1}{2} (g_{\alpha\rho} \partial_{\beta} F^{\rho} + g_{\beta\rho} \cdot \partial_{\alpha} F^{\rho})$$

Lemme. - Dans un système de coordonnées locales arbitraires, les composantes du tenseur de Ricci peuvent se mettre sous la forme

(3.4) $$R_{\alpha\beta} = R^{(h)}_{\alpha\beta} + L_{\alpha\beta}$$

où

(3.5) $$R^{(h)}_{\alpha\beta} = -\frac{1}{2} g^{\lambda\mu} \partial_{\lambda\mu} g_{\alpha\beta} + F_{\alpha\beta} (g_{\lambda\mu}, \partial_{\gamma} g_{\lambda\mu})$$

les $F_{\alpha\beta}$ étant des fonctions régulières.

Pham

Il suffit pour démontrer le lemme de chercher à mettre en évidence les dérivées du second ordre de $g_{\lambda\mu}$ dans $L_{\alpha\beta}$ et $R_{\alpha\beta}$. On a modulo les termes en $g_{\lambda\mu}$ et $\partial_\sigma g_{\lambda\mu}$

$$L_{\alpha\beta} \simeq \frac{1}{2}\left\{\partial_\beta (g_{\alpha\rho}\, g^{\lambda\mu}\, \Gamma^\rho_{\lambda\mu}) + \partial_\alpha (g_{\beta\rho}\, g^{\lambda\mu}\, \Gamma^\rho_{\lambda\mu})\right\}$$

$$\simeq \frac{1}{2}\, g^{\lambda\mu}\, (\partial_\beta[\lambda\mu, \alpha] + \partial_\alpha[\lambda\mu, \beta]\,)$$

soit

$$L_{\alpha\beta} \simeq \frac{1}{2} g^{\lambda\mu} (\partial_{\alpha\lambda}\, g_{\beta\mu} + \partial_{\beta\lambda} g_{\alpha\mu} - \partial_{\alpha\beta}\, g_{\lambda\mu})$$

D'autre part

$$R_{\alpha\beta} \simeq \partial_\lambda \Gamma^\lambda_{\alpha\beta} - \partial_\alpha \Gamma^\lambda_{\lambda\beta}$$

$$\simeq g^{\lambda\mu} (\partial_\lambda[\alpha\beta, \mu] - \partial_\alpha[\lambda\beta, \mu]\,)$$

soit

$$R_{\alpha\beta} \simeq \frac{1}{2} g^{\lambda\mu} (\partial_{\alpha\lambda}\, g_{\beta\mu} + \partial_{\beta\lambda}\, g_{\alpha\mu} - \partial_{\alpha\beta}\, g_{\lambda\mu} - \partial_{\lambda\mu}\, g_{\alpha\beta})$$

On en déduit

$$R_{\alpha\beta} \simeq -\frac{1}{2} g^{\lambda\mu}\, \partial_{\alpha\beta}\, g_{\lambda\mu} + L_{\alpha\beta}$$

puis le lemme.

Corollaire. - En coordonnées locales quelconques, on a

$$S_{\alpha\beta} = S_{\alpha\beta}^{(h)} + K_{\alpha\beta} \tag{3.6}$$

où

$$S_{\alpha\beta}^{(h)} = R_{\alpha\beta}^{(h)} - \frac{1}{2} R^{(h)} g_{\alpha\beta} \tag{3.7}$$

$$K_{\alpha\beta} = L_{\alpha\beta} - \frac{1}{2} L \, g_{\alpha\beta} \tag{3.8}$$

avec $R^{(h)} = g^{\lambda\mu} R_{\lambda\mu}^{(h)}$ et $L = g^{\lambda\mu} L_{\lambda\mu}$. Si les coordonnées sont harmoniques $K_{\alpha\beta} = 0$.

4. Application à l'étude des solutions des équations d'Einstein.

Théorème. - Toute solution du système des équations d'Einstein

$$S_{\alpha\beta} = \chi T_{\alpha\beta} \tag{4.1}$$

est une solution en coordonnées harmoniques du système

$$S_{\alpha\beta}^{(h)} = \chi T_{\alpha\beta} \tag{4.2}$$

$$\nabla_{\alpha} T^{\alpha}_{\beta} = 0 \tag{4.3}$$

Inversement toute solution du système (4. 2), (4. 3) satisfaisant sur une hypersurface Σ orientée dans l'espace aux conditions

(4. 4) $$F^{\rho} = 0$$

(4. 5) $$S^{0}_{\alpha} = \chi T^{0}_{\alpha}$$

est une solution du problème de Cauchy correspondant pour le système (4. 1) des équations d'Einstein.

En effet toute solution de (4. 2) écrite en coordonnées harmoniques vérifie (4. 2) car $L_{\alpha\beta} = 0$ et comme solution de (4. 1), elle doit vérifier les conditions de conservation $\nabla_{\alpha} S^{\alpha}_{\beta} = 0$ i. e. (4. 3).

Réciproquement considérons une solution de (4. 2), (4. 3) correspondant à des données de Cauchy satisfaisant sur Σ aux conditions (4. 4) et (4. 5). Sur Σ on a $S^{0}_{\alpha} = S^{(h)0}_{\alpha} + K^{0}_{\alpha}$ et en vertu de (4. 2) et (4. 5) on a

$$K^{0}_{\alpha} = 0 \qquad \text{sur } \Sigma$$

En explicitant l'expression de K^{0}_{α} , on a

$$K^{0}_{\alpha} = \frac{1}{2} g_{\alpha\rho}\, g^{0\beta}\, \partial_{\beta}\, F^{\rho} + \frac{1}{2} \partial_{\alpha} F^{0} - \frac{1}{2} g^{0}_{\alpha} \partial_{\rho} F^{\rho}$$

soit en tenant compte de la condition (4. 4) qui entraine sur Σ , $\partial_{i} F^{\rho} = 0$

$$K^0_\alpha = \frac{1}{2} g_{\alpha\rho}\, g^{00} \partial_0 F^\rho = 0$$

Comme Σ est orienté dans l'espace, $g^{00} \neq 0$, on a nécessairement $\partial_0 F^\rho = 0$.

Ainsi la solution considérée de (4.2), (4.3) satisfait sur Σ à

$$(4.6) \qquad \partial_0 F^\rho = 0$$

Pour cette solution on a encore

$$\nabla_\lambda K^{\lambda\mu} = \nabla_\lambda S^{\lambda\mu} - \nabla_\lambda S^{(h)\lambda\mu}$$

$$= - \chi \nabla_\lambda T^{\lambda\mu}$$

On en déduit en vertu de (4.3)

$$(4.7) \qquad \nabla_\lambda K^{\lambda\mu} = 0$$

Or

$$K^{\lambda\mu} = L^{\lambda\mu} - \frac{1}{2} L g^{\lambda\mu} = \frac{1}{2} (g^{\lambda\rho} \partial_\rho F^\mu + g^{\mu\rho} \partial_\rho F^\lambda) - \frac{1}{2} g^{\lambda\mu} \partial_\rho F^\rho$$

donne par dérivation

$$\partial_\lambda K^{\lambda\mu} = \frac{1}{2} (g^{\lambda\rho} \partial_{\lambda\rho} F^\mu + g^{\mu\rho} \partial_{\lambda\rho} F^\lambda - g^{\lambda\mu} \partial_{\lambda\rho} F^\rho) + \text{termes linéaires}$$

en $\partial_\rho F^\lambda$

Il en résulte que l'équation (4.7) peut s'écrire

(4.8) $$g^{\lambda\rho}\partial_{\lambda\rho}F^{\mu} + A^{\lambda\mu}{}_{\rho}\partial_{\lambda}F^{\rho} = 0$$

où $A^{\lambda\mu}{}_{\rho}$ sont des fonctions régulières.

Ainsi F^{ρ} satisfait à un système linéaire hyperbolique qui admet un théorème d'existence et d'unicité. Si Σ est orienté dans l'espace, la seule solution de (4.8) qui satisfait sur Σ à $F^{\rho} = 0$ et $\partial_0 F^{\rho} = 0$ est la solution nulle. Il en résulte que la solution considérée est une solution écrite en coordonnées harmoniques des équations d'Einstein (4.D).

5. Analyse formelle du système fondamental de l'hydrodynamique.

Le système fondamental des équations de l'hydrodynamique des fluides parfaits adiabatiques est formé des équations

(5.1) $$S_{\alpha\beta} = \chi\,(rfu_{\alpha}u_{\beta} - pg_{\alpha\beta})$$

(5.2) $$u^{\alpha}\partial_{\alpha}S = 0$$

(5.3) $$g_{\alpha\beta}\,u^{\alpha}u^{\beta} = +1$$

(5.4) $$dp = rdf - r\theta\, dS$$

On prendra f et S comme variables thermodynamiques r = = r(f, S) et p=p(f, S) sont alors des fonctions connues de f et S.

C'est un système de 16 équations aux dérivées partielles aux 16 fonctions inconnues $g_{\alpha\beta}$, f, S, u^{α}. On va d'abord faire une analyse formelle du problème de Cauchy. Pour cela on se donne sur une hypersurface Σ d'équation locale $x^0 = 0$, les valeurs des $g_{\alpha\beta}$, $\partial_0 g_{\alpha\beta}$, S et on cherche à déterminer la solution au voisinage de Σ. Nous supposons que Σ n'est pas tangente aux cônes élémentaires i.e.

(5.5) $$g^{00} \neq 0$$

et que sur Σ on a

(5.6) $$F^{?} = 0.$$

Une étude classique montre que si $g^{00} \neq 0$, les quantités S^0_{α} sont connues en fonction des données de Cauchy $g_{\alpha\beta}$, $\partial_0 g_{\alpha\beta}$. Les données de Cauchy ($g_{\alpha\beta}$, $\partial_0 g_{\alpha\beta}$, S) doivent alors vérifier les conditions de compatibilité

$$S^0_{\alpha} = \chi\,(rfu^0 u_{\alpha} - pg^0_{\alpha})$$

Supposons provisoirement connus les valeurs sur Σ de f. Les équations précédentes s'écrivent

(5.7) $$\chi rfu^0 u_{\alpha} = S^0_{\alpha} + \chi pg^0_{\alpha}$$

En tenant compte du caractère unitaire (5.3) de u^{α}, on tire

$$(\chi r f u^{0})^{2} = \left[\Omega^{0}(p)\right]^{2} = g^{\alpha\beta}(S^{0}_{\alpha} + \chi p g^{0}_{\alpha})(S^{0}_{\beta} + \chi p g^{0}_{\beta})$$

(5.7) donne alors la valeur de u^{0}

$$u^{0} = \frac{S^{00} + \chi p g^{00}}{\Omega^{0}(p)} \tag{5.8}$$

pui celle de

$$\chi r f = \frac{\left[\Omega^{0}(p)\right]^{2}}{S^{00} + \chi g^{00} p} \tag{5.9}$$

L'équation (5.9) définit implicitement la ou les valeurs possibles de f. On va l'écrire

$$F(f) \equiv \chi r f(S^{00} + \chi p g^{00}) - g^{\alpha\beta}(S^{0}_{\alpha} + \chi p g^{0}_{\alpha})(S^{0}_{\beta} + \chi p g^{0}_{\beta}) \tag{5.9'}$$

En dérivant par rapport à f, il vient en tenant compte de (5.4)

$$F'_{f} = \chi(r + r'_{f} f)(S^{00} + \chi p g^{00}) + \chi^{2} r^{2} f g^{00} - 2 g^{0\alpha} r (S^{0}_{\alpha} + \chi p g^{0}_{\alpha})$$

soit d'après (5.7)

$$F'_{f} = \chi^{2} r^{2} f \left[g^{00} - \left(1 - \frac{f r'_{f}}{r}\right) u^{0} u^{0} \right.$$

On voit que f est connue sur Σ si $F'_{f} \neq 0$ i.e.

$$g^{00} - \left(1 - \frac{f r_f'}{r}\right) u^0 u^0 \neq 0$$

Une fois f connue sur Σ_1, on en déduit u^α à l'aide de (5.7)

$$u^\alpha = \frac{S + \chi p g^{0\alpha}}{\Omega^0(p)}$$

si

$$u^0 \neq 0$$

Ceci étant en vertu des raisonnements faits au paragraphe 4 précédent, on peut substituer au système (5.1), (5.2), (5.3), (5.4) le système

(5.10) $$R^{(h)}_{\alpha\beta} = \chi\left[rfu_\alpha u_\beta - \frac{1}{2}(rf - 2p)\, g_{\alpha\beta}\right]$$

(5.11) $$u^\alpha \partial_\alpha S = 0$$

(5.12) $$\nabla_\alpha (ru^\alpha) = 0$$

(5.13) $$fru^\alpha \nabla_\alpha u^\beta - (g^{\alpha\beta} - u^\alpha u^\beta)(\partial_\alpha f - \theta \partial_\alpha S) = 0$$

où (5.12), (5.13) proviennent des conditions de conservation $\nabla_\alpha T^{\alpha\beta} = 0$ et (5.3). On remarque que (5.13) entraine

$$u^\alpha \nabla_\alpha u^\beta u_\beta = 0$$

ce qui montre que si u^α est unitaire sur Σ , il le reste au voisinage de Σ .

Supposons que les données de Cauchy sont données en termes de series formelles par rapport aux coordonnées locales et cherchons les solutions formelles du système (5.10), (5.11), (5.12), (5.13).

En mettant en évidence les dérivées $\partial_{00} g_{\alpha\beta}$, $\partial_0 S$, $\partial_0 f$, $\partial_0 u^\beta$ dans ces équations, il vient

(5.14) $$-\frac{1}{2} g^{00} \partial_{00} g_{\alpha\beta} = (d.c.)$$

(5.15) $$u^0 \partial_0 S = (d.c.)$$

(5.16) $$r\partial_0 u^0 + r'_f u^0 \partial_0 f + r'_S u^0 \partial_0 S = (d.c.)$$

(5.17) $$f u^0 \partial_0 u^\beta - (g^{0\beta} - u^0 u^\beta) \partial_0 f + g^{0\beta} \partial_0 S = (d.c.)$$

où les seconds membres sont connus en fonction des données de Cauchy. (5.14) donne $\partial_{00} g_{\alpha\beta}$ si $g^{00} \neq 0$; (5.15) donne $\partial_0 S$ si $u^0 \neq 0$, (5.16) et (5.17) pour $\beta = 0$ déterminent alors $\partial_0 u^0$, $\partial_0 f$ si le determinant du système $g^{00} - (1 - \frac{f\, r'_f}{r}) u^0 u^0 \neq 0$; enfin (5.17) pour $\beta = i$ donnent $\partial_0 u^\beta$ si $u^0 \neq 0$.

Ainsi sous la condition que:

$$g^{00} \neq 0 \qquad u^0 \neq 0 \qquad g^{00} - \left(1 - \frac{f x_f'}{r}\right) u^0 u^0 \neq 0$$

on peut calculer $\partial_{00} g_{\alpha\beta}$, $\partial_0 S$, $\partial_0 f$ et $\partial_0 u^\beta$. Les mêmes conclusions s'étendent aux dérivées d'ordre supérieure qu'on obtient en dérivant par rapport à x^0 les differentes équations, de sorte que les séries formelles cherchées sont uniquement déterminées.

Si le problème de Cauchy est analytique, il résulte du théorème de Cauchy-Kowaleski le résultat suivant.

Théorème. - Dans le cas analytique; si les données de Cauchy $g_{\alpha\beta}$, $\partial_0 g_{\alpha\beta}$, S satisfont aux conditions

1) la forme $g_{\alpha\beta}$ $X^\alpha X^\beta$ est hyperbolique normale
2) l'hypersurface Σ portant les données de Cauchy et définie localement par $x^0 = 0$, est orientée dans l'espace
3) sur Σ on a $F^\rho = 0$ et $S^0_\alpha = \chi T^0_\alpha$

alors le problème de Cauchy pour le système des équations de l'hydrodynamique admet une solution analytique et une seule au voisinage de tout point $x \in \Sigma$.

Variétés caractéristiques. - L'étude précédente montre que les variétés caractéristiques du problème de Cauchy sont définies par les équations suivantes

$$g^{\alpha\beta} \partial_\alpha \phi \partial_\beta \phi = 0$$

$$u^\alpha \partial_\alpha \phi = 0$$

$$(g^{\alpha\beta} - (1 - \frac{f r'_f}{r}) u^\alpha u^\beta) \partial_\alpha \phi \, \partial_\beta \phi = 0$$

Les premières définissent les ondes gravitationnelles , les secondes les ondes matérielles ou entropiques et les troisièmes les ondes hydrodynamiques. Un calcul classique montre que ces trois sortes d'ondes se propagent avec les vitesses 1, 0, v respectivement avec

$$v = \sqrt{\frac{r}{r'_{ff}}}$$

Sous l'exigence relativiste on doit avoir $\frac{f r'_f}{r} \geqslant 1$ pour que $v \leqslant 1$ ce qui entraine que les variétés caractéristiques définissant les ondes hydrodynamiques doivent être orientées dans le temps.

6. Théorème d'existence et d'unicité.

Reprenons le système étudié au paragraphe précédent

(6.1) $$R^{(h)}_{\lambda\mu} = \chi\left[rfu_\lambda u_\mu - \frac{1}{2}(rf - 2p) u_\lambda u_\mu\right]$$

(6.2) $$u^\alpha \partial_\alpha S = 0$$

(6.3) $$\nabla_\alpha (ru^\alpha) = 0$$

(6.4) $$fu^\alpha \nabla_\alpha u^\beta - (g^{\alpha\beta} - u^\alpha u^\beta)\partial_\alpha f - \theta g^{\alpha\beta} \partial_\alpha S = 0$$

Il ne se présente pas sous forme diagonale sauf pour (6.1) et (6.2). Nous le transformons de manière à obtenir un système diagonal.

Gardons pour le moment (6.1) et (6.2) tels quels.

Pour avoir l'équation en f, prenons la dérivée contractée ∇_β de (6.4) il vient

$$(g^{\alpha\beta} - u^\alpha u^\beta)\nabla_\alpha \nabla_\beta f - \theta g^{\alpha\beta}\nabla_\alpha\nabla_\beta S - f u^\alpha \nabla_\beta \nabla_\alpha u^\beta = h \, (1 \text{ en } g_{\alpha\beta},\ 1 \text{ en } S,$$

$$1 \text{ en } f,\ 1 \text{ en } u^\alpha)$$

où la notation au 2e membrree signifie que les termes non explicités contiennent des dérivées d'ordre maximum indiqué pour chaque inconnue. Pour avoir $u^\alpha \nabla_\beta \nabla_\alpha u^\beta$, considérons l'équation (6.3) qui se développe comme

$$\nabla_\beta u^\beta + \frac{r'_f}{r} u^\beta \nabla_\beta f = 0$$

en la dérivant suivant les lignes de courant et utilisant l'identité de Ricci $(\nabla_\alpha\nabla_\beta - \nabla_\beta\nabla_\alpha)\, u^\beta = - R_{\alpha\beta}\, u^\beta$, on a

$$u^\alpha \nabla_\beta \nabla_\alpha u^\beta + \frac{r'_f}{r} u^\alpha u^\beta \nabla_\alpha \nabla_\beta f = k \, (2 \text{ en } g_{\alpha\beta},\ 1 \text{ en } S,\ 1 \text{ en } f,\ 1 \text{ en } u^\alpha)$$

En portant dans l'équation en f, on obtient

$$(6.5) \qquad \left[g^{\alpha\beta} - \left(1 - \frac{f\, r'_f}{r}\right) u^\alpha u^\beta\right]\nabla_\alpha \nabla_\beta f - \theta g^{\alpha\beta}\nabla_\alpha\nabla_\beta S = \ell \, (2 \text{ en } g_{\alpha\beta},$$

$$1 \text{ en } S,\ 1 \text{ en } f,\ 1 \text{ en } u^\alpha)$$

Appliquons l'opérateur $u^{\gamma}\nabla_{\gamma}$ à cette équation

$$\left[g^{\alpha\beta} - (1 - \frac{f r'_f}{r})u^{\alpha}u^{\beta}\right] u^{\gamma}\nabla_{\gamma}\nabla_{\alpha}\nabla_{\beta} f - \theta g^{\alpha\beta} u^{\gamma}\nabla_{\gamma}\nabla_{\alpha}\nabla_{\beta} S = m\ (3 \text{ en } g_{\alpha\beta},$$

$$2 \text{ en } S,\ 2 \text{ en } f,\ 2 \text{ en } u^{\alpha})$$

Or l'identité de Ricci donne

$$(6.6) \qquad g^{\alpha\beta} u^{\gamma}\nabla_{\gamma}\nabla_{\alpha}\nabla_{\beta} S = g^{\alpha\beta} u^{\gamma}\nabla_{\alpha}\nabla_{\beta}\nabla_{\gamma} S - u^{\gamma} R_{\gamma}^{\ \rho}\nabla_{\rho} S$$

si on tient compte de $u^{\alpha}\partial_{\alpha} S = 0$, on a

$$(6.6) \qquad g^{\alpha\beta}u^{\gamma}\nabla_{\gamma}\nabla_{\alpha}\nabla_{\beta}S = n\ (2 \text{ en } g_{\alpha\beta},\ 2 \text{ en } S,\ 0 \text{ en } f,\ 2 \text{ en } u^{\alpha})$$

et l'équation en f s'écrit finalement

$$(6.7) \qquad \left[g^{\alpha\beta} - (1 - \frac{f r'_f}{r})u^{\alpha}u^{\beta}\right] u^{\gamma}\partial_{\alpha\beta\gamma} f = F\ (3 \text{ en } g_{\alpha\beta},\ 2 \text{ en } S,$$

$$2 \text{ en } f,\ 2 \text{ en } u^{\alpha})$$

Pour les inconnues u^{β} considérons le système (6.4) auquel nous appliquons l'opérateur $\left[g^{\alpha\beta} - (1 - \frac{f r'_f}{r})\ u^{\alpha}u^{\beta}\right]\nabla_{\alpha}\nabla_{\beta}$:

$$f\left[g^{\alpha\beta} - (1 - \frac{f r'_f}{r})\ u^{\alpha}u^{\beta}\right] u^{\gamma}\nabla_{\alpha}\nabla_{\beta}\nabla_{\gamma}u^{\lambda} - \left[g^{\alpha\beta} - (1 - \frac{f r'_f}{r})u^{\alpha}u^{\beta}\right](g^{\gamma\lambda} - u^{\gamma}u^{\lambda}).$$

$$\nabla_{\alpha}\nabla_{\beta}\nabla_{\gamma}f + \theta g^{\alpha\beta}g^{\gamma\lambda}\nabla_{\alpha}\nabla_{\beta}\nabla_{\gamma}S = p\ (3 \text{ en } g_{\alpha\beta},\ 2 \text{ en } S,\ 2 \text{ en } f,\ 2 \text{ en } u^{\alpha})$$

Pham

Appliquons l'opérateur $(g^{\gamma\lambda} - u^{\gamma}u^{\lambda})\nabla_{\gamma}$ à (6.5)

$$\left[g^{\alpha\beta} - (1 - \frac{f r'_f}{r}) u^{\alpha}u^{\beta}\right](g^{\gamma\lambda} - u^{\gamma}u^{\lambda})\nabla_{\alpha}\nabla_{\beta}\nabla_{\gamma} f - \theta g^{\alpha\beta}(g^{\gamma\lambda} - u^{\gamma}u^{\lambda})\nabla_{\alpha}\nabla_{\beta}\nabla_{\gamma} S$$

$$= q\ (3 \text{ en } g_{\alpha\beta},\ 2 \text{ en } S,\ 2 \text{ en } f,\ 2 \text{ en } u^{\alpha})$$

en tenant compte de cette relation l'équation en u^{λ} devient

$$\left[g^{\alpha\beta} - (1 - \frac{f r'_f}{r}) u^{\alpha}u^{\beta}\right]u^{\gamma}\nabla_{\alpha}\nabla_{\beta}\nabla_{\gamma}u^{\lambda} - \theta g^{\alpha\beta}(g^{\gamma\lambda} - u^{\gamma}u^{\lambda})\nabla_{\alpha}\nabla_{\beta}\nabla_{\gamma} S + \theta g^{\alpha\beta} g^{\gamma\lambda}\nabla_{\alpha}\nabla_{\beta}\nabla_{\gamma} S$$

$$= r(3 \text{ en } g_{\alpha\beta},\ 2 \text{ en } S,\ 2 \text{ en } f,\ 2 \text{ en } u^{\alpha})$$

Or $u^{\gamma}\partial_{\gamma}S = 0$ entraine (6.6) de sorte que on obtient finalement

(6.8) $$\left[g^{\alpha\beta} - (1 - \frac{f r'_f}{r}) u^{\alpha}u^{\beta}\right] u^{\gamma}\partial_{\alpha\beta\gamma} u^{\lambda} = U^{\lambda}(3 \text{ en } g_{\alpha\beta}, 2 \text{ en } S,\ 2 \text{ en } f, 2 \text{ en } u^{\alpha})$$

Nous avons ainsi transformé le système (6.1), (6.2), (6.3), (6.4) en le système diagonal suivant.

(6.9) $$-\frac{1}{2} g^{\alpha\beta}\partial_{\alpha\beta}\, g_{\lambda\mu} = G_{\lambda\mu}(1 \text{ en } g_{\alpha\beta},\ 0 \text{ en } S,\ 0 \text{ en } f,\ 0 \text{ en } u^{\alpha})$$

(6.10) $$u^{\alpha}\partial_{\alpha} S = 0$$

Pham

(6.11) $\left[g^{\alpha\beta} - (1 - \frac{f r'_f}{r}) u^\alpha u^\beta\right] u^\gamma \partial_{\alpha\beta\gamma} f = F$ (3 en $g_{\alpha\beta}$, 2 en S,

2 en f, 2 en u^α)

(6.11) $\left[g^{\alpha\beta} - (1 - \frac{f r'_f}{r}) u^\alpha u^\beta\right] u^\gamma \partial_{\alpha\beta\gamma} u^\lambda = U^\lambda$ (3 en $g_{\alpha\beta}$, 2 en S,

2 en f , 2 en u^α)

où (6.9) s'obtient à partir de (6.1) grâce à l'expression de $R^{(h)}_{\lambda\mu}$.

Ordonnons l'ensemble $\{g_{\alpha\beta}$, S, f, $u^\alpha\}$ en le numerotant de 1 à 16.

Nous avons avec les notations évidentes

$a(\alpha\beta) = g^{\alpha\beta} u^\gamma \partial_{\alpha\beta\gamma}$ $m(\alpha\beta) = 2$ $= 0, 1, 2, 3$

$a(11) = u^\alpha \partial_\alpha$ $m(11) = 1$

$a(12) = \left[g^{\alpha\beta} - (1 - \frac{f r'_f}{r}) u^\alpha u^\beta\right] u^\gamma \partial_{\alpha\beta\gamma}$ $m(12) = 3$

Associons aux inconnus et aux équations les indices suivants

$a(N) = \left[g^{\alpha\beta} - (1 - \frac{f r'_f}{r}) u^\alpha u^\beta\right] u^\gamma \partial_{\alpha\beta\gamma}$ $m(N) = 3$ N = 13, 14, 15, 16

Associons aux inconnus et aux équations les indices suivants

$s(\alpha\beta) = 4$ $s(11) = 3$ $s(12) = 3$ $s(N) = 3$

$t(\alpha\beta) = 3$ $t(11) = 3$ $t(12) = 1$ $t(N) = 1$

et dressons le tableau de l'ordre maximum des dérivations et celui des différences $s(i) - t(j)$:

1	0	0	0
	0		0
3	2	2	2
3	2	2	2

ordre maximum des dérivations

1	0	0	0
1	0	0	0
3	2	2	2
3	2	2	2

matrice ($s(i) - t(j)$)

Nous voyons que l'ordre maximum des dérivations ainsi que l'ordre des opérateurs différentiels $a(i)$ est compatible avec le choix des indices.

Nous avons ainsi montré que le système (6.9), (6.10), (6.11), (6.12) est un système quasi-linéaire au sens de Leray.

Montrons que c'est un système strictement hyperbolique.

Les données de Cauchy sur Σ sont par example les séries formelles calculées au § 5 (il suffit de prendre la somme des n premiers termes, pour n convenable). Leurs dérivées d'ordre $\leq s(i)+1$ sont manifestement de carrés localement intégrables. Nous supposons de plus que les données de Cauchy ($g_{\alpha\beta}$, $\partial_0 g_{\alpha\beta}$, S) sur Σ satisfont aux conditions:

1) la forme quadratique $g_{\alpha\beta} X^\alpha X^\beta$ est hyperbolique normale

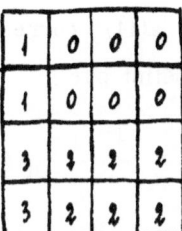

2) sur Σ on a $F^{\xi} = 0$, $S^{0}_{\alpha} = \chi T^{0}_{\alpha}$; la seconde relation défi nit une valeur admissible de f telle que $\frac{f r'_f}{r} \geqslant 1$.

Avec ces données on peut rechercher les demi-cônes $\Gamma^{+}_{x}(a_i)$.

A l'opérateur $a(\alpha\beta)$ correspond le cône $g^{\alpha\beta}u^{\gamma}\xi_{\alpha}\xi_{\beta}\xi_{\gamma} \neq 0$ et le demi-cône $\Gamma^{+}_{x}(\alpha\beta)$ défini par

$$g^{\alpha\beta}\xi_{\alpha}\xi_{\beta} \geqslant 0$$

A l'opérateur a(11) correspond le cône $u^{\alpha}\xi_{\alpha} = 0$ et le demi-cône $\Gamma^{+}_{x}(11)$ défini par

$$u^{\alpha}\xi_{\alpha} \geqslant 0$$

Aux opérateurs a(12) et a(N) correspond le cône

$\left[g^{\alpha\beta} - \left(1 - \frac{f r'_f}{r}\right)u^{\alpha}u^{\beta}\right] u^{\gamma}\xi_{\alpha}\xi_{\beta}\xi_{\gamma} = 0$ et le demi-cône $\Gamma^{+}_{x}(12) = \Gamma^{+}_{x}(N)$

défini par

$$\left[g^{\alpha\beta} - \left(1 - \frac{f r'_f}{r}\right) u^{\alpha}u^{\beta}\right]\xi_{\alpha}\xi_{\beta} \geqslant 0 \tag{6.12}$$

Sous l'hypothèse $\frac{f r'_f}{r} \geqslant 1$, nous voyons que l'intersections de ces trois demi-cônes $\Gamma^{+}_{x}(\alpha\beta)$, $\Gamma^{+}_{x}(11)$, $\Gamma^{+}_{x}(12) = \Gamma^{+}_{x}(N)$ n'est autre que le demi-cône (6.12) qui a un intérieur non vide.

Nous avons donc prouvé que l'opérateur différentiel $A = (a_i)$ est hyperbolique strict en chaque point x . Comme il est différentiable en x , il est strictement hyperbolique dans un voisinage ouvert connexe $\mathcal{U}$ de x . L'hypersurface initiale Σ a été choisie orientée dans l'espace relativement à A, il reste à préciser les valeurs sur Σ des dérivées d'indices 0 de $g_{\alpha\beta}$, S, f, u^α d'ordre $\leq s(i) - 1$, soit:

$$g_{\alpha\beta}\ , \partial_0 g_{\alpha\beta}, \partial_{00} g_{\alpha\beta}\ , \partial_{000} g_{\alpha\beta}$$

$$S\ , \partial_0 S, \partial_{00} S$$

$$f, \partial_0 f, \partial_{00} f$$

$$u^\alpha\ , \partial_0 u^\alpha, \partial_{00} u^\alpha$$

$g_{\alpha\beta}$, $\partial_0 g_{\alpha\beta}$, S ont été donnés dans le problème de Cauchy formel du § 5 $\partial_{00} g_{\alpha\beta}$, S , $\partial_0 S$, f, $\partial_0 f$, u^α, $\partial_0 u^\alpha$ sont obtenus par la résolution de ce problème, pour avoir $\partial_{000} g_{\alpha\beta}$, $\partial_{00} S$, $\partial_{00} f$, $\partial_{00} u^\alpha$, il suffit de dériver les équations (6.1), (6.2), (6.3) et (6.4).

Les hypothèses du théorème de Leray sont donc satisfaites. On en déduit que le problème de Cauchy posé pour le système non analytique (6.9), (6.10), (6.11), (6.12) admet une solution unique ($g_{\alpha\beta}$, S, f, u^α) au voisinage du point $x \in \Sigma$.

Il reste à montrer que cette solution est solution du système initial (6.1), (6.2), (6.3), (6.4) . Or si les données sont analytiques, la solution ($g_{\alpha\beta}$, f, S, u^α) est analytique et est nécessairement la solution analytique du premier problème.

Si les données ne sont pas analytiques, on approche le système par des systèmes analytiques et dans chaque cas, la solution du second problème de Cauchy est solution du premier problème de Cauchy. Par passage à la limite, la solution du problème de Cauchy de Leray est encore solution du système initial. Et nous avons démontré le théorème d'existence et d'unicité pour le problème de Cauchy non analytique relatif au sistème fondamental des équations de l'hydrodynamique relativiste des fluides parfaits.

Chapitre 3

Pham

L'HYDRODYNAMIQUE RELATIVISTE DES FLUIDES PARFAITS ISENTROPIQUES

§1. LA FORME DIFFERENTIELLE INVARIANTE

1. Système différentiel aux lignes de courant.

Le fluide parfait est dit isentropique si l'entropie S est constante. Dans ce cas (cf. Ch. 1, §4) le système différentiel aux lignes de courant s'écrit

$$\frac{dx^\alpha}{ds} = u^\alpha \tag{1.1}$$

$$u^\alpha \nabla_\alpha u_\beta - (g^\alpha{}_\beta - u^\alpha u_\beta) \frac{\partial_\alpha f}{f} = 0 \tag{1.2}$$

où s est l'abscisse curviligne sur les lignes de courant; f est l'indice du fluide. On a vu que

$$f = 1 + \epsilon + \frac{p}{r}$$

Nous nous proposons de mettre en évidence les propriétés géométriques du mouvement du fluide.

2. Variation d'une intégrale.

Soit V_n une variété différentiable de dimension n, $\pi : T(V_n) \rightarrow V_n$ le fibré des vecteurs tangents en tous les points de V_n et $D(V_n) \rightarrow V_n$

le fibré des directions tangentes. $T(V_n)$ est de dimension 2n et $D(V_n)$ de dimension 2n - 1. Comme $T(V_n)$ et $D(V_n)$ sont localement triviaux et qu'on peut choisir comme cartes locales des cartes locales induites par celles de V_n, en coordonnées locales un point de $T(V_n)$ est défini par l'ensemble (x^α, X^α) où (x^α) est un point de l'ouvert U de coordonnées locales de V_n et (X^α) les composantes d'un vecteur tangent en $(x^\alpha) \in U$ relativement à ces coordonnées (x^α) ; un point de $D(V_n)$ est défini par l'ensemble (x^α, u^α) où u^α sont les paramètres directeurs de la direction.

Soit $C: [t_0, t_1] \to V_n$ une courbe différentiable dans V_n d'origine $x_0 = x(t_0)$ et d'extrémité $x_1 = x(t_1)$. En coordonnées locales, elle est définie par la représentation paramétrique

(2.1) $$x^\alpha = x^\alpha(t) .$$

On posera

(2.2) $$\dot{x}^\alpha = \frac{dx^\alpha}{dt} .$$

Cette courbe se relève dans $T(V_n)$ en la courbe $L : t \to (x^\alpha(t), \dot{x}^\alpha(t))$ et si $\dot{x}^\alpha(t)$ n'est pas nulle pour tout t, en la courbe Γ dans $D(V_n)$.

Soit $F: \omega(V_n) \to R$ une fonction à valeurs scalaires donnée sur $T(V_n)$ positivement homogène de degré 1 par rapport à X i.e. pour tout x fixé, $F(x, \lambda X) = \lambda F(x, X)$. A toute courbe $C: [t_0, t_1] \to V_n$, $F(x^\alpha, \dot{x}^\alpha)$ est une fonction de t et on peut calculer l'intégrale

(2.3) $$I = \int_{t_0}^{t_1} F\left(x^\alpha(t), \dot{x}^\alpha(t)\right) dt$$

Cette intégrale est en fait intrinsèquement attachée à F et C et ne dépend pas de la représentation paramétrique.

Calculons la variation de cette intégrale pour une variation quelconque à extrémités non fixes de C . En supposant que C est dans un ouvert de coordonnées locales, on a

$$\delta I = F_{t_1} \cdot \delta t_1 - F_{t_0} \cdot \delta t_0 + \int_{t_0}^{t_1} \delta F \, dt$$

soit d'après un raisonnement classique du calcul des variations

(2.4) $$\delta I = \langle \omega, \delta x \rangle_{x_1} - \langle \omega, \delta x \rangle_{x_0} - \int_{t_0}^{t_1} \langle P, \delta x \rangle \, dt$$

où ω est la 1 forme définie sur $T(V_n)$ par

(2.5) $$\omega = \frac{\partial F}{\partial \dot{x}^\alpha} dx^\alpha$$

et P un covecteur défini en composantes par

(2.6) $$P_\alpha = \frac{d}{dt} \frac{\partial F}{\partial \dot{x}^\alpha} - \frac{\partial F}{\partial x^\alpha}$$

les P_α ne sont autres que les premiers membres des équations d'Euler du calcul des variations. δx est un vecteur.

Si C n'est pas dans un ouvert de coordonnées locales, on la recouvre par un nombre fini de cartes locales et on peut étudier la variation de l'intégrale I qui est la même.

3. Principe d'extrémum pour les lignes de courant.

Appliquons les résultats précédents au cas où V est l'espace-temps et

(3.1) $$F = f\sqrt{g_{\alpha\beta}\,\dot{x}^{\alpha}\,\dot{x}^{\beta}}$$

On a à calculer les variations de l'intégrale

(3.2) $$S = \int_{t_0}^{t_1} f\sqrt{g_{\alpha\beta}\,\dot{x}^{\alpha}\,\dot{x}^{\beta}}\;dt$$

supposée évaluée le long d'un courbe C orientée dans le temps.

On a

$$\frac{\partial F}{\partial \dot{x}^{\alpha}} = \frac{f g_{\alpha\beta}\,\dot{x}^{\beta}}{\sqrt{g_{\lambda\mu}\,\dot{x}^{\lambda}\,\dot{x}^{\mu}}}\,,\quad \frac{\partial F}{\partial x^{\alpha}} = \frac{\partial_{\alpha} f g_{\beta\gamma}\,\dot{x}^{\beta}\,\dot{x}^{\gamma} + \frac{1}{2} f\,\partial_{\alpha} g_{\beta\gamma}\,\dot{x}^{\beta}\,\dot{x}^{\gamma}}{\sqrt{g_{\lambda\mu}\,\dot{x}^{\lambda}\,\dot{x}^{\mu}}}$$

$$P_{\alpha} = \frac{d}{dt}\,\frac{f g_{\alpha\beta}\,\dot{x}^{\beta}}{\sqrt{g_{\lambda\mu}\,\dot{x}^{\lambda}\dot{x}^{\mu}}} - \frac{\partial_{\alpha} f g_{\beta\gamma}\,\dot{x}^{\beta}\,\dot{x}^{\gamma} + \frac{1}{2} f\,\partial_{\alpha} g_{\beta\gamma}\,\dot{x}^{\beta}\,\dot{x}^{\gamma}}{\sqrt{g_{\lambda\mu}\,\dot{x}^{\lambda}\,\dot{x}^{\mu}}}$$

AU lieu d'un paramètre t quelconque, prenons l'arc s de courbe comme paramètre. Le vecteur $\dot{x}^{\alpha} = \frac{dx^{\alpha}}{ds}$ est alors le vecteur vitesse unitaire, de sorte que l'on ait par un calcul facile

(3.3) $$P_{\alpha} = f\left[u^{\alpha}\nabla_{\alpha}u_{\beta} - (g^{\alpha}_{\beta} - u^{\alpha}u_{\beta})\;\frac{\partial_{\alpha} f}{f}\right]$$

(3.4) $$\omega = f u_{\alpha}\,dx^{\alpha}\,.$$

On obtient ainsi la formule

(3.5) $$\delta S = \langle \omega, \delta x \rangle_{x_1} - \langle \omega, \delta x \rangle_{x_0} - \int_{s_0}^{s_1} \langle P, \delta x \rangle \, ds$$

qui donne la variation de l'intégrale

(3.6) $$S = \int_{s_0}^{s_1} f ds$$

pour des extrémités non fixes.

Si les variations sont à extrémités fixes, $\delta x_0 = \delta x_1 = 0$, il vient

(3.7) $$\delta S = - \int_{s_0}^{s_1} \langle P, \delta x \rangle \, ds$$

Pour que S est extrémum, il faut et il suffit que $P = 0$, c'est-à-dire

$$u^\alpha \nabla_\alpha u_\beta - (g^\alpha_\beta - u^\alpha u_\beta) \frac{\partial_\alpha f}{f} = 0$$

équations formellement identiques à (1.1). D'où

Théorème. - Dans tout mouvement d'un fluide parfait isentropique, les lignes de courant sont localement des lignes orientées dans le temps extrémales de l'intégrale (3.6) pour des variations à extrémités fixes.

Introduisons la métrique conforme $\bar{g} = f^2 g$. On a

(3.8) $$\bar{g}_{\alpha\beta} = f^2 g_{\alpha\beta} \qquad \bar{g}^{\alpha\beta} = f^{-2} g^{\alpha\beta}$$

Vis à vis de cette métrique, l'arc de courbe est défini par $d\bar{s} = f ds$, de sorte que les lignes de courant sont définies comme extrémales de

$$(3.9) \qquad \bar{S} = \int_{\bar{s}_0}^{\bar{s}_1} d\bar{s}$$

Ces extrémales sont des géodesiques de $(V_4, \bar{g})$. En posant

$$(3.10) \qquad C_\alpha = f u_\alpha$$

on voit que $\bar{C}_\alpha = f u_\alpha$ et $\bar{C}^\alpha = f^{-1} u^\alpha$, de sorte que $\bar{g}_{\alpha\beta}\, \bar{C}^\alpha \bar{C}^\beta = 1$ et ces géodésiques ont pour équations

$$(3.11) \qquad \bar{C}^\alpha \nabla_\alpha \bar{C}_\beta = 0$$

Corollaire. - Lesllignes de courant du fluide parfait isentropique sont géodésiques orientées dans le temps de $(V, \bar{g})$.

4. L'invariant intégral de l'hydrodynamique.

Considérons un mouvement du fluide, défini par example par un pro blème de Cauchy. Soit $\mathcal{T}$ un tube de courant s'appuyant sur un cycle Γ_0 de dimension tracé sur l'hypersurface initiale Σ (non tangente aux lignes de courant) et soit Γ_1 un cycle tracé sur $\mathcal{T}$, homotope à Γ_0. Chaque ligne de courant de $\mathcal{T}$ est limité en $x_0 \in \Gamma_0$ et $x_1 \in \Gamma_1$. Nous pouvons appliquer la formule (3.5) à chacune de ces lignes de courant, $P = 0$, la variation totale de S est nulle quand x_0 décrit le cycle Γ_0, il vient

$$(4.2) \qquad \int_{\Gamma_0} \omega = \int_{\Gamma_1} \omega$$

La 1-forme ω a pour expression:

$$\omega = C_\alpha \, dx^\alpha \tag{4.2}$$

la propriété (4.1) se traduit par l'énoncé suivant qui généralise un théorème classique sur la conservation de la circulation.

Théorème. - Etant donné un cycle Γ à une dimension non tangent aux lignes de courant, la circulation du vecteur courant C_α le long de Γ reste invariante quand Γ se déforme sur le tube de courant défini par Γ .

Si D est une variété différentiable à 2 dimensions, à bord ∂D, la formule de Stokes donne

$$\int_{\partial D} \omega = \int_D d\omega \tag{4.3}$$

L'intégrale de la forme

$$\Omega = d\omega$$

sur la sous-variété D se conserve quand elle se déforme de manière que chaque point reste sur la même ligne de courant.

Dans le langage de H. Poincaré, Ω définit un invariant intégral pour le système différentiel aux lignes de courant

$$\frac{dx^\alpha}{ds} = u^\alpha$$

et ω définit un invariant intégral relatif. La 2-forme Ω joue un rôle fondamental dans la description du mouvement. Elle admet l'expression locale

(4.4) $$\Omega_{\alpha\beta} = \partial_\alpha C_\beta - \partial_\beta C_\alpha$$

Théorème. - La 2-forme Ω est une forme invariante pour le système différentiel aux lignes de courant i.e.

(4.5) $$\mathcal{L}_c \Omega = 0$$

$\mathcal{L}_c$ est la dérivée de Lie suivant C.

En effet on a en utilisant l'identité du calcul des variations

$$\mathcal{L}_c \Omega = (di_c + i_c d)\Omega$$

comme $\Omega = d\omega$, $d\Omega = 0$, il reste $di_c\Omega$; or

$$(i_c\Omega)_\beta = C^\alpha \Omega_{\alpha\beta} = C^\alpha (\partial_\alpha C_\beta - \partial_\beta C_\alpha)$$

En introduisant la connexion riemannienne associée à la métrique conforme $\bar{g}$, on obtient

$$\mathcal{L}_c \Omega = \bar{C}^\alpha (\bar{\nabla}_\alpha \bar{C}_\beta - \bar{\nabla}_\beta \bar{C}_\alpha)$$

Comme $\bar{C}_\alpha$ est unitaire, $\bar{C}^\alpha \bar{\nabla}_\beta \bar{C}_\alpha = 0$ et d'autre part $\bar{C}_\alpha$ est un champ géodésique d'après (3.11), on a bien $\mathcal{L}_c \Omega = 0$.

La 2-forme Ω est une forme invariante par le système différentiel aux lignes de courant. Nous allons rechercher tous les systèmes diffé rentiels qui la laissent invariante. Il nous faut déterminer tous les champs de vecteur X tels que

$$\Omega_{\alpha\beta} X^{\alpha} = 0$$

L'existence de X dépend du rang du système précédent, comme $\Omega_{\alpha\beta}$ est antisymétrique, il vient :

1. - si Ω est de rang 2 , en chaque point x les vecteurs caractéristiques forment un 2-plan π_x . Le champ de 2-plan π admet des variétés intégrales de dimension 2 engendries par les lignes de courant.
2. - si Ω est de rang 0 , $\Omega = 0$. Comme $\Omega = d\omega$, ω est une 1-forme fermé, il existe une fonction ϕ telle que $\omega = d\phi$. Par suite $C_\alpha = \partial_\alpha \phi$: les lignes de courant sont trajectoires orthogonales à la famille d'hypersurfaces $\phi = const.$

Ces résultats sont importants pour l'étude des mouvements rotationnels et irrotationnels du fluide.

§2. MOUVEMENTS ROTATIONNELS ET IRROTATIONNELS

5. Tenseur tourbillon et équations de Helmholtz.

Définition. - On appelle tenseur tourbillon le tenseur antisymétrique d'ordre 2 défini par la 2-forme invariante Ω .

Il constitue la véritable extension relativiste du rotationnel des vitesses introduit en mécanique classique. Si l'on se rappelle l'expression de $f = 1 + \varepsilon c^{-2} + pr^{-1} . c^{-2}$ où c est la vitesse de la lu -

mière, on voit que $C_\alpha = f u_\alpha$ diffère de u_α par des termes en c^{-2}, $\Omega_{\alpha\beta} = \partial_\alpha C_\beta - \partial_\beta C_\alpha$ diffère de $\overset{\circ}{\Omega}_{\alpha\beta} = \partial_\alpha u_\beta - \partial_\beta u_\alpha$ par des termes en c^{-2}.

Théorème. - Le tenseur tourbillon satisfait aux équations de Helmholtz

$$C^\rho \nabla_\rho \Omega_{\alpha\beta} + \nabla_\alpha C^\rho \Omega_{\rho\beta} + \nabla_\beta C^\rho \Omega_{\alpha\rho} = 0 \tag{5.1}$$

En effet un calcul simple montre que ces équations sont une conséquence de l'équation $\mathcal{L}_C \Omega = 0$ qui exprime que Ω est une forme invariante. On fait le calcul en métrique initiale (V, g).

Définition. - On dit qu'un mouvement du fluide est rotationnel si $\Omega \neq 0$ et irrotationnel si $\Omega = 0$.

Théorème. - Pour qu'un mouvement du fluide parfait isentropique soit irrotationnel, il faut et il suffit que les lignes de courant soient orthogonales à une même hypersurface (locale).

En effet soit Σ une hypersurface orientée dans l'espace telle que $\Omega_{\alpha\beta} = 0$ sur Σ. On peut choisir des coordonnées locales telles que Σ soit représentée par $x^0 = 0$ et que les lignes de courant par $x^i = \text{const.}$ (coordonnées de Gauss). Les équations de Helmholtz montrent alors que $\partial_0 \Omega_{\alpha\beta} = 0$. Il en résulte que $\Omega_{\alpha\beta} = 0$ au voisinage de Σ.

6. Vecteur tourbillon.

On suppose que le mouvement est irrotationnel. Au point $x \in V_4$ étudions le 2-plan Π_x formé des vecteurs X^α caractéristiques i.e. tels que

$$\Omega_{\alpha\beta} X^\beta = 0 \tag{6.1}$$

On aurait dejà dans π_x le vecteur u^α tangent à la ligne de courant passant par x . Pour achever de determiner π_x , il nous suffit de rechercher un second vecteur non colinéaire à u^α . Nous choi - sissons un tel vecteur θ orthogonal au premier. Ce vecteur est défini par les équations

(6.2) $$\Omega_{\alpha\beta}\theta^\beta = 0 \qquad \theta^\alpha u^\alpha = 0$$

Le vecteur θ^α n'est défini qu'à un facteur près, on a par un calcul algébrique

(6.3) $$\theta^\alpha = \frac{1}{2}\eta^{\alpha\beta\gamma\delta} u_\beta \Omega_{\gamma\delta}$$

où $\eta_{\alpha\beta\gamma\delta}$ est la forme élément de volume riemannien de (V, g). On remarque que $\theta^\alpha = 0$ entraine $\Omega_{\alpha\beta} = 0$.

Définition. - Au vecteur θ^α défini par (6.3) on donne le nom de vecteur tourbillon, à ses trajectoires le nom de lignes de tourbillon.

D'après la définition les lignes de tourbillon sont orthogonales aux lignes de courant. D'autre part le système différentiel aux lignes de tourbillon

$$\frac{dx^\alpha}{dt} = \theta^\alpha$$

admet la 2-forme Ω comme forme invariante. On en déduit immédiatement les propriétés suivantes:

Théorème. - Etant donné un cycle Γ à une dimension non tangent aux lignes de tourbillon, la circulation du vecteur tourbillon le long de Γ reste invariante quand on déforme Γ sur le tube de tourbillons défini par Γ .

Soit $\mathcal{T}$ un tube de lignes de courant, Γ et Γ' deux cycles homotopes sur $\mathcal{T}$. Chacun de ces cycles définit un tube de tourbillons, soit Θ et Θ' . Soit Γ_1 un cycle sur Θ homotope à Γ . Alors le tube de courant passant par Γ_1 coupe le tube de tourbillons Θ' suivant un cycle Γ_1' homotope à Γ' . Comme ω est un invariant intégral relatif pour les lignes de courant et aussi pour les lignes de tourbillons, on a

$$\int_{\Gamma_1'} \omega = \int_{\Gamma_1} \omega$$

Cette propriété constitue la généralisation relativiste d'un théorème de Helmholtz en dynamique classique.

Enfin le champ de 2-plans $x \to \Pi_x$ défini par le système caractéristique de la forme Ω

$$\Omega_{\alpha\beta} X^\beta = 0$$

est un champ complétement intégrable. Aux variétés intégrales à deux dimensions W_2 on donne le nom de variétés caractéristiques de Ω . Ces variétés peuvent être engendrées par des lignes de courant et par des lignes de tourbillon qui sont orthogonales sur W_2 . Si donc on mène les lignes de courant passant par les points d'un ligne de tourbillon, les trajectoires orthogonales de ces lignes de courant sur W_2 sont lignes de tourbillon. Cela veut dire que si une ligne fluide est de tourbillon à un instant, elle reste de tourbillon à tout instant.

Pham

§3. MOUVEMENTS PERMANENTS

7. Espace-temps stationnaire.

On dit qu'un espace-temps (V_4, g) est stationnaire s'il existe un groupe connexe d'isométries globales à un paramètre ne laissant invariant aucun point de V_4, à trajectoires orientées dans le temps et tel que

1) chaque trajectoire z est homéomorphe à R

2) il existe une variété différentiable V_3 à trois dimensions et un difféomorphisme $V_4 \rightarrow V_3 \times R$ appliquant les trajectoires z sur la droite facteur R.

V_4 apparait comme une variété fibrée triviale de base V_3 de fibre type R. Les fibres sont des trajectoires d'isométries. On les appelle les lignes de temps. On appelle espace la variété de base V_3. Celle-ci est difféomorphe à la variété quotient de V_4 par la relation d'équivalence définie par le groupe d'isométries.

Si ξ est le vecteur générateur infinitésimal du groupe d'isométries il satisfait aux équations de Killing

$$(\mathcal{L}_\xi g)_{\alpha\beta} = \nabla_\alpha \xi_\beta + \nabla_\beta \xi_\alpha \tag{7.1}$$

Il résulte de la définition qu'il existe des systèmes de coordonnées locales (x^0, x^i) tel que les x^i soient un système de coordonnées locales sur V_3 et que x^0 définisse les points sur les trajectoires de ξ, de sorte que les sections d'espace x^0 = const sont globalement définies et difféomorphes à V_3. On dira que ces coordonnées locales

(x^0, x^i) sont localement adaptées au groupe d'isométries si le générateur infinitésimal ξ y admet les composantes contravariantes

$$(7.2) \qquad \xi^0 = 1 \qquad \xi^i = 0$$

Si $g_{\alpha\beta}$ sont les composantes du tenseur métrique dans ces systèmes de coordonnées, les composantes covariantes de ξ sont

$$\xi_\alpha = g_{0\alpha}$$

Les équations de Killing (7.1) se traduisent comme

$$\nabla_\alpha \xi_\beta = \nabla_\alpha g_{0\beta} + \Gamma^\rho_{\alpha 0} \cdot g_{\rho\beta} = [\alpha 0, \beta]$$

spit

$$\nabla_\alpha \xi_\beta + \nabla_\beta \xi_\alpha = \partial_0 g_{\alpha\beta} = 0$$

Ainsi dans les coordonnées adaptées les $g_{\alpha\beta}$ sont indépendantes de x^0.

En décomposant la forme métrique suivant la variable directrice x^0 on a

$$(7.3) \qquad g = \frac{g_{0\alpha}\, g_{0\beta}\, dx^\alpha \otimes dx^\beta}{g_{00}} + \hat{g}_{ij}\, dx^i \otimes dx^j$$

où

$$(7.4) \qquad \hat{g}_{ij} = g_{ij} - \frac{g_{0i}\, g_{0j}}{g_{00}}$$

définit une métrique définie négative sur les sections d'espace. Elle est invariante par tout changement de système de coordonnées adaptée de la forme

$$x^{0'} = x^0 + \psi(x^i) \qquad x^{i'} = x^i$$

On munira V_3 de cette métrique $\hat{g}$.

8. Mouvement permanent.

On dit que le mouvement du fluide parfait isentropique est permanent si l'espace-temps stationnaire et si le groupe d'isométries laisse invariants l'indice f et le vecteur vitesse unitaire i.e.

(8.1) $$\mathcal{L}_\xi f = 0 \qquad \mathcal{L}_\xi u = 0$$

Si les coordonnées sont adaptées ces conditions (8.1) se traduisent par

(8.2) $$\partial_0 f = 0 \qquad \partial_0 u_\alpha = 0$$

Théorème. - Pour que le mouvement du fluide parfait isentropique soit permanent, il faut et il suffit que l'espace-temps soit stationnaire.

Choisissons des coordonnées locales adaptées (x^0, x^i) et soit Σ l'hypersurface d'équation $x^0 = 0$. Σ est orientée dans le temps. Il résulte du problème de Cauchy que sur Σ et sur les hypersurfaces voisines $x^0 =$ const , on a (chp II, 5, (5.9))

$$\chi r f\,(S^{00} + \chi p g^{00}) - g^{\alpha\beta}(S^0_\alpha + \chi p g^0_\alpha)\,(S^0_\beta + \chi p g^0_\beta) = 0$$

En coordonnées adaptées, $\partial_0 S^0_\alpha = 0$, $\partial_0 g_{\alpha\beta} = 0$. Il en résulte qu'en dérivant par rapport à x^0 on obtient compte tenu de l'équation thermodynamique $dp = r df - \theta\, dS$ $(dS = 0)$

$$\chi (f r'_f + r)\,(S^{00} + \chi p g^{00})\,\partial_0 f + \chi^2 r^2 f\,\partial_0 f g^{00} - 2 g^{\alpha 0}(S^0_\alpha + \chi p g^0_\alpha)\chi\, r\,\partial_0 f = 0$$

soit en tenant compte du Ch. II, 5, (5.7)

$$\left[g^{00} - \left(1 - \frac{f r'_f}{r}\right) u^0 u^0\right] \cdot \partial_0 f = 0$$

Ou en déduit $\partial_0 f = 0$ sur Σ , puis $\partial_0 u^\alpha = 0$. Le mouvement est donc permanent.

Théorème. - Dans tout mouvement permanent du fluide, la fonction scalaire

(8.3) $$H = \xi^\alpha C_\alpha$$

conserve une valeur constante le long de chaque ligne de courant.

Il nous suffit de montrer que $\mathcal{L}_C(i_\xi \omega) = 0$ où $\omega = C_\alpha dx^\alpha$. Or on a en utilisant l'identité du calcul des variations

$$\mathcal{L}_C(i_\xi \omega) = (i_C\, d + d\, i_C)\, i_\xi \omega$$

$$= i_C\, d\, i_\xi \omega$$

$$= i_C (\mathcal{L}_\xi - i_\xi d) \omega$$

Or il est manifeste que $\mathcal{L}_\xi \omega = 0$ ce qui veut dir que le système différentiel aux lignes de courant admet la transformée infinitésimale ξ, on en déduit également $\mathcal{L}_\xi \omega = 0$. Il vient

$$\mathcal{L}_C (i_\xi \omega) = - i_C i_\xi d\omega$$

$$= i_\xi i_C \Omega = 0$$

Rémarque.- Le système différentiel aux lignes de courant admet la forme invariante Ω et la transformée infinitésimale ξ . Comme $\mathcal{L}_\xi \theta = 0$, on voit que le système différentiel aux lignes de tourbillon possède la même propriété. On en déduit que $H = C^\alpha \xi_\alpha$ est également constant le long des lignes de tourbillons. H est donc constant sur chaque variété caractéristique W_2 de Ω .

On a

$$dH = \Omega_{\alpha\beta} \xi^\beta dx^\alpha \tag{8.4}$$

formule qui rend les résultats précédents évidents.

9. Le théorème de Bernouilli.

Introduisons la grandeur d'espace du vecteur u^α relativement à la direction de temps ξ . Soit

$$- v^2 = \hat{g}_{ij} u^i u^j$$

En vertu du caractère unitaire de u, on a

$$g_{\alpha\beta}\ u^{\alpha}u^{\beta} = \frac{1}{g_{00}}\ (g_{0\alpha}u^{\alpha})^2 + \hat{g}_{ij}\ u^i\ u^j = 1$$

d'où

$$(u_0\ )^2 = g_{00}(1 + v^2) \tag{9.1}$$

L'intégrale première H a pour valeur en coordonnées adaptées $C_0 = fu_0$. On en déduit

$$H^2 = f^2 g_{00}\ (1 + v^2)$$

En posant $U = g_{00}$, il vient

Théorème. - Le mouvement permanent d'un fluide parfait isentropique satisfaits le long de chaque ligne de courant à

$$f^2\ U\ (1 + v^2) = \text{const}\ . \tag{9.2}$$

où U est le potentiel principal de gravitation.

Ce théorème généralise le théorème de Bernouilli.. En effet de l'équation thermodynamique, il vient

$$f = 1 + \int_{p_0}^{p} \frac{dp}{c^2 r}\ .$$

On en déduit aux termes en c^{-2} près

$$\frac{1}{2}c^2 U \ + \ U\left(\frac{1}{2}\ v^2 + \int_{p_0}^{p} \frac{dp}{r}\ \right) = \text{const.}$$

Pham

§4. PROJECTIONS DANS L'ESPACE

10. Un problème du calcul des variations.

On se propose d'étudier le mouvement permanent dans l'espace V_3. Pour cela il nous faut étudier les projections des géodésiques de $(V_4 \; \bar{g})$ sur l'espace quotient $(V_3 \; \hat{\bar{g}})$.

Un tel problème a été résolu dans le cas plus général d'une variété fuislerienne $(V_{n+1} \;, \mathcal{L})$ définie par une variété différentiable V_{n+1} munie d'une fonction $\mathcal{L}(x, X)$ positivement homogène de degré 1 sur le fibré des direetions $D(V_{n+1})$. On supposera que $(V_{n+1} \;, \mathcal{L})$ admet un groupe connexe d'isométries globales définies par un champ de vecteurs ξ tel que

$$\mathcal{L}_\xi \mathcal{L} = 0$$

On rapportera $(V_{n+1} \;, \mathcal{L})$ à des coordonnées locales $(x^i \;, x^0)$ adaptées à son groupe d'isométries et on désignera par V_n la variété quotient.

Le système différentiel aux extıemales de $\mathcal{L}$ admet l'invariant intégral relatif

$$\omega = \partial_{\dot{\alpha}} \mathcal{L} \, dx^\alpha \tag{10.1}$$

où $\partial_{\dot{\alpha}} = \frac{\partial}{\partial \dot{x}^\alpha}$. $\mathcal{L}$ ne dépend pas de x^0 , $\partial_0 \mathcal{L} = 0$, on a l'intégrale première provenant de l'équation d'Euler en x^0

$$\partial_{\dot{0}} \mathcal{L} = h \tag{10.2}$$

de sorte que $\partial_{\dot 0}\mathcal{L}\, dx^0 = K\, dx^0$ constitue un invariant intégral pour la famille (E_k) des extrèmales correspondant à la valeur h. Il en résulte que

(10. 3) $$\pi = \partial_{\dot k}\mathcal{L}\, dx^k$$

est un invariant intégral relatif pour la famille (E_h).

Si $\partial_{\dot 0\dot 0}\mathcal{L} \neq 0$, on peut resoudre (10. 2) par rapport à $\dot x^0$, soit

(10. 4) $$\dot x^0 = \varphi(x^i, \dot x^j, h)$$

où φ est une fonction homogène de degré 1 en $\dot x^j$. D'autre part en vertu de l'homogénéité de $\mathcal{L}$ on a

$$\dot x^i \partial_{\dot i}\mathcal{L} + \dot x^0 \partial_{\dot 0}\mathcal{L} = \mathcal{L}$$

Par suite la forme $\dot\pi = \dot x^i \partial_{\dot i}\mathcal{L}$ peut s'exprimer par une fonction L des variables x^i, $\dot x^j$, k, soit

(10. 5) $$L(x^i, \dot x^j, h) = \mathcal{L}(x^i, \dot x^j, \varphi(x^i, \dot x^j, h)) - h\,\varphi(x^i, \dot x^j, h)$$

et l'on a

$$\partial_{\dot i}L = \partial_{\dot i}\mathcal{L} + \partial_{\dot 0}\mathcal{L}\,\partial_{\dot i}\varphi - h\,\partial_{\dot i}\varphi = \partial_{\dot i}\mathcal{L} + h\,\partial_{\dot i}\varphi - h\,\partial_{\dot i}\varphi = \partial_{\dot i}\mathcal{L}$$

On a démontré le théorème.

Théorème. - Les projections sur V_n des extrèmales (E_h) pour une valeur h donné sont les extremales de la fonction L. Elles sont définies pour un système différentiel qui admet l'invariant intégral relatif

$$\pi = \partial_{\dot{k}} L \, dx^k.$$

11. Cas d'une métrique riemannienne.

Considérons le cas où la fonction $\mathcal{L}$ est définie par

$$\mathcal{L}^2 = g_{\alpha\beta} \dot{x}^\alpha \dot{x}^\beta$$

$\alpha, \beta = 0, 1, \ldots, n$. On suppose que $g_{00} \neq 0$.

Le procédé de descente conduit à former l'équation

$$(11.1) \qquad \frac{1}{2} \partial_{\dot{0}} \mathcal{L}^2 = g_{00} \dot{x}^0 + g_{0i} \dot{x}^i = h\mathcal{L}$$

et à éliminer $\dot{x}^0$ entre cette équation et

$$(11.2) \qquad L = \mathcal{L} - h\dot{x}^0$$

L'élimination donne

$$(11.3) \qquad L = \sqrt{\left(1 - \frac{h^2}{g_{00}}\right) g_{ij} \dot{x}^i \dot{x}^j} + h \frac{g_{0i} \dot{x}^i}{g_{00}}$$

Si $g_{00} = 0$, on a

$$(11.4) \qquad \mathcal{L}^2 = 2g_{0i} \dot{x}^i \dot{x}^0 + g_{ij} \dot{x}^i \dot{x}^j$$

On supposera $g_{0i} \dot{x}^i \neq 0$. Le procédé de descente conduit à éliminer $\dot{x}^0$ entre (11.2) et les relations

$$g_{0i} \dot{x}^i = h\mathcal{L}$$

$$L = \mathcal{L} - h \dot{x}^0$$

L'élimination donne

$$(11.5) \qquad L = \frac{g_{0i} \dot{x}^i}{2h} + h \frac{g_{ij} \dot{x}^i \dot{x}^j}{2g_{0i} \dot{x}^i}$$

Application aux mouvements permanents. - Il suffit de remplacer dans les formules précédents $g_{\alpha\beta}$ par $f^2 g_{\alpha\beta}$ et on obtient la fonction L dont les extrémales donnent le mouvement dans l'espace. Il y a un seul cas car $g_{00} \neq 0$. On obtient ainsi:

$$(11.6) \qquad L = \sqrt{\left(1 - \frac{h^2}{f^2 g_{00}}\right) f^2 g_{ij} \dot{x}^i \dot{x}^j} + h \frac{g_{0i} \dot{x}^i}{g_{00}}$$

Il serait intéressant de développer les calculs.

12. Projection des géodésiques de longueur nulle.

On les considère comme limites des géodésiques orientées dans le temps. Dans notre problème

$$\mathcal{L} = \sqrt{g_{\alpha\beta} \dot{x}^\alpha \dot{x}^\beta}$$

en dérivant par rapport à $\dot{x}^0$, on a $h\mathcal{L} = g_{0\alpha} \dot{x}^\alpha$, ce qui montre que $h \to \infty$ lorsque $\mathcal{L} \to 0$, h garde le signe de $g_{0\alpha}\, \dot{x}^{\alpha}$. Les extremales de L coincident avec celles de L/h .

Par suite les extrémales cherchées qui définissent les projections des géodésiques isotropes de (V_4, g) sont les extrémales de la fonction

$$\Lambda = \lim_{h \to \infty} \frac{1}{h} L(x^i, \dot{x}^j, h) \tag{12.1}$$

1er cas : $g_{00} \neq 0$.- Le passage à la limite donne

$$\Lambda = \varepsilon\varepsilon' \sqrt{-\frac{1}{g_{00}} \hat{g}_{ij}\, \dot{x}^i \dot{x}^j} - \frac{g_{0i}\, \dot{x}^i}{g_{00}} \tag{12.2}$$

où ε' est le signe de $g_{0\alpha}\, \dot{x}^{\alpha}$ et ε le signe de g_{00} , puis

$$\dot{x}^0 = \varepsilon\varepsilon' \sqrt{-\frac{1}{g_{00}} \hat{g}_{ij}\, \dot{x}^i \dot{x}^j} - \frac{g_{0i}\, \dot{x}^i}{g_{00}} \tag{12.3}$$

2e cas : $g_{00} = 0$. Le passage à la limite donne

$$L = \frac{g_{ij}\, \dot{x}^i \dot{x}^j}{2 g_{0i}\, \dot{x}^i} \tag{12.4}$$

$$\dot{x}^0 = -\frac{g_{ij}\, \dot{x}^i \dot{x}^j}{2 g_{0i}\, \dot{x}^i} \tag{12.5}$$

Nous appliquons ces résultats à l'étude du principe de Fermat.

13. Le principe de Fermat.

On sait que les rayons lumineux [16] sont géodésiques isotropes de la variété riemannienne $(V_4, \bar{g})$ définie par l'espace-temps V_4 muni de la métrique

(13.1) $$\bar{g}_{\alpha\beta} = g_{\alpha\beta} - (1 - \frac{1}{\lambda\mu}) u_\alpha u_\beta$$

Supposons que le mouvement est permanent. Si λ et μ sont constants, le groupe d'isométries de (V_4, g) induit un groupe d'isométries sur $(V_4, \bar{g})$

On choisira des coordonnées adaptées.

Mais alors que les trajectoires d'isométries de (V_4, g) sont orientées dans le temps, les trajectoires d'isométries induites sur $(V_4, \bar{g})$ peuvent être orientées dans le temps, dans l'espace or isotropes . En effet si ζ est le générateur infinitésimal du groupe d'isométries de $(V_4, \bar{g})$ on a pour les composantes contravariantes

$$\zeta^0 = \xi^0 = 1 \qquad \zeta^i = \xi^i = 0$$

le carré de ce vecteur a pour valeur

(13.2) $$\bar{g}_{00} = g_{00} - (1 - w^2) u_0 u_0$$

où $w^2 = \frac{1}{\lambda\mu}$ est le carré de la vitesse de propagation de la lumière dans le fluide.

Si nous introduisons la grandeur d'espace du vecteur vitesse unitaire u^α relativement à la direction de temps ζ , $v^2 = - \hat{g}_{ij} u^i u^j$, on a vu que

$$(u_0)^2 \neq g_{00}(1 + v^2)$$

En portant cette valeur dans (13.2)$_b$, il vient

$$\bar{g}_{00} = g_{00}\,(v^2 w^2 + w^2 - v^2) \tag{13.3}$$

$\bar{g}_{00}$ peut changer de signe.

En appliquant les formules du paragraphe précédent, on obtient le théorème suivant qui donne la loi de propagation de la lumière dans l'espace.

Théorème. - Si le mouvement du fluide est permanent et tel que $\bar{g}_{00} \neq 0$, les rayons lumineux dans l'espace sont les extrémales de l'intégrale

$$\int_{x_0}^{x_1} \Lambda\, du = \int_{x_0}^{x_1} \left[\varepsilon\varepsilon' \sqrt{- \frac{1}{\bar{g}_{00}}\, \hat{g}_{ij}\, \dot{x}^i \dot{x}^j} - \frac{\bar{g}_{0i}\, \dot{x}^i}{\bar{g}_{00}} \right] du \tag{13.4}$$

où $\dot{x}^i = \frac{dx^i}{du}$, pour des variations à extrémités fixes dans V_3. Le temps mis par un rayon pour aller du point x_0 au point x_1 est donné par

$$\int_{x_0}^{x_1} dt = \int_{x_0}^{x_1} \left[\varepsilon\varepsilon' \sqrt{- \frac{1}{\bar{g}_{00}}\, \hat{g}_{ij}\, \dot{x}^i \dot{x}^j} - \frac{\bar{g}_{0i}\, \dot{x}^i}{\bar{g}_{00}} \right] du \tag{13.5}$$

Il est extrémum .

Dans le cas $\bar{g}_{00} = 0$, on a

$$\int_{x_0}^{x_1} \Lambda\, du = \int_{x_0}^{x_1} - \frac{\bar{g}_{ij}\, \dot{x}^i \dot{x}^j}{2\bar{g}_{0i}\, \dot{x}^i}\, du \tag{13.6}$$

$$\int_{x_0}^{x_1} dt = \int_{x_0}^{x_1} - \frac{\bar{g}_{ij}\, \dot{x}^i \dot{x}^j}{2\bar{g}_{0i}\, \dot{x}^i}\, du \tag{13.7}$$

Il est clair que les résultats ne dépendent pas de la variable auxiliaire u . D'autre part si l'espace-temps est statique orthogonal et si les lignes de courant coincident avec les lignes de temps, on a la métrique d'univers

$$g = U dx^0 \otimes dx^0 + g_{ij} \, dx^i \otimes dx^j$$

et la métrique associée

$$\bar{g} = \frac{U}{n^2} dx^0 \otimes dx^0 + g_{ij} \, dx^i \otimes dx^j$$

où $n^2 = \lambda \mu$. On peut alors mettre (13.5) sous la forme

$$\int_{x_0}^{x_1} dt = \int_{x_0}^{x_1} n \sqrt{U} \, d\sigma$$

où $d\sigma^2 = g_{ij} \, dx^i \, dx^j$ est l'élément linéaire de $(V_3, \bar{g})$. Dans le cas d'un espace temps plat $U = 1$, le théorème précédent se traduit par

$$\delta \int_{x_0}^{x_1} n d\sigma = 0$$

C'est l'énoncé du principe de Fermat en optique classique. Ce théorème que nous avons démontré constitue l'énoncé du principe de Fermat en relativité générale , dans le cas où le fluide est en mouvement. Par ce théorème se trouve également démontrée l'équivalence entre le principe d'action et le principe du moindre temps.

14. Application: loi relativiste de la composition des vitesses.

Plaçons nous dans l'espace temps de Minkowski rapporté à des coordonnées orthonormales. u^{α} est le vecteur vitesse unitaire d'univers dont les composantes sont déterminées classiquement à partir de la vitesse d'espace $\vec{\beta}$ si c est prise comme unité. Un calcul facile donne la métrique associée que nous écrivons sous la forme

$$(14.1) \qquad d\bar{s}^2 = \frac{V^2 - \beta^2}{1 - \beta^2}(dx^0)^2 + 2\,\frac{1 - V^2}{1 - \beta^2}\,\beta_i\, dx^0 dx^i - \sum_i (dx^i)^2 + \frac{1 - V^2}{1 - \beta^2}\,(\beta_i dx^i)^2$$

Cette métrique est du type hyperbolique normal. Il est à noter un changement de l'ordre dans la signature au passage de $V^2 = \beta^2$. On le met en évidence en choisissant l'axe x^1 parallèle à $\vec{\beta}$ (vitesse du fluide). On a ainsi

$$(14.2) \qquad d\bar{s}^2 = \frac{V^2 - \beta^2}{1 - \beta^2}(dx^0)^2 + 2\,\frac{(1 - V^2)}{1 - \beta^2}\,dx^0 dx^1 - \frac{1 - V^2}{1 - \beta^2}(dx^1)^2 - (dx^2)^2 - (dx^3)^2.$$

que l'on peut mettre sous la forme canonique par une décomposition en carrés. Si $V^2 \neq \beta^2$, on obtient

$$d\bar{s}^2 = \frac{1 - \beta^2}{V^2 - \beta^2}\left[\frac{V^2 - \beta^2}{1 - \beta^2}\,dx^0 + \frac{(1 - V^2)\beta}{1 - \beta^2}\,dx^1\right]^2 - \frac{(1 - \beta^2)V^2}{V^2 - \beta^2}(dx^1)^2 -$$

$+ (dx^2)^2 - (dx^3)^2$.

et l'on voit que pour $V^2 > \beta^2$ on a la signature + - - - et pour $V^2 < \beta^2$ la signature - + - - . Pour $V^2 = \beta^2$, on obtient

$$d\bar{s}^2 = 2V dx^0 dx^1 - (1+V^2)(dx^1)^2 - (dx^2)^2 - (dx^3)^2$$

qui reste de signature + - - -

A partir de la métrique associée (14. 1) cherchons à exprimer le théorème en prenant llarc σ du rayon comme paramètre. On a à remplacer dans (13. 5) $\dot{x}^i$ par $\lambda^i = dx^i / d\sigma$ où $d\sigma^2 = (dx^1)^2 + (dx^2)^2 + (dx^3)^2$. On en tire ainsi:

$$\frac{dt}{d} = \frac{1}{W} = \varepsilon\varepsilon' \sqrt{\frac{1-\beta^2}{V^2-\beta^2} \quad V^2-\beta^2 + (1-V^2)\,(\beta_i\lambda^i)^2} - \frac{(1-V^2)\,(\beta_i\lambda^i)}{V^2-\beta^2}$$

Si $V^2-\beta^2 \neq 0$, cette relation donne

$$1-\beta^2 - (1-\beta^2)\,W^2 - (1-V^2)\,(1-W\beta_i\lambda^i)^2 = 0$$

Si on interprète $\vec{V}$ comme vitesse absolue et $\vec{W}$ comme vitesse relative de propagation de la lumière , on a manifestement

$$(14.3) \qquad \vec{V}^2 = \frac{1}{(1+\vec{W}\cdot\vec{\beta})^2}\left[\vec{W}^2 + \vec{\beta}^2 + 2\vec{W}\cdot\vec{\beta} + (\vec{W}\cdot\vec{\beta})^2 - \vec{W}^2\vec{\beta}^2\right]$$

On vérifie par un calcul direct que cette relation reste valable dans le cas $V^2 = \beta^2$. C'est la formule relativiste de la composition des vitesses. Il est aisé de vérifier qu'on peut la mettre sous la forme

Pham

$$\vec{V} = \frac{1}{1+\vec{W_0}\vec{\beta}} \left[\left(1+ \frac{\vec{W_0}\vec{\beta}}{\beta^2}\right)\vec{\beta} + \sqrt{1-\beta^2}\left(\vec{W} - \frac{\vec{W}\cdot\vec{\beta}}{\vec{\beta}^2}\vec{\beta}\right)\right]$$

Nous obtenons ainsi à partir du principe de Fermat une démonstra - tion de la loi relativiste de composition des vitesses.

Pham

BIBLIOGRAPHIE

[1] . M. ABRAHAM - R.C. Circ. Mat. Palermo 22 (1909), 27-35

[2] . N.L. BALAZS - The propagation of light rays in moving media. Jour. Optical Soc. Amer. 45 (1955)

[3] . C. CATTANEO - C.R. Ac. Sc. Paris, 247 (1958), 431-433

[4] . Y. CHOQUET-BRUHAT - Fluides relativistes de conductivité in finie, Astronautica Acta, 6 (1960), 354-365

[5] . Y. CHOQUET-BRUHAT - Etude des équations des fluides chargés relativistes inductifs et conducteurs, Comm. Math. Phys. 3 (1966), 334-357

[6] .. M. KRANYS - Il Nuovo Cimento, 50B, (1967), 48-63

[7] . LANDAU - LIFCHITZ - Fluid mechanics , Pergamon, London (1958)

[8] . J. LERAY - Hyperbolic differential equations - Inst. for Advanced Studies, Princeton (1953) , notes miméographinés

[9] . A. LICHNEROWICZ - Théories relativistes de la gravitation et de l'électromagnétisme, Masson, Paris (1955)

[10] . A. LICHNEROWICZ - Relativistic hydrodynamics and magnetohydro dynamics, Benjamin (1967)

[11] B. MAHJOUB - CR. Ac. Sc. , Paris, 247, (1968), 668-671 et 268A (1969), 1440-1442.

[12] . C. MARLE - Sur l'établissement des équations de l'hydrodynamique des fluides relativistes dissipatifs. Ann. Inst. Henri Poincaré vol. X, n. 2, (1969), 127-194

[13] . C. ECKART - The thermodynamics of irreversible process, Phys. Rev. 58, (1940)

[14] .. PHAM MAU QUAN - Sur une théorie relativiste des fluides thermodynamiques, Ann. Mat. Pura e appl. (4), 38 (1955)

[15] . PHAM MAU QUAN - Etude électromagnétique et thermodynamique d'un fluide relativiste chargé, J. Mech. Anal. 5 (1956), 473-538

[16] . PHAM MAU QUAN - Inductions électromagnétiques en relativité générale et principe de Fermat, Arch. Rat. Mech. Anal., 1, (1957)

[17] . PHAM MAU QUAN - Thermodynamique d'un fluide relativiste, Boll. U. M. I. (1960) (3) vol 15, 105-118.

[18] . PHAM MAU QUAN - C. R. Ac. Sc. , Paris, 261 (1965) 3049-3052

[19] . PHAM MAU QUAN - Magnétohydrodynamique relativiste, Ann. Inst. Henri Poincaré 2 (1965), 21-85

[20] . PHAM MAU QUAN - Sur les équations des fluides chargés inductifs en Relativité générale - Rendiconti di Matematica, 1. 2; vol. 2 Serie VI (1969)

[21] . G. PICHON - Etude relativiste des fluides visqueux et chargés, Ann. . Inst. Henri Poincaré, 2 (1965). 21-85

[22] . AH. TAUB - Relativistic hydrodynamics, Arch. Rat. Mech. Anal. 3 (1959) et in Lectures in Applied Mathematics, 8 (1967)

CENTRO INTERNAZIONALE MATEMATICO ESTIVO

(C. I. M. E.)

ONDES DES CHOC, ONDES INFINITESIMALES ET RAYONS EN HYDRODYNAMIQUE ET MAGNETOHYDRODYNAMIQUE RELATIVISTES

A. LICHNEROWICZ

Corso tenuto a Bressanone dal 7 al 16 Giugno 1970

ONDES DES CHOC, ONDES INFINITESIMALES ET RAYONS EN HYDRODYNAMIQUE ET MAGNETOHYDRODYNAMIQUE RELATIVISTES

Introduction.

On sait l'importance mathématique et physique prise récemment par la magnetohydrodynamique relativiste. Dans ce cadre, on s'est proposé d'étudier en détail les ondes infinitesimales et les ondes de choc. L'extension relativiste des conditions de compressibilité de Hermann Weil joue dans la théorie développée un rôle important.

On a introduit systématiquement un instrument mathématique commode (les tenseurs-distributions) par l'analyse des différentes ondes. Cet instrument est d'abord appliqué en hydrodynamique, à l'étude des ondes soniques et des rayons correspondants.

En passant au cadre de la magnétohydrodynamique, on étudie ensuite successivement le système différentiel fondamental, la structure des ondes magnétosoniques et des ondes d'Alfven, les rayons correspondants, enfin les ondes de choc qui font l'objet d'une analyse détaillée. On établit en particulier grâce à l'introduction d'une fonction d'Hugoniot convenable, un important théorème d'existence et d'unicité par les solutions non triviales des équations de choc.

En traitant les ondes magnétosoniques par le formalisme des ondes de choc, on met en évidence la non-invariance de la direction des rayons par l'opérateur de discontinuité infinitésimale.

Ces leçons qui se suffisent à elles-mêmes représent une synthèse de certains de mes travaux durant la période 1966-69

I - Tenseurs-distributions

II - Hydrodynamique relativiste

III - Les équations de la magnétohydrodynamique relativiste

Lichnerowicz

IV - Ondes de choc en magnétohydrodynamique

V - Fonction d'Hugoniot et orientation des ondes de choc

VI - Ondes de choc et ondes d'Alfven

VII - Vitesses des ondes de choc et théorèmes fondamentaux

VIII - Retour aux rayons magnétosoniques.

I. TENSEURS-DISTRIBUTIONS ET DISCONTINUITES

1. Tenseurs-distributions sur une variété riemannienne.

a) Soit V_n une variété différentiable orientée, de dimension n et classe C^{h+1} $(h \geqslant 0)$; nous disposons sur V_n d'une métrique riemannienne ds^2 de signature arbitraire et de classe cC^h $(0 \leqslant k \leqslant h)$. Localement:

$$ds^2 = g_{\alpha\beta}\, dx^\alpha\, dx^\beta \qquad (\alpha, \beta = 0, 1, \dots, n-1).$$

Si T et U sont deux p-tenseurs, nous notons $(T, U)_x$ le produit scalaire de T et U au point x de V_n. En coordonnées locales:

$$(T, U)_x = T_{\alpha_1 \dots \alpha_p}(x) U^{\alpha_1 \dots \alpha_p}(x)$$

Soit $D(p, V_n)$ l'espace des p-tenseurs à support compact de classe C^h sur V_n. Si T est un p-tenseur localement sommable arbitraire, nous pouvons poser pour $U \in D(p, V_n)$:

(1.1) $$\langle T, U \rangle = \int_{V_n} (T, U)_x\, \eta(x)$$

où y est l'élément de volume riemannien de la variété. Localement:

(1.2) $$\eta = \sqrt{|g|}\, dx^0 \wedge \dots \wedge dx^{n-1}.$$

Un p-tenseur-distribution T de V_n est une forme linéaire continue, à valeur scalaires, sur l'espace $D(p, V_n)$. Continu est ici entendu au sens usuel en théorie des distributions. Si $U \in D(p, V_n)$, nous désignons par $T[U]$

ou $\langle T, U \rangle$ la valeur pour U du tenseur-distribution T.

Un p-tenseur ordinaire localement sommable T de V_n peut être identifié avec un tenseur-distribution au moyen de la formule (1.1). Le tenseur-distribution est noté T^D, ou parfois T par abus de notation, quand aucune confusion n'est possible.

b) Si T est un scalaire-distribution et V un p-tenseyr ordinaire, TV est le p-tenseur distribution défini naturellement par:

$$TV[U] = T[(V, U)]$$

Cela posé, soit Ω le domaine d'un système de coordonnées locales (x^α) et donnons-nous dans Ω, n^p scalaires distributions $T_{\alpha_1 \dots \alpha_p}$. L'expression

$$T = T_{\alpha_1 \dots \alpha_p} dx^{\alpha_1} \otimes \dots \otimes dx^{\alpha_p} \tag{1.3}$$

définit un p-tenseur-distribution: si (e_α) est le repère dual du corepère dx^α:

$$U = U^{\alpha_1 \dots \alpha_p} e_{\alpha_1} \otimes \dots \otimes e_{\alpha_p}$$

et $T[U]$ est donné par la somme:

$$T[U] = T_{\alpha_1 \dots \alpha_p} [U^{\alpha_1 \dots \alpha_p}]$$

Inversement tout p-tenseur distribution T dans Ω peut être représenté par une telle expression: désignons par $T_{\alpha_1 \dots \alpha_p}$ les scalaires-distributions définis par:

$$T_{\alpha_1 \ldots \alpha_p}(f) = T\left[f\, e_{\alpha_1} \otimes \ldots \otimes e_{\alpha_p}\right] \qquad f \in D(0, \Omega)$$

On a:

$$T[U] = T\left[U^{\alpha_1 \ldots \alpha_p}\right] e_{\alpha_1} \ldots e_{\alpha_p} = T_{\alpha_1 \ldots \alpha_p} U^{\alpha_1 \ldots \alpha_p}$$

et T admet bien l'expression (1. 3). Ainsi, dans le domaine d'un système de coordonnées locales, tout p-tenseur-distribution peut être, comme un p-tenseur ordinaire, rapporté à ces coordonnées, les composantes étant des scalaires-distributions. Nous supposons maintenant $h, k \geqslant 2$.

c) Soit T un p-tenseur ordinaire, ∇T sa dérivée covariante dans la connexion riemannienne. Si $U \in D(p+1, V_n)$

$$\langle \nabla T, U \rangle = \int_{V_n} \nabla_\rho T_{\alpha_1 \ldots \alpha_p} U^{\rho\alpha_1 \ldots \alpha_p} \eta$$

soit, par intégration par parties:

$$\langle \nabla T, U \rangle = \int_{V_n} \nabla_\rho (T_{\alpha_1 \ldots \alpha_p} U^{\rho\alpha_1 \ldots \alpha_p}) \eta -$$

$$- \int_{V_n} T_{\alpha_1 \ldots \alpha_p} \nabla_\rho U^{\rho\alpha_1 \ldots \alpha_p} \eta$$

Le premier terme du second membre est nul. On introduit ainsi l'opérateur $\underline{\delta}$ de codérivation sur les (p+1)-tenseurs, définis par:

$$\underline{\delta} : U_{\beta_1 \ldots \beta_{p+1}} \longrightarrow - \nabla_\rho U^\rho{}_{\alpha_1 \ldots \alpha_p}$$

Ainsi la formule précédente s'écrit:

(1.4) $$\langle \nabla T, U\rangle = \langle T, \underline{\delta} U\rangle$$

Cela posé, on définit naturellement la dérivée covariante d'un p-tenseur distribution T comme le (p+1)-tenseur distribution ∇T déterminé par la relation (1.4) où $U \in D(p, V_n)$. Il est aisé de voir que toutes les propriétés classiques de la dérivée covariante dans une connexion riemannienne, ainsi que les formules correspondantes, demeurent valables pour les tenseurs-distributions.

2. Les distributions Y^+, Y^- et $\bar{\delta}$ relatives à une hypersurface.

a) Nous considérons maintenant exclusivement un domaine Ω de V_n correspondant à des coordonnées locales. Soit Σ une hypersurface régulière définie par l'équation locale $\varphi = 0$ (φ de classe C^2) qui partage Ω en deux domaines Ω_o et Ω_1 correspondant respectivement à $\varphi < 0$ et $\varphi > 0$. Nous notons par $l \neq 0$ le gradient de φ.

Soit Y^o_φ, ou plus brièvement Y^o (resp Y^1_φ ou (Y^1) la fonction sur Ω qui vaut 1 (resp 0) dans Ω_o et 0 (resp 1) dans Ω_1; ces fonctions définissent dans Ω des distributions désignées par les mêmes notations, par l'intermédiaire des formules:

(2.1) $$\langle Y^o, f\rangle = \int_\Omega Y^o f\eta = \int_{\Omega_o} f\eta \qquad \langle Y^1, f\rangle = \int_\Omega Y^1 f\eta = \int_{\Omega_1} f\eta \quad (f \in D(0, \Omega))$$

b) Considérons la classe des (n-1)-formes covérifiant la relation:

(2.2) $$\eta = d\varphi \wedge \omega = l \wedge \omega$$

Si ω et ω' sont deux formes de cette classe, il existe une (n-2)-forme μ telle que $\omega' = \omega + d\varphi \wedge \mu$. Soit $\partial\Omega_0$ et $\partial\Omega_1$ les bords orientés sur Σ de Ω_0 et Ω_1 ($\partial\Omega_0 = -\partial\Omega_1$). D'après la remarque précédente, l'intégrale:

$$\int_{\partial\Omega_0} f\omega = -\int_{\partial\Omega_1} f\omega$$

a une valeur bien déterminée, independante du choix de ω dans la classe envisagée. Nous pouvons ainsi définir un scalaire-distribution $\bar{\delta}_\varphi$, ou plus bièrèvement $\bar{\delta}$, par la relation:

$$\text{(2.3)} \qquad \langle \bar{\delta}, f \rangle = \int_{\partial\Omega_0} f\omega = -\int_{\partial\Omega_1} f\omega \qquad f \in D(0,\Omega)$$

$\bar{\delta}_\varphi$ est la <u>mesure de Dirac</u> relative à φ ; son support est porté par Σ.

c) Proposons-nous d'évaluer les tenseurs-distributions dérivés des scalaires-distributions Y^0 et Y^1. A cet effet, introduisons dans Ω un système de coordonnées (y^α) tel que $y^0 = \varphi$; dans ce système l a pour composantes $l'_0 = 1$, $l_i = 0$ (i, j = 1, 2, ..., n-1). De l'expression de η on déduit que la (n-1)-forme

$$\omega = \sqrt{|g|}\, dy^1 \wedge \ldots \wedge dy^{n-1}$$

vérifie (2.2). Si $U \in D(1,\Omega)$, on a:

$$\langle \nabla Y^1, U \rangle = \langle Y^1, \underline{\delta} U \rangle$$

soit

$$\langle \nabla Y^1, U \rangle = -\langle Y^1, \frac{1}{\sqrt{|g|}} \partial_\alpha (U^\alpha \sqrt{|g|}) \rangle =$$

$$= -\int_{\Omega_1} \partial_\alpha (U^\alpha \sqrt{|g|}) dy^o \wedge dy^1 \wedge \ldots \wedge dy^{n-1}$$

soit encore, d'après la formule de Stockes:

$$\langle \nabla Y^1, U \rangle = -\int_{\partial \Omega_1} U^o \sqrt{|g|}\, dy^1 \wedge \ldots \wedge dy^{n-1} = -\int_{\partial \Omega_1} 1_\alpha U^\alpha \omega = \langle 1 \bar{\delta}, U \rangle$$

En raisonnant sur Y^o, on obtient ainsi:

(2.4) $$\nabla Y^1 = 1 \bar{\delta} \qquad \nabla Y^o = -1 \bar{\delta}$$

On vérifie de même qu'il existe un scalaire-distribution $\bar{\delta}'$ de support Σ tel que:

(2.5) $$\nabla \bar{\delta} = 1 \bar{\delta}'$$

3. Tenseurs-discontinuités à la traversée d'une hypersurface.

a) Considérons un p-tenseur T sur Ω satisfaisant les hypothèses suivantes:

A_1) Sur chacun des domaines Ω_o et Ω_1, le tenseur T est un tenseur ordinaire de classe C^1.

A_2) Quand φ tend vers zéro par valeurs négatives (resp. positives), T et ∇T convergent uniformément vers des fonctions à valeurs tensorielles définies sur Σ et notées T_o, $(\nabla T)_o$ (resp. T_1, $(\nabla T)_1$).

Nous introduisons les tenseurs-discontinuités sur Σ :

$$[T] = T_1 - T_o \qquad [\nabla T] = (\nabla T)_1 - (\nabla T)_o$$

Dans Ω , se trouvent définis de manière naturelle les tenseurs distributions Y^oT, $Y^o\nabla T$, Y^1T, $Y^1\nabla T$. Si T^D est le tenseur-distribution défini par le tenseur T défini presque partout dans Ω , on a en termes de distributions:

(3.1) $$T^D = Y^oT + Y^1T$$

Le tenseur-distribution ∇T^D, dérivée au sens des distributions de T^D, s'écrit:

$$\nabla T^D = \nabla(Y^oT) + \nabla(Y^1T)$$

avec:

$$\nabla(Y^oT) = -1\bar{\delta} \otimes T_o + Y^o\nabla T \qquad \nabla(Y^1T) = 1\bar{\delta} \otimes T_1 + Y^1\nabla T$$

On en déduit, compte tenu de (3.1):

(3.2) $$\nabla T^D = 1\bar{\delta} \otimes [T] + (\nabla T)^D$$

où $(\nabla T)^D$ est le tenseur distribution défini par le tenseer ordinaire défini presque partout ∇T, dérivée covariante usuelle du tenseur T.

b) Etudions la dérivée du tenseur-distribution $\bar{\delta}[T]$ de Ω . En coordonnées (y^α), on a:

$$\nabla_i(\bar{\delta}[T]) = \nabla_i \bar{\delta}\,[T] + \bar{\delta}\,\nabla_i[T]$$

où $\nabla_i \bar{\delta} = l_i \bar{\delta}' = 0$. D'après les hypothèses de convergence uniforme $\nabla_i[T] = [\nabla_i T]$. Ainsi l'on a $\bar{\delta}[\nabla_i T] = \nabla_i(\bar{\delta}[T])$. Il en résulte qu'il existe un p-tenseur distribution δT, à support sur Σ, tel que l'on ait la formule:

$$\bar{\delta}[\nabla T] = \nabla(\bar{\delta}[T]) + l \otimes \delta T \tag{3.3}$$

c) Nous considérons maintenant des tenseurs satisfaisant toujours aux hypothèses A_1, A_2 mais qui sont supposés continus dans Ω. Ces tenseurs définissent d'une manière naturelle une algèbre $\mathcal{A}$ de tenseurs. La formule (3. 3) devient alors:

$$\bar{\delta}[\nabla_\alpha T] = l_\alpha \nabla T \tag{3.4}$$

Considérons l'application:

$$\delta : T \in \mathcal{A} \rightarrow \delta T$$

où δT est un tenseur-distribution à support sur Σ. On déduit de (3. 4) que l'application δ est une dérivation: si a et b sont deux réels et si $T, U \in \mathcal{A}$ sont deux p-tenseurs, il résulte de (3. 4):

$$\delta(aT + bU) = a\,\delta T + b\,\delta U$$

Si $T, U \in \mathcal{A}$ sont respectivement un p-tenseur et un q-tenseur, on a:

$$\delta(T \otimes U) = \delta T \otimes U + T \otimes \delta U$$

δT est appelé la <u>discontinuité infinitésimale</u> de T et δ <u>l'opérateur de discontinuité infinitésimale.</u>

4. <u>Formule concernant les dérivées secondes.</u>

Plaçons-nous maintenant dans les hypothèses suivantes:

B_1) <u>Le tenseur T est continu sur Ω. Sur chacun des domaines Ω_0 et Ω_1, T est un tenseur de classe C^2</u>

B_2) <u>Quand φ tend vers zéro par valeur négatives (resp. positives), ∇T et $\nabla\nabla T$ convergent uniformément vers des fonctions à valeurs tensorielles définies sur Σ et notées $(\nabla T)_0$ $(\nabla\nabla T)_0$ (resp. $(\nabla T)_1$, $(\nabla\nabla T)_1$).</u>

Ainsi le tenseur ∇T satisfait lui-même aux hypothèses A_1, A_2.

De (3.3) appliqué à ∇T, il résulte:

$$[\nabla_\alpha \nabla_\beta T] = \nabla_\alpha(\delta [\nabla_\beta T])+ 1 \delta \nabla_\beta T$$

On en déduit d'après (3.4):

$$[\nabla_\alpha \nabla_\beta T] = \nabla_\alpha(1_\beta \delta T)+1_\alpha \delta \nabla_\beta T$$

soit:

$$[\nabla_\alpha \nabla_\beta T] = \nabla_\alpha 1_\beta \delta T + 1_\beta \nabla_\alpha \delta T + 1_\alpha \delta \nabla_\beta T \tag{4.1}$$

La métrique étant C^2, le tenseur de courbure est continue à la traversée de Σ, Il en résulte d'après l'identité de Bianchi

$$[\nabla_\alpha \nabla_\beta T] = [\nabla_\beta \nabla_\alpha T]$$

Comme $\nabla_\alpha l_\beta = \nabla_\beta l_\alpha = \nabla_\alpha \nabla_\beta \varphi$, on déduit de (4.1)

$$l_\beta \nabla_\alpha \delta T + l_\alpha \delta \nabla_\beta T = l_\alpha \nabla_\beta \delta T + l_\beta \delta \nabla_\alpha T$$

soit:

$$(4.2) \qquad l_\alpha (\delta \nabla_\beta T - \nabla_\beta \delta T) - l_\beta (\delta \nabla_\alpha T - \nabla_\alpha \delta T) = 0$$

On déduit de (4.2) qu'il existe un p-tenseur-distribution T a support sur Σ tel que:

$$\delta . \nabla_\beta T - \nabla_\beta \delta T = l_\beta \bar{T}$$

En substituant dans (4.1), on obtient une formule utile:

$$(4.3) \qquad [\nabla_\alpha \nabla_\beta T] = \nabla_\alpha l_\beta \delta T + l_\alpha \nabla_\beta \delta T + l_\beta \nabla_\alpha \delta T + l_\alpha l_\beta \bar{T}$$

II. HYDRODYNAMIQUE RELATIVISTE ET HYPOTHESES DE COMPRESSIBILITE.

5. Fluide parfait thermodynamique.

a) Soit V_4 un espace-temps muni d'une métrique hyperbolique ds^2, de signature +---, satisfaisant aux hypothèses usuelles de différentiabilité. En coordonnées locales $ds^2 = g_{\alpha\beta} dx^\alpha dx^\beta$ ($\alpha, \beta = 0, 1, 2, 3$). Dans un domaine de

V_4, un fluide parfait est décrit par le tenseur d'énergie:

(5.1) $$T^{(f)}_{\alpha\beta} = (\rho + p) u_\alpha u_\beta - p g_{\alpha\beta}$$

où ρ est la densité d'énergie propre, p la pression et u_α le vecteur-vitesse unitaire du fluide, orienté vers le futur; ρ contient une densité de matière et une densité d'énergie interne. Nous posons:

$$\rho = c^2 r \left(1 + \frac{\varepsilon}{c^2}\right) \qquad r > 0$$

où r est la densité de matière du fluide et ε son énergie interne spécifique. Considérons le scalaire

$$\rho + p = c^2 r \left(1 + \frac{\varepsilon}{c^2} + \frac{p}{c^2 r}\right)$$

Nous posons

$$i = \varepsilon + \frac{p}{r} = \varepsilon + pV \qquad (\text{où } V = 1/r)$$

V est le volume spécifique et i l'enthalpie spécifique. A cette variable nous substituons la variable thermodynamique équivalente, appelée l'indice du fluide, défini par:

$$f = 1 + \frac{i}{c^2}$$

Avec ces notations, le tenseur d'énergie s'écrit:

(5.2) $$T^{(f)}_{\alpha\beta} = c^2 r f u_\alpha u_\beta - p g_{\alpha\beta}$$

b) La température propre Θ du fluide et son eutropie spécifique S peuvent être définies, comme en hydrodynamique classique, par la relation différentielle

$$\Theta \, dS = d\varepsilon + p\,dV = di - Vdp = c^2 df - Vdp \qquad (\Theta > 0)$$

Ainsi

(5. 3) $$c^2 df = Vdp + \Theta\, dS$$

En relativité la variable thermodynamique $\tau = fV$ (volume dynamique) joue un rôle important et se substitue le plus souvent au volume spécifique classique. Il est commode d'adopter p et S comme variables thermodynamiques de bases Nous considérons $\tau = \tau(p, S)$ comme une fonction donnée définissante, par le fluide envisagé une équation d'état

c) Le système différentiel de l'hydrodynamique relativiste est fourni par les considérations suivantes: nous supposons d'abord que la densité de matière (qui correspond au nombre spécifique de particules) est conservative. Si ∇ est l'opérateur de dérivation covariante:

(5. 4) $$\nabla_\alpha (r u^\alpha) = 0$$

D'autre part les équations de la dynamique relativiste sont fournies par la conservation du tenseur d'énergie:

(5. 5) $$\nabla_\alpha T^{(f)\alpha\beta} = 0$$

Nous allons transformer le système (5 4), (5. 5) en un système équivalent; (5. 5) s'écrit explicitement:

(5. 6) $$\nabla_\alpha (c^2 r f u^\alpha) u_\beta + c^2 r f u^\alpha \nabla_\alpha u_\beta - \partial_\beta p = 0$$

Par produit par u^{β}, il vient, compte-tenu duc caractère unitaire du vecteur-vitesse:

(5.7) $$\nabla_{\alpha}(c^2 rfu^{\alpha}) - u^{\alpha}\partial_{\alpha}p = 0$$

soit:

$$c^2 f\,\nabla_{\alpha}(ru^{\alpha}) + c^2 ru^{\alpha}\,\partial_{\alpha}f - u^{\alpha}\partial_{\alpha}p = 0$$

D'après (5.3) cette relation peut s'écrire:

x $$c^2 f\,\nabla_{\alpha}(ru^{\alpha}) + r\,\Theta\,u^{\alpha}\partial_{\alpha}S = 0$$

Ainsi (5.4) entraine l'équation dite de flot adiabatique:

(5.8) $$u^{\alpha}\partial_{\alpha}S = 0$$

En reportant (5.7) dans (5.6) on obtient le systeme différentiel aux lignes de courant:

(5.9) $$c^2 rfu^{\alpha}\nabla_{\alpha}u^{\beta} - (g^{\alpha\beta} - u^{\alpha}u^{\beta})\partial_{\alpha}p = 0$$

Le système (5.4), (5.5) est équivalent au système formé par (5.4), (5.8) et le système différentiel (5.9) aux lignes de courant.

6. Vitesse d'une hypersurface par rapport au fluide et ondes soniques.

a) Soit Σ une hypersurface régulière dans un domaine de V_4, d'équation $\varphi = 0$ (avec $l = d\varphi$). La vitesse v^{Σ} de l'hypersurface Σ par rapport au fluide, c'est-à-dire par rapport à la direction temporelle u, est donnée classiquement par la formule:

$$(6.1) \qquad \frac{(v^\Sigma)^2}{c^2} = y^\Sigma \qquad \text{avec} \qquad y^\Sigma = \frac{(u^\alpha l_\alpha)^2}{(u^\alpha l_\alpha)^2 - l^\alpha l_\alpha}$$

On voit que, quel que soit l, y^Σ est positif et que l'on a:

$$l^\alpha l_\alpha > 0 \Longleftrightarrow y^\Sigma > 1 \qquad l^\alpha l_\alpha = 0 \Longleftrightarrow y^\Sigma = 1$$

$$l^\alpha l_\alpha < 0 \Longleftrightarrow y^\Sigma < 1$$

b) Dans un domaine Ω où les variables thermodynamiques p, S et le vecteur-vitesse u sont continues, supposons les dérivées premières discontinues à la traversée de Σ, de façon que p, S, u^α vérifient les hypothèses du § 3, c. D'après (3.4), nous posons dans la suite:

$$\bar{\delta}[\nabla_\alpha p] = l_\alpha \delta p \qquad \bar{\delta}[\nabla_\alpha S] = l_\alpha \delta S \qquad \bar{\delta}[\nabla_\alpha u^\beta] = l_\alpha \delta u^\beta$$

Etudions à quelle condition l'une au moins des distributions δp, δS, δu^β est non nulle.

La relation (5.4) peut s'écrire, compte-tenu de (5.8),

$$(6.2) \qquad r \nabla_\alpha u^\alpha + r'_p u^\alpha \partial_\alpha p = 0 \qquad (\text{où } r = r(p, S))$$

De $\tau = fV$, on déduit en dérivant en p à S constant:

$$\tau'_p = f'_p V + f V'_p$$

De (5. 3) on déduit:

$$\frac{c^2\tau'_p}{V} = c^2 f'_p + c^2 f \frac{V'_p}{V} = V - c^2 f \frac{r'_p}{r} = V(1-c^2 fr'_p)$$

Nous introduisons dans la suite la quantité γ définie, à partir de l'équation d'état, par la relation:

$$c^2\tau'_p = - V^2(\gamma -1) \tag{6.3}$$

de telle sorte que $\gamma = c^2 fr'_p$. La relation (6. 2) peut s'écrire:

$$c^2 rf\ \nabla_\alpha u^\alpha + \gamma u^\alpha \partial_\alpha p = 0 \tag{6.2'}$$

En écrivant cette relation de part et d'autre de Σ et retranchant, il vient, après produit par $\bar{\delta}$:

$$c^2 rf\ \bar{\delta}\left[\nabla_\alpha u^\alpha\right] + \gamma u^\alpha\ \bar{\delta}\left[\nabla_\alpha\ p\right] = 0$$

ce qui s'écrit:

$$c^2 rf\ l_\alpha \delta u^\alpha + \gamma (u^\alpha l_\alpha) \delta p = 0 \tag{6.4}$$

On déduit de même de la relation (5. 8):

$$(u^\alpha l_\alpha)\delta S = 0 \tag{6.5}$$

Enfin le système (5. 9) donne:

$$c^2 rf(u^\alpha l_\alpha)\delta u^\beta - (g^{\alpha\beta} - u^\alpha u^\beta) l_\alpha \delta p = 0 \tag{6.6}$$

Pour $u^{\alpha} l_{\alpha}=0$, δS peut être non-nulle. On obtient ainsi les hypersurfaces engendrées par des lignes de courant, dites ondes d'eutropie (ou de matière). Leur vitesse par rapport au fluide est nulle.

Supposons $u^{\alpha} l_{\alpha} \neq 0$; on a $\delta S=0$ de (6.6) il résulte que si $\delta p=0$, on a $\delta u^{\beta}=0$. Nous sommes ainsi conduits d'après (6.4) à multiplier (6.6) par l_{β}. Il vient

$$c^2 rf(u^{\alpha} l_{\alpha}) l_{\beta} \delta u^{\beta} - (g^{\alpha\beta} - u^{\alpha} u^{\beta}) l_{\alpha} l_{\beta} \delta p = 0$$

soit d'après (6.4):

$$\left\{(g^{\alpha\beta} - u^{\alpha} u^{\beta}) l^{\alpha} l^{\beta} + \gamma (u^{\alpha} l_{\alpha})^2\right\} \delta p = 0$$

Ainsi si $\delta p \neq 0$, l'hypersurface Σ vérifie l'équation:

$$P(l) \equiv (g^{\alpha\beta} - u^{\alpha} u^{\beta}) l_{\alpha} l_{\beta} + \gamma (u^{\alpha} l_{\alpha})^2 = 0 \qquad (l_{\alpha} = \partial_{\alpha} \varphi) \tag{6.7}$$

qui est l'équation aux ondes soniques du fluide; (6.7) exprime que des ondes, qui sont naturellement des caractéristiques du système différentiel (5.4), (5.5) sont les hypersurfaces tangentes aux cônes du second degré définis par dualité à partir du tenseur:

$$h^{\alpha\beta} = g^{\alpha\beta} + (\gamma - 1) u^{\alpha} u^{\beta}$$

Si v est la vitesse des ondes soniques par rapport au fluide, on déduit de (6.1) et (6.7)

$$\frac{v^2}{c^2} = \frac{1}{\gamma}$$

Nous postulons dans la suite que $v < c$ (ou $\gamma > 1$), ce qui revient à postuler que les ondes soniques sont orientées dans le temps. Pour que v soit $< c$, il faut et il suffit d'après (6. 3) que τ'_p soit < 0 .

c) Pour $\gamma > 1$, le tenseur $h^{\alpha\beta}$ définit une forme quadratique de type hyperbolique normal; $h_{\alpha\beta}$ désigne le tenseur covariant inverse de $h^{\alpha\beta}$. La génératrice de contact de Σ, onde sonique, avec le cône est définie par le vecteur:

$$(6.8) \qquad N^{\beta} = \frac{1}{2}\,\frac{\partial P(l)}{\partial l_{\beta}} = h^{\alpha\beta} l_{\alpha} = l^{\beta} + (\gamma - 1)(u^{\alpha} l_{\alpha}) u^{\beta}$$

Soit v^{β} la composante tangente à Σ de la vitesse u^{β} du fluide:

$$u^{\beta} = v^{\beta} + \frac{u^{\alpha} l_{\alpha}}{l^{\alpha} l_{\alpha}}\, l^{\beta} \qquad (v^{\beta} l_{\beta} = 0)$$

Notons que, N^{β} étant tangent à Σ, <u>la direction de N^{β} est celle de v^{β}</u>:

$$(6.9) \qquad N^{\beta} = (\gamma - 1)(u^{\alpha} l_{\alpha}) v^{\beta}$$

7. <u>Propriété fondamentale des rayons en hydrodynamique.</u>

a) Les bicaractéristiques ou <u>rayons</u> associés aux ondes soniques (et qui les engendrent) sont les trajectoires du champ sur Σ des vecteurs $N^{\beta} = h^{\alpha\beta} l_{\alpha}$. Ce sont des géodésiques isotropes relatives à la métrique hyperbolique normale définie par $h_{\alpha\beta}$.

Il résulte de (6. 6) que, par une onde sonique Σ :

$$(7.1) \qquad \delta u^{\beta} = \frac{(g^{\alpha\beta} - u^{\alpha} u^{\beta}) l_{\alpha}}{c^2 r f u^{\alpha} l_{\alpha}}\, \delta p$$

s'exprime simplement en fonction de δp. Nous nous proposons de montrer que δp (et par suite les δu^{β}) se propage le long des rayons, c'est-à-dire que δp verifie un sistème differentiel de la forme:

$$N^{\beta} \nabla_{\beta} \delta p + A \, \delta p = 0$$

Nous postulons dans la suite les hypothèses B_1 et B_2 du § 4 pour p, S, u^{λ}. De la formule (4. 3) il résulte qu'il existe des distributions $\bar{p}, \bar{S}, \bar{u}^{\lambda}$ telles que:

$$\bar{\delta}\left[\nabla_{\alpha} \nabla_{\beta} p\right] = \nabla_{\alpha} l_{\beta} \delta p + l_{\alpha} \nabla_{\beta} \delta p + l_{\beta} \nabla_{\alpha} \delta p + l_{\alpha} l_{\beta} \bar{p} \tag{7.2}$$

$$\bar{\delta}\left[\nabla_{\alpha} \nabla_{\beta} S\right] = l_{\alpha} l_{\beta} \bar{S} \tag{7.3}$$

$$\bar{\delta}\left[\nabla_{\alpha} \nabla_{\beta} u\right] = \nabla_{\alpha} l_{\beta} \delta u^{\lambda} + l_{\alpha} \nabla_{\beta} \delta u^{\lambda} + l_{\beta} \nabla_{\alpha} \delta u^{\lambda} + l_{\alpha} l_{\beta} \bar{u}^{\lambda} \tag{7.4}$$

b) Partons de la relation (6. 2'), divisons la part $c^2 rf$ et dérivons-la. Il vient

$$\nabla_{\beta} \nabla_{\alpha} u^{\alpha} + \frac{\gamma}{c^2 rf} u^{\alpha} \nabla_{\beta} \nabla_{\alpha} p + \nabla_{\beta} \left(\frac{\gamma}{c^2 rf} u^{\alpha}\right) \nabla_{\alpha} p = 0$$

En écrivant cette relation de part et d'autre de Σ et retranchant, on obtient:

$$\left[\nabla_{\beta} \nabla_{\alpha} u^{\alpha}\right] + \frac{\gamma}{c^2 rf} u^{\alpha} \bar{\delta}\left[\nabla_{\beta} \nabla_{\alpha} p\right] + \bar{\delta}\left[\nabla_{\beta}\left(\frac{\gamma}{c^2 rf} u^{\alpha}\right) \nabla_{\alpha} p\right] = 0 \tag{7.5}$$

Or:

$$\bar{\delta}\left[\nabla_{\beta}\left(\frac{\gamma}{c^2 rf} u^{\alpha}\right) \nabla_{\alpha} p\right] = \left\{\nabla_{\alpha} p\right\}_1 \bar{\delta}\left[\nabla_{\beta}\left(\frac{\gamma}{c^2 rf} u^{\alpha}\right)\right] + \left\{\nabla_{\beta}\left(\frac{\gamma}{c^2 rf} u^{\alpha}\right)\right\}_0 \bar{\delta}\left[\nabla_{\alpha} p\right]$$

est une combinaison linéaire de termes en δp et δu^λ et est donc proportionnel à δp. Nous écrivons (7.5) sous la forme

(7.6) $$\bar{\delta}\left[\nabla_\beta \nabla_\alpha u^\alpha\right] + \frac{\gamma}{c^2 rf} u^\alpha \bar{\delta}\left[\nabla_\beta \nabla_\alpha p\right] \simeq 0$$

où $\simeq$ signifie modulo des termes proportionnels à δp.

En procédant de même sur la relation (5.8) relative à l'entropie, on a:

$$u^\alpha \bar{\delta}\left[\nabla_\alpha \nabla_\beta S\right] \simeq 0$$

soit d'après (7.3):

$$u^\alpha l_\alpha l_\beta \bar{S} \simeq 0$$

Ainsi $\bar{S}$ est $\simeq 0$ et il vient:

(7.7) $$\bar{\delta}\left[\nabla_\beta \nabla_\alpha S\right] \simeq 0$$

Enfin en se livrant au même raisonnement sur (5.9), il vient.

(7.8) $$c^2 rf\, u^\alpha \bar{\delta}\left[\nabla_\alpha \nabla_\beta u^\alpha\right] - (g^{\alpha\beta} - u^\alpha u^\beta)\, \bar{\delta}\left[\nabla_\alpha \nabla_\beta p\right] \simeq 0$$

En tenant compte de (7.6) dans (7.8) on obtient:

$$\gamma\, u^\alpha u^\beta\, \bar{\delta}\left[\nabla_\alpha \nabla_\beta p\right] + (g^{\alpha\beta} - u^\alpha u^\beta)\, \bar{\delta}\left[\nabla_\alpha \nabla_\beta p\right] \simeq 0$$

soit d'après (7.2):

$$h^{\alpha\beta}(l_\alpha \nabla_\beta \delta p + l_\beta \nabla_\alpha \delta p + l_\alpha l_\beta \overline{p}) \simeq 0$$

Or Σ étant sonique vérifie P(l)=0 et le terme en $\overline{p}$ disparait. Il reste:

$$h^{\alpha\beta} \; l_\alpha \nabla_\beta \delta p \simeq 0$$

soit:

(7.9) $$N^\beta \nabla_\beta \delta p \simeq 0$$

On a le théorème qui traduit la <u>propriété fondamentale des rayons</u>:

<u>Théorème</u>. <u>Sous les hypothèses du § 4, les discontinuités infinitésimales δp, δu^λ relatives à une onde sonique Σ si propagent le long des rayons associés selon les systèmes différentiels:</u>

$$N^\beta \nabla_\beta \delta p \simeq 0 \qquad\qquad N^\beta \nabla_\beta \delta u^\lambda \simeq 0$$

c) Nous avons vu que l'opérateur δ définit une dérivation; la métrique et Σ étant supposées de classe C^2 dans Ω , on a:

$$\delta g^{\alpha\beta} = 0 \qquad\qquad \delta l_\alpha = 0$$

On peut mettre les équations fondamentales relatives aux ondes soniques sous une forme commode.

En écrivant (5.4) de part et d'autre de Σ et retranchant, il vient immédiatement

$$\delta(ru^\alpha l_\alpha) = 0$$

Ainsi le scalaire:

$$a = ru^{\alpha} l_{\alpha} \tag{7.10}$$

est <u>invariant par la dérivation δ</u>

De même, en écrivant (5.5) de part et d'autre de Σ et retranchant, on obtient

$$\delta(T^{(f)\alpha\beta} l_{\alpha}) = 0$$

et le vecteur,:

$$W^{\beta} = c^2 rf(u^{\alpha} l_{\alpha}) u^{\beta} - pl^{\beta} = c^2 afu^{\beta} - pl^{\beta} \tag{7.11}$$

est <u>invariant par δ</u>. Ce vecteur peut être décomposé en la somme:

$$W^{\beta} = c^2 afv^{\beta} + \left(c^2 af \frac{u^{\alpha} l_{\alpha}}{l^{\alpha} l_{\alpha}} - p\right) l^{\beta} \tag{7.12}$$

et ses composantes tangentielles et normale par rapport à Σ sont invariantes. Ainsi:

$$\delta(fv^{\beta}) = 0$$

<u>En particulier la direction du rayon, qui est d'après (6.9) celle de v^{β}, est invariante par l'opérateur de discontinuité infinitésimale δ.</u>

8. Hypothèses de compressibilité.

a) J'ai été conduit à adopter par les fluides parfaits relativistes les hypothèses de compressibilité suivantes portant sur la fonction τ (p, S). Dans le domaine envisagé des variables p et S, on suppose que l'on a:

$$(H_1) \qquad \tau'_p < 0 \qquad \tau'_S > 0$$

et la condition de convexité:

$$(H_2) \qquad \tau''_{p^2} > 0$$

L'inégalité $\tau'_p < 0$ exprime que v est $< c$ ou que les ondes soniques sont orientées dans le temps. Les hypothèses (H_1) et (H_2) se réduisent à l'approximation classique aux hypothèses usuelles dites de Hermann-Weil et nous mõntrérons qu'elles jouent le même rôle qu'elles, pour la thermodynamique des ondes de chac en hydrodynamique et magnétohydrodynamique (relativistes). On vérifie facilement que $\tau'_p < 0$ implique les deux autres conditions par les gaz polytropiques relativistes. Dans des travaux encore partiellement inédits Israel et Lucquiaud en ont donné des justifications du point de vue de la mécanique statistique

b) En inversant la fonction $\tau = \tau(p, S)$, on obtient une fonction $p=p(\tau, S)$ exprimant la pression en fonction des variables τ et S. On a identiquement en τ et S

$$(8.1) \qquad p = p\left\{\tau(p, S), S\right\}$$

Nous nous proposons d'évaluer les dérivées partielles de $p(\tau, S)$. Par dérivation de (8.1) par rapport à p, à S constant, il vient:

(8.2) $$p'_{\tau}\, \tau'_{p} = 1$$

On en déduit:

(8.3) $$p'_{\tau} = \frac{1}{\tau'_{p}}$$

De même, par dérivation de (8.1) par rapport à S, à p constant, on a:

$$p'_{\tau}\, \tau'_{S} + p'_{S} = 0$$

Il en résulte:

(8.4) $$p'_{S} = -\frac{\tau'_{S}}{\tau'_{p}}$$

Les hypothèses de compressibilité (H_1) se traduisent denc par les inégalités $p'_{\tau} < 0$, $p'_{S} > 0$.

c) En dérivant (8.2) par rapport à p, à S constant, il vient:

$$p''_{\tau^2}\, (\tau'_{p})^2 + p'_{\tau}\, \tau''_{p^2} = 0$$

On en déduit:

(8.5) $$p''_{\tau^2} = -\frac{\tau''_{p^2}}{(\tau'_{p})^3}$$

et (H_2) se traduit, modulo (H_1), par $p''_{\tau^2} > 0$.

Nous n'étudierons pas, pour elles-mêmes, les ondes de choc de l'hydrodynamique relativiste, mais nous considérerons une telle étude comme un cas particulier de l'étude complète des ondes de choc de la magnétohydrodynamique, à laquelle nous procéderons.

III. LES EQUATIONS DE LA MAGNETOHYDRODYNAMIQUE RELATIVISTE.

9. Le tenseur d'énergie de la magnétohydrodynamique.

a) Supposons le fluide envisagé soumis à un champ électromagnétique décrit par l'ensemble de deux 2-tenseurs antisymétriques H et G; H est ici le tenseur champ électrique-induction magnétique et vérifie le premier groupe des équations de Maxwell $dH=0$ (où d désigne la différentiation extérieure). Si $*$ est l'opérateur d'adjonction sur les tenseurs antisymétriques, les vecteurs orthogonaux à u, donc spatiaux

$$e_\beta = u^\alpha H_{\alpha\beta} \qquad b_\beta = u^\alpha ({*}H)_{\alpha\beta}$$

sont respectivement le vecteur champ électrique et le vecteur induction magnétique relatifs à la direction temporelle u. Soit μ, constante donnée, la perméabilité magnétique du fluide. Le vecteur champ magnétique h est supposé relié à l'induction magnétique b par la relation:

$$b_\beta = \mu h_\beta$$

Le courant électrique J est sensiblement la somme de deux termes:

$$J^{\beta} = \nu u^{\beta} + \sigma e^{\beta}$$

où ν est la densité propre de charge électrique du fluide et σ sa conductivité.

b) La magnétohydrodynamique est ici l'étude des propriétés d'un fluide idéal relativiste de conductivité infinie $\sigma = \infty$; J étant essentiellement fini, il en est de même pour σ, e, et l'on a nécessairement e=0.

Par rapport à la direction temporelle définie par le vecterr-vitesse u du fluide, le champ électromagnérique est réduit au champ magnétique h. D'après des résultats classiques, ce champ admet le tenseur d'énergie:

$$\tau_{\alpha\beta} = \mu \left\{ |h|^2 \left(u_{\alpha} u_{\beta} - \frac{1}{2} g_{\alpha\beta}\right) - h_{\alpha} h_{\beta} \right\}$$

où $|h|^2 = -h^{\rho} h_{\rho}$ est strictement positif pour $h_{\rho} \neq 0$. Le tenseur d'énergie total fluide-champs s'en déduit:

$$T_{\alpha\beta} = (c^2 rf + \mu |h|^2) u_{\alpha} u_{\beta} - q g_{\alpha\beta} - \mu h_{\alpha} h_{\beta} \tag{9. 1}$$

où l'on a posé:

$$q = p + \frac{1}{2} \mu |h|^2$$

c) Le système différentiel fondamental de la magnétohydrodynamique relativiste est constitué par les équations suivantes: l'équation de conservation de la densité de matière:

$$\nabla_{\alpha}(r u^{\alpha}) = 0 \tag{9. 2}$$

les équations de Maxwell (dH=0) qui peuvent s'écrire ici:

(9.3) $$\nabla_\alpha (u^\alpha h^\beta - h^\alpha u^\beta) = 0$$

et les équations de la dynamique relativiste:

(9.4) $$\nabla_\alpha T^{\alpha\beta} = 0$$

où $T^{\alpha\beta}$ est donné par (q-1).

10. Conséquences du système fondamental.

Nous utiliserons dans la suite un certain nombre de relations, conséquences du système (9.2), (9.3), (9.4).

a) Partons des équations de Maxwell explicitées sous la forme:

(10.1) $$h^\beta \nabla_\alpha u^\alpha + u^\alpha \nabla_\alpha h^\beta - h^\alpha \nabla_\alpha u^\beta - u^\beta \nabla_\alpha h^\alpha = 0$$

et projetons-les successivement sur les directions définies par les deux vecteurs orthogonaux u et h. Par produit scalaire par u il vient, compte-tenu du caractère unitaire de u:

(10.2) $$u^\alpha u^\beta \nabla_\alpha h_\beta - \nabla_\alpha h^\alpha = 0$$

Par produit scalaire par h, on obtient:

$$\frac{1}{2} u^\alpha \nabla_\alpha |h|^2 + |h|^2 \nabla_\alpha u^\alpha + h^\alpha h^\beta \nabla_\alpha u_\beta = 0$$

soit, compte-tenu de l'orthogonalité de u et h:

(10.3) $$\frac{1}{2} u^\alpha \nabla_\alpha |h|^2 + |h|^2 \nabla_\alpha u^\alpha - h^\alpha u^\beta \nabla_\alpha h_\beta = 0$$

b) En explicitant (9.4), on a:

(10.4) $$\nabla_\alpha\left\{(c^2rf+\mu|h|^2)u^\alpha\right\} u_\beta +(c^2rf+\mu|h|^2)u^\alpha\nabla_\alpha u_\beta -$$
$$-\partial_\beta(p+\frac{1}{2}\mu|h|^2)-\mu\nabla_\alpha h^\alpha h_\beta -\mu h^\alpha\nabla_\alpha h_\beta =0$$

Par produit par u^β, on en déduit:

(10.5) $$\nabla_\alpha\left\{(c^2rf+\mu|h|^2)u^\alpha\right\} -u^\alpha\partial_\alpha(p+\frac{1}{2}\mu|h|^2)-\mu h^\alpha u^\beta\nabla_\alpha h_\beta =0$$

c'est-à-dire:

$$\nabla_\alpha(c^2rfu^\alpha)-u^\alpha\partial_\alpha p+\mu\left\{\frac{1}{2}u^\alpha\nabla_\alpha|h|^2+|h|^2\nabla_\alpha u^\alpha-h^\alpha u^\beta\nabla_\alpha h_\beta\right\} =0$$

qui, compte-tenu de (10.3), peut s'écrire

$$\nabla_\alpha(c^2rfu^\alpha)-u^\alpha\partial_\alpha p =0$$

relation indentique à (5.7). En faisant intervenir Θ et S, il vient encore:

$$(c^2f\nabla_\alpha(ru^\alpha)+r\,\Theta\, u^\alpha\partial_\alpha S =0$$

Ainsi (9.2) entraine toujours l'équation de flot adiabatique:

(10.6) $$u^\alpha\partial_\alpha S =0$$

En utilisant (10.6) paur transformer (10.4), on obtient le système différentiel aux lignes de courant:

$$(c^2 rf + \mu |h|^2) u^\alpha \nabla_\alpha u^\beta - (g^{\alpha\beta} - u^\alpha u^\beta)\, \partial_\alpha (p + \frac{1}{2}\mu\, |h|^2) +$$
$$+ \mu\, h^\lambda u^\mu \nabla_\lambda h_\mu u^\beta - \mu \nabla_\alpha h^\alpha h^\beta - \mu h^\alpha \nabla_\alpha h^\beta = 0$$

En développant, il vient:

$$(c^2 rf + \mu |h|^2) u^\alpha \nabla_\alpha u^\beta - (g^{\alpha\beta} - u^\alpha u^\beta)\, \partial_\alpha p - \frac{1}{2}\mu\, g^{\alpha\beta} \nabla_\alpha |h|^2 +$$
$$+ \frac{1}{2}\mu u^\alpha u^\beta \nabla_\alpha |h|^2 + \mu h^\lambda u^\mu \nabla_\lambda h_\mu u^\beta - \mu \nabla_\alpha h^\alpha h^\beta - \mu h^\alpha \nabla_\alpha h^\beta = 0$$

En tenant compte de (10. 3), on obtient:

$$(10.7) \qquad (c^2 rf + \mu |h|^2) u^\alpha \nabla_\alpha u^\beta - (g^{\alpha\beta} - u^\alpha u^\beta)\, \partial_\alpha p - \frac{1}{2}\mu g^{\alpha\beta} \nabla_\alpha |h|^2 +$$
$$+ \mu u^\alpha u^\beta \nabla_\alpha |h|^2 + \mu |h|^2 \nabla_\alpha u^\alpha u^\beta - \mu \nabla_\alpha h^\alpha h^\beta - \mu h^\alpha \nabla_\alpha h^\beta = 0$$

Par produit par h_β, on déduit de (10. 7):

$$(c^2 rf + \mu |h|^2) u^\alpha h^\beta \nabla_\alpha u_\beta - h^\alpha \partial_\alpha p - \frac{1}{2}\mu h^\alpha \nabla_\alpha |h|^2 + \mu |h|^2 \nabla_\alpha h^\alpha +$$
$$+ \frac{1}{2}\mu\, h^\alpha \nabla_\alpha |h|^2 = 0$$

soit, après simplification,

$$(c^2 rf + \mu |h|^2) u^\alpha u^\beta \nabla_\alpha h_\beta + h^\alpha \partial_\alpha p - \mu |h|^2 \nabla_\alpha h^\alpha = 0$$

Compte-tenu de (10. 2), on obtient la relation simple:

(10.8) $$c^2 rf \nabla_\alpha h^\alpha + h^\alpha \partial_\alpha p = 0$$

11. Ondes magnétosoniques et ondes d'Alfven.

a) Dans un domaine Ω où les variables thermodynamiques p, S et les vecteurs u^α, h^α sont continus, soit Σ une hypersurface d'équation $\varphi = 0$ (φ de classe C^2; $l = d\varphi$), à la traversée de laquelle les dérivées premières sont discontinues, de façon que p, S, u^α, h^α vérifient les hypothèses du § 3, c. Comme précédemment nous posons:

$$\bar{\delta}[\nabla_\alpha p] = l_\alpha \delta p \qquad \bar{\delta}[\nabla_\alpha S] = l_\alpha \delta S \qquad \bar{\delta}[\nabla_\alpha u^\beta] = l_\alpha \delta u^\beta$$

$$[\nabla_\alpha h^\beta] = l_\alpha \delta h^\beta$$

Etudions à quelle continion l'une au moins des distributions δp, δS, δu^β, δh^β est non nulle.

Le système différentiel (9.2), (9.3), (9.4) est équivalent au système différentiel formé par (9.2), (9.3), (10.6), (10.7). Les relations (9.2) et (10.6) donnent d'abord, exactement comme dans le cas de l'hydrodynamique:

(11.1) $$c^2 rf\, l_\alpha \delta u^\alpha + \gamma\, (u^\alpha l_\alpha) \delta p = 0$$

et

(11.2) $$(u^\alpha l_\alpha) \delta S = 0$$

Des équations de Maxwell (9.3) ou (10.1), on déduit par un raisonnement identique à celui du § 6:

$$u^{\beta} l_{\alpha} \, \delta h^{\alpha} + (h^{\alpha} l_{\alpha}) \delta u^{\beta} - (u^{\alpha} l_{\alpha}) \delta h^{\beta} - h^{\beta} l_{\alpha} \delta u^{\alpha} = 0 \tag{11.3}$$

Enfin le système différentiel aux lignes de courant donne de manière analogue:

$$(c^2 rf + \mu |h|^2)(u^{\alpha} l_{\alpha}) \, \delta u^{\beta} - (g^{\alpha\beta} - u^{\alpha} u^{\beta}) l_{\alpha} \delta p - \frac{1}{2} \mu l^{\beta} \delta |h|^2 + \tag{11.4}$$

$$+ \mu (u^{\alpha} l_{\alpha}) u^{\beta} \delta |h|^2 + \mu |h|^2 u^{\beta} l_{\alpha} \delta u^{\alpha} - \mu h^{\beta} l_{\alpha} \delta h^{\alpha} - \mu (h^{\alpha} l_{\alpha}) \delta h^{\beta} = 0$$

Pour $u^{\alpha} l_{\alpha} = 0$, δS peut être nulle et on obtient de nouveau les ondes de matière. Nous supposons dans la suite $u^{\alpha} l_{\alpha} \neq 0$ et par suite $\delta S = 0$.

b) Des relations (10. 2) et (10. 3), conséquences des équations de Maxwell, on déduit les relations suivantes conséquences de (11. 3):

$$l_{\alpha} \, \delta h^{\alpha} - (u^{\alpha} l_{\alpha}) u^{\beta} \, \delta \, h_{\beta} = 0$$

$$\frac{1}{2} (u^{\alpha} l_{\alpha}) \, \delta |h|^2 + |h|^2 l_{\alpha} \, \delta u^{\alpha} - (h^{\alpha} l_{\alpha}) \, u^{\beta} \, \delta \, h_{\beta} = 0$$

Par élimination de $u^{\beta} \delta h_{\beta}$ entre ces deux relations, il vient:

$$\frac{1}{2} (u^{\alpha} l_{\alpha})^2 \, \delta |h|^2 + |h|^2 (u^{\alpha} l_{\alpha}) l_{\beta} \, \delta u^{\beta} - (h^{\alpha} l_{\alpha}) l_{\beta} \delta h^{\beta} = 0 \tag{11.5}$$

Considérons maintenant la relation (10. 8), conséquence de (10. 7). On en déduit:

$$c^2 rf \, l_{\alpha} \delta h^{\alpha} + h^{\alpha} l_{\alpha} \delta p = 0 \tag{11.6}$$

Notons que comme $\delta S=0$, il résulte de (5.3) $\delta p=c^2 r\,\delta f$. Par suite (11.6) peut s'écrire:

$$f l_\alpha \delta h^\alpha + h^\alpha l_\alpha \delta f = 0\,0$$

Ainsi le scalaire $b=fh^\alpha l_\alpha$ est invariant par la dérivation δ

Enfin multiplions (11.4) par l_β. Il vient:

$$(c^2 rf+\mu|h|^2)(u^\alpha l_\alpha)\, l_\beta \delta u^\beta -(g^{\alpha\beta}-u^\alpha u^\beta) l_\alpha l_\beta \delta p - \frac{1}{2}\, l^\beta l_\beta \delta|h|^2 +$$

$$+\mu(u^\alpha l_\alpha)^2 \delta |h|^2 + \mu|h|^2(u^\alpha l_\alpha) l_\beta \delta u^\beta - 2\mu(h^\alpha l_\alpha) l_\beta \delta h^\beta = 0$$

c'est-à-dire:

$$c^2 rf(u^\alpha l_\alpha) l_\beta \delta u^\beta -(g^{\alpha\beta}-u^\alpha u^\beta) l_\alpha l_\beta \delta p - \frac{1}{2}\mu l^\beta l_\beta \delta|h|^2 +$$

$$+\mu\left\{2|h|^2(u^\alpha l_\alpha) l_\beta \delta u^\beta +(u^\alpha l_\alpha)^2 \delta|h|^2 - 2(h^\alpha l_\alpha) l_\beta \delta h^\beta\right\} = 0$$

D'après (11.5), on obtient:

$$c^2 rf(u^\alpha l_\alpha) l_\beta \delta u^\beta -(g^{\alpha\beta}-u^\alpha u^\beta) l_\alpha l_\beta \delta p - \frac{1}{2}\mu l^\beta l_\beta \delta|h|^2 = 0 \tag{11.7}$$

Eliminons $\delta|h|^2$ entre (11.5) et (11.7). Par produit de (11.5) par $\mu l^\rho l_\rho$, de (11.7) par $(u^\alpha l_\alpha)^2$ et addition, on obtient:

$$c^2 rf(u^\alpha l_\alpha)^{3\,3} l_\beta \delta u^\beta -(u^\alpha l_\alpha)^2 (g^{\alpha\beta}-u^\alpha u^\beta) l_\alpha l_\beta \delta p + \text{ll} \; 1$$

$$+\mu l^\rho l_\rho\left\{|h|^2(u^\alpha l_\alpha) l_\beta \delta u^\beta -(h^\alpha l_\alpha) l_\beta \delta h^\beta\right\} = 0$$

c'est-à-dire:

$$\left\{c^2 rf(u^\alpha l_\alpha)^2 + \mu|h|^2 l^\rho l_\rho\right\}(u^\alpha l_\alpha) l_\beta \delta u^\beta - (u^\alpha l_\alpha)^2 (l^\beta l_\beta - (u^\beta l_\beta)^2)\delta p -$$

$$- \mu h^\alpha l_\alpha l^\rho l_\rho l_\beta \delta h^\beta = 0$$

A cette relation, nous pouvons adjoindre (11.1), soit:

$$c^2 rf l_\beta \delta u^\beta + \gamma (u^\alpha l_\alpha) \delta p = 0$$

et (11.6) soit:

$$(h^\alpha l_\alpha) \delta p + c^2 rf l_\beta \delta h^\beta = 0$$

Le déterminant de ces trois équations linéaires aux inconnues $l_\beta \delta u^\beta$, $\delta p, l_\beta \delta h^\beta$ s'écrit:

$$H = \begin{vmatrix} \left\{c^2 rf(u^\alpha l_\alpha)^2 + \mu|h|^2 l^\rho l_\rho\right\}(u^\alpha l_\alpha) & -(u^\alpha l_\alpha)^2 (l^\beta l_\beta - (u^\beta l_\beta)^2) & -\mu h^\alpha l_\alpha l^\rho l_\rho \\ c^2 rf & \gamma (u^\alpha l_\alpha) & 0 \\ 0 & h^\alpha l_\alpha & c^2 rf \end{vmatrix}$$

On obtient par développement:

$$\frac{H}{c^2 rf} = \left\{c^2 rf(u^\alpha l_\alpha)^2 + \mu|h|^2 l^\rho l_\rho\right\}(u^\alpha l_\alpha)^2 \gamma + c^2 rf(u^\alpha l_\alpha)^2 (l^\beta l_\beta - (u^\beta l_\beta)^2) -$$

$$- \mu (h^\alpha l_\alpha)^2 l^\rho l_\rho$$

Pour que le système considéré admette des solutions autres quel la solution nulle, il faut et il suffit qhe H=0. S'il en est ainsi, les relations (11.1), (11.5) et (11.6) fournissent $l_\beta \delta u^\beta$, $l_\beta \delta h^\beta$, $\delta|h|^2$ en fonction de δp.

c) L'étude précédente concernait δp et les composantes normales à Σ de u^β et h^β. Décomposons ces vecteurs selon leurs composantes tangentielles et normales à Σ. Il vient:

$$u^\beta = v^\beta + \frac{u^\alpha l_\alpha}{l^\alpha l_\alpha} l^\beta \qquad h^\beta = t^\beta + \frac{h^\alpha l_\alpha}{l^\alpha l_\alpha} l^\beta$$

où $v^\beta l_\beta = t^\beta l_\beta = 0$. Compte-tenu du b, les formules (11.3) et (11.4) fournissent relations de la forme:

$$(11.9) \qquad (h^\alpha l_\alpha)\,\delta v^\beta - (u^\alpha l_\alpha)\,\delta t^\beta \simeq 0$$

$$(11.10) \qquad (c^2 rf + \mu |h|^2)(u^\alpha l_\alpha)\,\delta v^\beta - \mu (h^\alpha l_\alpha)\,\delta t^\beta \simeq 0$$

où le symbole $\simeq$ signifie encore modulo des termes proportionnels à δp. Le déterminant des équations (11.9), (11.10) aux inconnues δv^β, δt^β s'écrit:

$$D(l) = (c^2 rf + \mu|h|^2)(u^\alpha l_\alpha)^2 - \mu (h^\alpha l_\alpha)^2$$

d) Nous avons ainsi ~~mis~~ en évidence trois types d'ondes ou d'hypersurfaces caractéristiques du système fondamentale de la magnétohydrodynamique

1) Les <u>ondes d'entropie</u> ou hypersurfaces engendrées par des lignes

de courant d'équation

$$u^\alpha l_\alpha = 0$$

à la traversée desquelles δS peut être $\neq 0$.

2) Les <u>ondes magnétosoniques,</u> hypersurfaces telles que $l = d\varphi$ vérifie:

$$(11.11) \quad P(l) \equiv c^2 rf(\gamma-1)(u^\alpha l_\alpha)^4 + (c^2 rf + \mu |h|^2 \gamma)(u^\alpha l_\alpha)^2 l^\beta l_\beta - \\ -\mu (h^\alpha l_\alpha)^2 l^\beta l_\beta = 0$$

A la traversée d'une onde magnétosonique δp et par suite $l_\beta \delta u^\beta$, $l_\beta \delta h^\beta$, $\delta |h|^2$ peuvent être $\neq 0$

3) Les <u>ondes d'Alfven</u> , hypersurfaces vérifiant:

$$D(l) \equiv (c^2 rf + \mu |h|^2)(u^\alpha l_\alpha)^2 - \mu (h^\alpha l_\alpha)^2 = 0$$

A la traversée diune onde d'Alfven, il peut y avoir discontinuité effective des dérivées des composantes tangentielles de la vitesse et du champ magnétique: δv^β et δt^β peuvente être $\neq 0$ pour $\delta p = 0$.

Posons pour abréger $\beta = \sqrt{(c^2 rf + \mu |h|^2)/\mu}$. L'équation $D(l)=0$ aux ondes d'Alfven peut s'écrire:

$$\left\{(\beta u^\alpha + h^\alpha) l_\alpha\right\} \left\{(\beta u^\beta - h^\beta) l_\beta\right\} = 0$$

et l'on voit que les ondes d'Alfven sont engendrées par les trajectoires des champt de vecteurs:

Lichnerowicz

$$A^{\alpha} = \beta u^{\alpha} + h^{\alpha} \qquad B^{\alpha} = \beta u^{\alpha} - h^{\alpha}$$

ce qui définit deux types d'ondes dites ondes A ou ondes B. On note que:

$$A^{\alpha} A_{\alpha} = B^{\alpha} B_{\alpha} = \beta^2 - |h|^2 > 0$$

Les vecteurs A et B sont temporels et orientés vers le futur.

12. Vitesses des ondes magnétosoniques et des ondes d'Alfven.

a) Une hypersurface Σ (d'équation $\varphi = 0$, avec $l = d\varphi$) étant donnée, nous avons associé à u et l (voir (6.1)) le paramètre:

$$y^{\Sigma} = \frac{(u^{\alpha} l_{\alpha})^2}{(u^{\alpha} l_{\alpha})^2 - l^{\alpha} l_{\alpha}} \geqslant 0$$

La définition précédente est équivalente à la relation:

$$(1 - y^{\Sigma})(u^{\alpha} l_{\alpha})^2 = -y^{\Sigma}(l^{\alpha} l_{\alpha}) \tag{12.1}$$

De manière analogue, associons au vecteur h et à l, une composante h_n définie par:

$$h_n^2 = \frac{(h^{\alpha} l_{\alpha})^2}{(u^{\alpha} l_{\alpha})^2 - l^{\alpha} l_{\alpha}} \geqslant 0$$

h_n^2 vérifie le lemme suivant.

Lemme 1°) On a toujours $h_n^2 \leq |h|^2$

2°) Pour que $h_n^2 = |h|^2$, il faut et il suffit que l appartienne au 2-plan Π défini par u et h.

En effet considérons au point x de Σ un repère orthonormé (e_o, e_i) (i, j=1, 2, 3) tel que $e_o = u$. On a $h^o = 0$ et il vient d'après l'inégalité de Schwarz:

$$(h^i l_i)^2 \leq \sum_i (h^i)^2 \sum_i (l^i)^2$$

c'est-à-dire

$$\frac{(h^i l_i)^2}{\sum (l^i)^2} \leq \sum_i (h^i)^2 = |h|^2$$

On obtient

$$h_n^2 \leq |h|^2$$

ce qui établit le 1°. Pour que l'égalité ait lieu il faut et il suffit que si nous décomposons l selon un vecteur colinéaire à u et un vecteur k orthogonal à u, k soit colinéaire à h, ce qui démontre le 2°.

b) En introduisant h_n^2, le polynome P(l) peut s'écrire:

$$P(l) \equiv c^2 rf(\gamma - 1)(u^\alpha l_\alpha)^4 + (c^2 rf + \mu |h|^2 \gamma - \mu h_n^2)(u^\alpha l_\alpha)^2 l^\beta l_\beta +$$
$$+ \mu h_n^2 (l^\beta l_\beta)^2$$

D'après (12. 1), P(l) peut s'exprimer en termes de y^Σ par:

$$\frac{(1-y^{\Sigma})^2\, P(1)}{(1^\beta : 1_\beta)^2} = \Pi(y^{\Sigma}) \tag{12.2}$$

où $\Pi(y)$ désigne le trinome du second degré:

$$\Pi(y) = c^2 rf(\gamma-1)y^2 - (c^2 rf + \mu|h|^2\gamma - \mu h_n^2)y(1-y) + \mu h_n^2(1-y)^2$$

c'est-à-dire en développant et ordannant en y:

$$\Pi(y) = (c^2 rf + \mu|h|^2)\gamma\, y^2 - (c^2 rf + \mu|h|^2\gamma + \mu h_n^2)y + \mu h_n^2 \tag{12.3}$$

<u>Sous la seule hypothèse $\tau'_p < 0$</u> (soit $\gamma > 1$), $\Pi(y)$ a les propriétés suivantes

$$\Pi(0) = \mu h_n^2 \geqslant 0 \qquad \Pi(1) = c^2 rf(\gamma-1) > 0$$

D'autre part:

$$\Pi\left(\frac{v^2}{c^2}\right) = \Pi\left(\frac{1}{\gamma}\right) = (c^2 rf + \mu|h|^2)\frac{1}{\gamma} - (c^2 rf + \mu|h|^2\gamma + \mu h_n^2)\frac{1}{\gamma} + \mu h_n^2$$

Ce qui peut s'écrire:

$$\Pi\left(\frac{v^2}{c^2}\right) = \mu(|h|^2 - h_n^2)\left(\frac{1}{\gamma} - 1\right) \leqslant 0$$

Ainsi $\Pi(y)$ a deux zéros entre 0 et 1 que nous désignons par y^{ML} et y^{MR}. Il existe donc pour les ondes magnétosoniques et par rapport au fluide deux vitesses v^{ML} et v^{MR}, définies par $(v^{ML})^2/c^2 = y^{ML}$ et $(v^{MR})^2/c^2 = y^{MR}$, satisfaisant les inégalités:

(12. 4) $$v^{ML} \leq v \leq v^{MR} < c$$

v^{ML} est dite vitesse des ondes magnétosoniques lentes et v^{MR} vitesse des ondes magnétosoniques rapides.

c) En introduisant h_n^2 dans D(l), on a:

(12. 5) $$D(l) = (c^2 rf + \mu|h|^2)((u^\alpha l_\alpha)^2 - l^\alpha l_\alpha)(y^\Sigma - \frac{\mu h_n^2}{c^2 rf + \mu|h|^2})$$

Ainsi les ondes d'Alfven admettent par rapport au fluide une vitesse v^A donné par:

(12. 6) $$\frac{(v^A)^2}{c^2} = \frac{\mu h_n^2}{c^2 rf + \mu|h|^2} < 1$$

Evaluons:

$$\Pi\left(\frac{(v^A)^2}{c^2}\right) = \frac{(v^A)^2}{c^2}\left\{\mu h_n^2 \gamma - (c^2 rf + \mu|h|^2 \gamma + \mu h_n^2) + c^2 rf + \mu|h|^2\right\}$$

soit, après semplifications,

$$\Pi\left(\frac{(v^A)^2}{c^2}\right) = \frac{(v^A)^2}{c^2}\,\mu(h_n^2 - |h|^2)(\gamma - 1) \leq 0$$

On établit ainsi que:

$$v^{ML} \leq v^A \leq v^{MR} \tag{12.7}$$

Soit Σ une onde d'alfven de vitesse $v^A \neq 0$ ($h_n^2 \neq 0$); elle vérifie $D(l)=0$. Notons que paur que $P(l)=0$ il faut et il suffit que $\Pi\left(\frac{(v^A)^2}{c^2}\right)=0$, c'est-à-dire que $h_n^2 = |h|^2$. D'après le lemme 1 doit appartenir au 2-plan Π défini par u et h((puur $u^\alpha l_\alpha \neq 0$).

13. Représentation des cônes d'ondes dans R^3.

a) En un point x du fluide, soit T_x^*, l'espace des covecteurs tangents à l'espace-temps. Le cône fondamental de l'espace-temps peut-être défini par le cône Γ des covecteurs l vérifiant:

$$\Gamma : g^{\alpha\beta} l_\alpha l_\beta = 0$$

D'après l'étude du § 11, les caractéristiqaes ou ondes de la magnétohydrodynamique sont définies par les hypersurfaces tangentes à l'un des trois cônes d'équations:

$$\Gamma_E : u^\alpha l_\alpha = 0, \qquad \Gamma_H : P(l)=0, \qquad \Gamma_A : D(l)=0$$

Introduisons en x un repère orthonormé (e_0, e_i) ($i=1,2,3$) tel que l'on ait:

$$e_o = u \qquad e_3 = \frac{h}{|h|}$$

Nous posons provisoirement dans ce paragraphe:

$$l_o = t \quad l_1 = x \quad l_2 = y \quad l_3 = z$$

Avec ces notations, nous obtenons pour les différents cônes les équations explicites:

$$\Gamma : \quad t^2-x^2-y^2-z^2 = 0$$

$$\Gamma_E : \quad t = 0$$

$$\Gamma_H : \quad c^2 rf(\gamma-1)t^4+(c^2 rf+\mu|h|^2\gamma)t^2(t^2-x^2-y^2-z^2)- \quad h^2 z^2$$

$$7- \mu|h|^2 z^2(t^2-x^2-y^2-z^2)=0$$

$$\Gamma_A : \quad (c^2 rf+\mu|h|^2)t^2-\mu|h|^2 z^2=0$$

avec $\gamma > 1$. En introduisant β, on obtient pour équations de Γ, Γ_H, Γ_A (après avoir posé $\bar\beta^2 = \beta^2 / |h|^2$)

$$\Gamma : \quad t^2-x^2-y^2-z^2 = 0$$

$$\Gamma_H : \quad (\bar\beta^2-1)(\gamma-1)t^4+(\bar\beta^2+\gamma-1)t^2(t^2-x^2-y^2-z^2)-z^2(t^2-x^2-y^2-z^2)=0$$

$$\Gamma_A : \quad \bar\beta^2 t^2-z^2 = 0$$

Nous appelons <u>indicatrices dans R^3</u> les sections de ces cônes par l'hyperplan t=1. Nous obtenons ainsi les trois indicatrices suivantes:

$$\begin{cases} S & : 1-x^2-y^2-z^2=0 \\ S_H & : \bar{\beta}^2\gamma-(\bar{\beta}^2+\gamma-1)(x^2+y^2+z^2)-z^2(1-x^2-y^2-z^2)=0 \\ S_A & : z=\pm\bar{\beta} \end{cases}$$

Il est aisé de discuter la forme de ces indicatrices qui admettent 0z comme axe de rotation. Il suffit de couper les indicatrices par le plan x=0, ce qui conduit au cercle

$$y^2+z^2=1$$

aux deux droites

$$z=\pm\bar{\beta}$$

et par S_H à la quartique:

$$C_H: \bar{\beta}^2\gamma-(\bar{\beta}^2+\gamma-1)(y^2+z^2)-z^2(1-y^2-z^2)=0$$

admettant 0y et 0z pour axes de symétrie. L'équation de C_H peut s'écrire:

$$y^2=-\frac{(z^2-\gamma)(z^2-\bar{\beta}^2)}{z^2(\bar{\beta}^2+\gamma-1)}$$

Pour $z^2=\gamma$ et $z^2=\bar{\beta}^2$, on a $y^2=0$ et C_H admet les asymptotes $z^2=\bar{\beta}^2+\gamma-1$.

Nous appelons cas général le cas où $\gamma \neq \bar{\beta}^2$. Par exemple pour $\gamma < \bar{\beta}^2$, on a pour C_H la forme de la fig. 1; Γ_H est décomposée topologiquement en deux parties Γ_{H_1} et Γ_{H_2}; la partie extérieure Γ_{H_1} correspond aux ondes lentes et la partie intérieure Γ_{H_2} aux ondes rapides.

Nous appelons <u>cas singulier</u> le cas où $\gamma = \bar{\beta}^2$, c'est-à-dire où:

(13.1)
$$\frac{v^2}{c^2} = \frac{\mu |h|^2}{c^2 rf + \mu |h|^2}$$

C_H admet des points singuliers sur 0z et a la forme del la fig. 2.

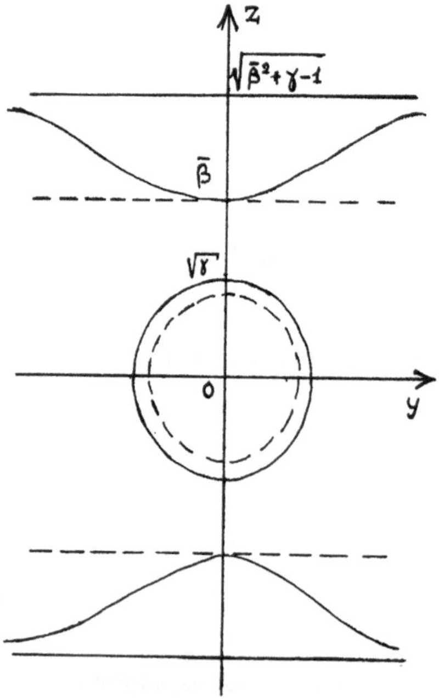

fig. 1

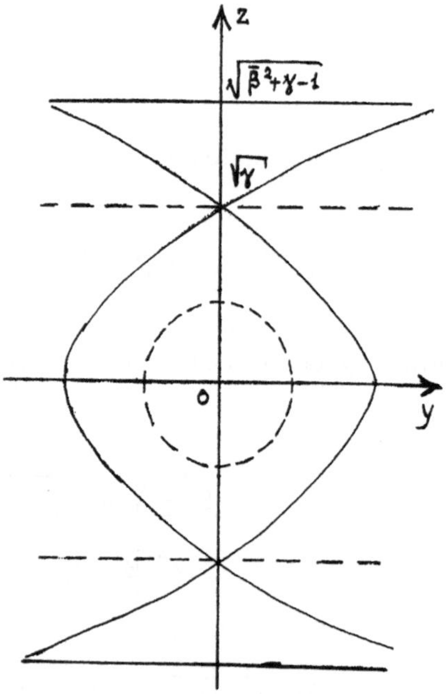

fig. 2

Le cône Γ_H admet deux génératrices singulières $x=0$, $y=0$, $z^2=\gamma t^2$ contenues dans le 2-plan π defini par ru et h.

Dans tous les cas, Γ_A est l'ensemble des deux hyperplans $z=\pm \beta t$; leur intersection est le 2-plan π' defini par $z=0$, $k=0$, c'est-à-dire l'orthocomplément du 2-plan π. Dans le cas général les hyperplans Γ_A sont tangents à Γ_H le long des génératrices contenues dans π et définies par $z^2=\bar{\beta}^2 t^2$. Dans le cas singulier, ils passent par ces génératrices qui sont les génératrices singulières.

L'intersection de l'hyperplan Γ $(t=0)$ soit avec $\Gamma_{H'}$, soit avec chacun des hyperplans de . Γ_A n'est autre que le 2-plan π'.

b) Etudions l'intersection $\Gamma_H \cap \Gamma_A$. Si nous posons $z^2=\bar{\beta}^2 t^2$ dans l'équation de Γ_A il vient

$$(\bar{\beta}^2-1)(\gamma-1)t^4+(\bar{\beta}^2+\gamma-1)t^2\left\{t^2(1-\bar{\beta}^2)-x^2-y^2\right\}-$$

$$-\bar{\beta}^2 t^2\left\{t^2(1-\bar{\beta}^2)-x^2-y^2\right\}=0$$

soit après simplifications:

$$(\gamma-1)t^2(x^2+y^2)=0$$

L'intersection se compose donc du 2-plan π' (d'équations $z=0$, $t=0$) et des deux génératrices de Γ_H contenues dans le 2-plan π ($x=0$, $y=0$) et définies par $z=\pm \beta t$. Dans le cas géneral ($\gamma \neq \bar{\beta}^2$), les hyperplans de Γ_A sont tangents à Γ_H le long de ces génératrices; dans le cas singulier ($\gamma=\bar{\beta}^2$), nous avons vu que ces génératrices sont singulières.

c) Désignons par $a(x,\partial)$ $(x\in\Omega)$ un opérateur différentiel homogène d'ordre m. Si $l \in T^*_x$, $a(x,l)$ est un polynome homogène en l de degré m. Soit $V_x(a)$ le cône de T^*_x défini par l'équation $a(x,l)=0$, où x est fixé. BNous considérons $V_x(a)$ comme un cône projectif. L'opérateur a est dit strictement hyperbolique en x, si $V_x(a)$ vérifie l'hypothèse suivante:

Il existe dans T^*_x des points l telle que toute droite issue de l et ne passant pas par le sommet du cône coupe la surface du cône en m points réels et distincts.

Etudions les opérateurs P(l) et D(l) qui apparaissent dans le système différentiel de la magnétohydrodynamique. Plaçons-nous dans le cas général: l'opérateur P(l) est strictement hyperbolique, mais il n'est est manifestement pas de même pour l'opérateur D(l) qui n'est que le produit de deux opérateurs strictement hyperboliques du 1e ordre correspondant aux deux hyperplans. Le système différentiel étudié n'est donc pas strictement hyperbolique.

Il en est a fortiori de même dans le cas singulier où l'opérateur P(l) lui-même n'est pas strictement hyperbolique.

En me plaçant dans le cas général, j'ai établi cependant ailleurs (Relativistic hydrodynamics and Magnetohydrodynamics, Benjamin, New York (1967)) un théorème d'existence et d'unicité pour le problème de Cauchy relatif au système de la magnétohydrodynamique, sur une classe convenable de fonctions C^{∞}.

14. Propriété fondamentale des rayons associés aux ondes magnétosoniques.

a) Soit φ une solution de l'équation aux ondes magnétosoniques $P(l)=0$. Nous supposons que pour les ondes Σ envisagées D(l) est

$\neq 0$. Il résulte du § 11 (en particulier (11.9) et (11.10) que les "discontinuités infinitésimales" δu^β, δh^β peuvent s'exprimer de manière unique en termes de δp. Nous aurons besoin en particulier de l'expression de $\delta |h|^{22}$. D'après (11.7):

$$\frac{1}{2}\mu\, l^\rho l_\rho\, \delta|h|^2 = c^2 r f (u^\alpha l_\alpha) l_\beta\, \delta u^\beta - (l^\alpha l_\alpha - (u^\alpha l_\alpha)^2)\delta p$$

On en déduit en exprimant $l_\beta \delta u^\beta$ à partir de (11.1):

$$\frac{1}{2}\mu\, l^\rho l_\rho\, \delta|h|^2 = -\gamma (u^\alpha l_\alpha)^2 \delta p - (l^\alpha l_\alpha - (u^\alpha l_\alpha)^2)\, \delta p$$

c'est-à-dire

$$\text{(14.1)} \qquad \frac{1}{2}\mu\, l^\rho l_\rho\, \delta|h|^2 = -\left\{(\gamma - 1)(u^\alpha l_\alpha)^2 + l^\alpha l_\alpha\right\}\delta p$$

b) Les ondes magnétosoniques sont les shypersurfaces tangentes aux cônes $P(l)=0$. La génératrice de contact avec le cône est définie par le cvecteur:

$$N^\beta = \frac{1}{2}\frac{\partial P(l)}{\partial l_\beta}$$

soit:

$$\text{(14.2)} \qquad nN^\beta = 2c^2 r f(\gamma - 1)(u^\alpha l_\alpha)^3 u^\beta + (c^2 r f + \mu |h|^2 \gamma)u^\alpha l_\alpha (l^\rho l_\rho u^\beta + (u^\alpha l_\alpha) l^\beta)$$
$$- \mu h^\alpha l_\alpha (l^\rho l_\rho h^\beta + (h^\alpha l_\alpha) l^\beta)$$

Ce vecteur étant tangent à l'onde magnétosonique Σ, (14.2) peut encore s'écrire en prenant les composantes tangentielles des différents termes:

(14.3) $$N^{\beta} = 2c^2 rf(\gamma - 1)(u^{\alpha} l_{\alpha})^3 v^{\beta} + (c^2 rf + \mu |h|^2 \gamma)(u^{\alpha} l_{\alpha})(l^{\rho} l_{\rho}) v^{\beta} - \mu (h^{\alpha} l_{\alpha})(l^{\rho} l_{\rho}) t^{\beta}$$

Les bicaracteristiques ou rayons associés à ces ondes Σ sont les trajectoires sur Σ du champ des vecteurs N^{β}. Nous nous proposons de montrer que δp (et par suite les δu^{β}, δh^{β}) se propagent le long des rayons, c'est-à-dire que δp vérifie un système différentiel de la forme:

$$N^{\beta} \nabla_{\beta} \delta p \simeq 0$$

où le symbole $\simeq$ signifie modulo des termes proportionnels à δp. Nous postulons dans la suite les hypothèses B_1 et B_2 du § 4 pour $p, S, u^{\lambda}, h^{\lambda}$. D'après (4.3), il existe des distributions $\bar{p}, \bar{S}, \bar{u}^{\lambda}, \bar{h}^{\lambda}$ telles que:

(14.4) $$\bar{\delta} [\nabla_{\alpha} \nabla_{\beta} p] = \nabla_{\alpha} l_{\beta} \delta p + l_{\alpha} \nabla_{\beta} \delta p + l_{\beta} \nabla_{\alpha} \delta p + l_{\alpha} l_{\beta} \bar{p}$$

(14.5) $$\bar{\delta} [\nabla_{\alpha} \nabla_{\beta} S] = l_{\alpha} l_{\beta} \bar{S}$$

(14.6) $$\bar{\delta} [\nabla_{\alpha} \nabla_{\beta} u^{\lambda}] = \nabla_{\alpha} l_{\beta} \delta u^{\lambda} + l_{\alpha} \nabla_{\beta} \delta u^{\lambda} + l_{\beta} \nabla_{\alpha} \delta u^{\lambda} + l_{\alpha} l_{\beta} \bar{u}^{\lambda}$$

(14.7) $$\bar{\delta} [\nabla_{\alpha} \nabla_{\beta} h^{\lambda}] = \nabla_{\alpha} l_{\beta} \delta h^{\lambda} + l_{\alpha} \nabla_{\beta} \delta h^{\lambda} + l_{\beta} \nabla_{\alpha} \delta h^{\lambda} + l_{\alpha} l_{\beta} \bar{h}^{\lambda}$$

c) Des relations (11.1) et (11.2), on déduit comme précedemment:

(14.8) $$c^2 rf \,\bar{\delta}\left[\nabla_\beta \nabla_\alpha u^\alpha\right] + \gamma u^\alpha \bar{\delta}\left[\nabla_\beta \nabla_\alpha p\right] \simeq 0$$

(14.9) $$\bar{\delta}\left[\nabla_\beta \nabla_\alpha S\right] \simeq 0$$

Des relations (10.2), (10.3) conséquences des équations de Maxwell, on déduit par dérivation, en raisonnant comme au § 7:

$$\bar{\delta}\left[\nabla_\beta \nabla_\alpha h^\alpha\right] - u^\alpha u^\lambda \bar{\delta}\left[\nabla_\beta \nabla_\alpha h_\lambda\right] \simeq 0$$

et

$$\frac{1}{2} u^\alpha \bar{\delta}\left[\nabla_\alpha \nabla_\beta |h|^2\right] + |h|^2 \bar{\delta}\left[\nabla_\beta \nabla_\alpha u^\alpha\right] - h^\alpha u^\lambda \bar{\delta}\left[\nabla_\beta \nabla_\alpha h_\lambda\right] \simeq 0$$

En multipliant la première de ces relations par h^β, la seconde par u^β, et retranchant, il vient:

(14.10) $$\frac{1}{2} u^\alpha u^\beta \bar{\delta}\left[\nabla_\alpha \nabla_\beta |h|^2\right] + |h|^2 u^\beta \bar{\delta}\left[\nabla_\beta \nabla_\alpha u^\alpha\right] - h^\beta \bar{\delta}\left[\nabla_\beta \nabla_\alpha h^\alpha\right] \simeq 0$$

De (10.8) il vient de même:

(14.11) $$c^2 rf \bar{\delta}\left[\nabla_\beta \nabla_\alpha h^\alpha\right] + h^\alpha \bar{\delta}\left[\nabla_\beta \nabla_\alpha p\right] \simeq 0$$

De (14.10) on tire, compte-tenu de (14.11) et (14.8):

(14.12) $$\frac{1}{2} c^2 rf\, u^\alpha u^\beta \bar{\delta}\left[\nabla_\alpha \nabla_\beta |h|^2\right] \simeq (|h|^2 \gamma u^\alpha u^\beta - h^\alpha h^\beta)\, \bar{\delta}\left[\nabla_\alpha \nabla_\beta p\right].$$

d) Prenons enfin la dérivée covariante contractée de (10.7). On obtient, compte-tenu de (14.5):

$$(c^2 rf+\mu|h|^2)u^\beta \bar{\delta}[\nabla_\beta \nabla_\alpha u] -(g^{\alpha\beta}-u^\alpha u^\beta)\bar{\delta}[\nabla_\alpha \nabla_\beta p] - \frac{1}{2}\mu\bar{\delta}[\nabla_\alpha \nabla_\beta |h|^2]+$$

$$+\mu u^\alpha u^\beta \bar{\delta}[\nabla_\alpha \nabla_\beta |h|^2]+\mu|h|^2 u^\beta \bar{\delta}[\nabla_\beta \nabla_\alpha u^\alpha] -2\mu h^\beta \bar{\delta}[\nabla_\beta \nabla_\alpha h^\alpha] \simeq 0$$

Ce qui peut s'écrire:

$$c^2 rf\, u^\beta \bar{\delta}[\nabla_\beta \nabla_\alpha u^\alpha]-(g^{\alpha\beta}-u^\alpha u^\beta)\bar{\delta}[\nabla_\alpha \nabla_\beta p] - \frac{1}{2}\mu\bar{\delta}[\nabla_\beta \nabla^\beta |h|^2] +$$

$$+\mu\left\{2|h|^2 u^\beta \bar{\delta}[\nabla_\beta \nabla_\alpha u^\alpha] +u^\alpha u^\beta \bar{\delta}[\nabla_\alpha \nabla_\beta |h|^2] -2h^\beta \bar{\delta}[\nabla_\beta \nabla_\alpha h^\alpha]\right\} \simeq 0$$

où la seconde ligne disparait en vertu de (14.10). En utilisant (14.8) il vient:

$$\left\{(\gamma -1)u^\alpha u^\beta +g^{\alpha\beta}\right\}\bar{\delta}[\nabla_\alpha \nabla_\beta p] +\frac{1}{2}\mu\bar{\delta}[\nabla_\beta \nabla^\beta |h|^2] \simeq 0 \tag{14.13}$$

Introduisons l'expression:

$$\bar{\delta}[\nabla_\alpha \nabla_\beta |h|^2] = \nabla_\alpha l_\beta\, \delta |h|^2 + l_\alpha \nabla_\beta\, \delta |h|^2 + l_\beta \nabla_\alpha\, \delta |h|^2 + l_\alpha l_\beta \overline{|h|}^2$$

On a:

$$\frac{1}{2}\bar{\delta}[\nabla_\beta \nabla^\beta |h|^2] \simeq l^\beta \nabla_\beta\, \delta |h|^2 + \frac{1}{2} l^\rho l_\rho \overline{|h|}^2$$

et (14.12) peut s'écrire:

Lichnerowicz

$$c^2 rf(u^\alpha l_\alpha) u^\beta \nabla_\beta \delta |h|^2 + \frac{1}{2} c^2 rf(u^\alpha l_\alpha)^2 |\bar{h}|^2 \simeq (h^2 \quad u \; u \; -h \; h$$

$$(|h|^2 \gamma \; u^\alpha u^\beta - h^\alpha h^\beta) \bar{\delta} [\nabla_\alpha \nabla_\beta p]$$

Après produit par $c^2 rf(u^\rho l_\rho)^2$, la formule (14.13) prend la forme:

$$(14.14) \quad c^2 rf \left\{ (\gamma - 1) u^\alpha u^\beta + g^{\alpha\beta} \right\} (u^\rho l_\rho)^2 \bar{\delta} [\nabla_\alpha \nabla_\beta p] + \mu c^2 rf(u^\rho l_\rho)^2 l^\beta \nabla_\beta \delta |h|^2 +$$

$$+ \mu l^\rho l_\rho \left\{ -c^2 rf(u^\alpha l_\alpha) u^\beta \nabla_\beta \delta |h|^2 + (|h|^2 \gamma \; u^\alpha u^\beta - h^\alpha h^\beta) \bar{\delta} [\nabla_\alpha \nabla_\beta p] \right\} \simeq 0$$

soit en ordonnant:

$$(14.15) \quad \left\{ c^2 rf((\gamma - 1) u^\alpha u^\beta + g^{\alpha\beta})(u^\rho l_\rho)^2 + \mu |h|^2 \gamma \; l^\rho l_\rho u^\alpha u^\beta - \right.$$

$$\left. - \mu \; l^\rho l_\rho h^\alpha h^\beta \right\} \bar{\delta} [\nabla_\alpha \nabla_\beta p] + \mu c^2 rf(u^\rho l_\rho) \left\{ (u^\rho l_\rho) l^\beta - l^\rho l_\rho u^\beta \right\} \nabla_\beta \delta |h|^2 \simeq 0$$

Substituons à $\bar{\delta} [\nabla_\alpha \nabla_\beta p]$ sa valeur tirée de (14.4). Le coefficient de $\bar{p}$ est:

$$c^2 rf((\gamma - 1) u^\alpha u^\beta + g^{\alpha\beta})(u^\rho l_\rho)^2 l_\alpha l_\beta + \mu |h|^2 \gamma \, l^\rho l_\rho (u^\alpha l_\alpha)^2 - \mu l^\rho l_\rho (h^\alpha l_\alpha)^2$$

soit:

$$c^2 rf(\gamma - 1)(u^\alpha l_\alpha)^4 + (c^2 rf + \mu |h|^2 \gamma) l^\rho l_\rho (u^\alpha l_\alpha)^2 - \mu l^\rho l_\rho (h^\alpha l_\alpha)^2 = P(l)$$

Ainsi le coefficient de p est nul. Il vient ainsi à partir de (14.15):

$$(14.16)\quad 2\left\{c^2rf(\gamma-1)(u^\alpha l_\alpha)^3u^\beta + c^2rf(u^\alpha l_\alpha)^2l^\beta + \mu|h|^2\gamma\, l^\rho l_\rho(u^\alpha l_\alpha)u^\beta - \right.$$

$$\left. -\mu l^\rho l_\rho(h^\alpha h_\alpha)h^\beta\right\}\nabla_\beta\delta p + \mu c^2rfu^\rho l_\rho\left\{(u^\rho l_\rho)l^\beta - l^\rho l_\rho u^\beta\right\}\nabla_\beta\delta|h|^2 \simeq 0$$

D'après le calcul précédent concernant le coefficient de p, le vecteur coefficient de $\nabla_\beta\delta p$ dans (14.16) est tangent à Σ. Il en est de même manifestement pour le coefficient de $\nabla_\beta\delta|h|^2$. La relation précédente peut donc s'écrire après division par 2:

$$\left\{c^2rf(\gamma-1)(u^\alpha l_\alpha)^3v^\beta + \mu|h|^2\gamma\, l^\rho l_\rho(u^\alpha l_\alpha)v^\beta - \mu l^\rho l_\rho(h^\alpha l_\alpha)t^\beta\right\}\nabla_\beta\delta p$$

$$-\frac{1}{2}\mu c^2rfl^\rho l_\rho(u^\alpha l_\alpha)v^\beta\nabla_\beta\delta|h|^2 \simeq 0$$

soit d'après la relation (14.1):

$$\left\{2c^2rf(\gamma-1)(u^\alpha l_\alpha)^3v^\beta + (c^2rf+\mu|h|^2\gamma)l^\rho l_\rho(u^\alpha l_\alpha)v^\beta - \right.$$

$$\left. -\mu l^\rho l_\rho(h^\alpha l_\alpha)t^\beta\right\}\nabla_\beta\delta p \simeq 0$$

c'est-à-dire d'après (14.3):

$$N^\beta\nabla_\beta\delta p \simeq 0$$

Nous obtenons ainsi :

Théorème. Sous les hypothèses du § 4, les discontinuités infinitésimales δp, δu^λ, δh^λ relatives à une onde magnétosonique Σ se propagent le long des rayons associés selon les systèmes différentiels:

$$N^{\beta}\nabla_{\beta}\,\delta p \simeq 0 \qquad N^{\beta}\nabla_{\beta}\,\delta u^{\lambda} \simeq 0 \qquad N^{\beta}\nabla_{\beta}\,\delta h^{\lambda} = 0$$

15. Propriété des rayons associés aux ondes d'Alfven.

a) Soit φ une soluzion de l'équation aux ondes d'Alfen par example d'espèce A. On a:

$$A^{\alpha}\partial_{\alpha}\varphi = A^{\alpha}l_{\alpha} = (\beta u^{\alpha} + h^{\alpha})l_{\alpha} = 0 \tag{15.1}$$

Nous supposons que pour les ondes envisagées $P(l)\neq 0$. Par suite:

$$\delta p = 0 \qquad l_{\alpha}\,\delta u^{\alpha} = 0 \qquad l_{\alpha}\,\delta h^{\alpha} = 0 \qquad \delta|h|^{2} = 0$$

De plus (14.15) se réduit à $P(l)\bar{p}=0$; par suite $\bar{p}=0$ et l'on a:

$$\bar{\delta}\left[\nabla_{\alpha}\nabla_{\beta}p\right] = 0 \tag{15.2}$$

De (14.8), (14.11)(14.12) il résulte:

$$\bar{\delta}\left[\nabla_{\beta}\nabla_{\alpha}u^{\alpha}\right] = 0 \qquad \bar{\delta}\left[\nabla_{\beta}\nabla_{\alpha}h^{\alpha}\right] = 0 \qquad \bar{\delta}\left[\nabla_{\alpha}\nabla_{\beta}|h|^{2}\right] = 0 \tag{15.3}$$

b) Pour les ondes d'Alfven envisagées ici, seules les discontinuités δv^{A}, δt^{λ} peuvent être non nulles. Des relations (11.9), (11.10) où $u^{\alpha}l_{\alpha}\neq 0$ et où les seconds membres sont nuls, on déduit que δt^{λ} est proportionnel à δv^{λ}. Le symbole $\simeq$ signifie dans ce § modulo des termes linéaires par rapport aux δv^{λ} (ou aux δt^{λ}).

Lichnerowicz

Compte-tenu de (15.2), (15.3), les équations de Maywell est le système aux lignes de courant donnent par dérivation:

$$h^{\alpha}\,\bar{\delta}\left[\nabla_{\alpha}\nabla_{\beta}\,u^{\lambda}\right] - u^{\alpha}\,\bar{\delta}\left[\nabla_{\alpha}\nabla_{\beta}\,h^{\lambda}\right] \simeq 0$$

et:

$$\beta^{2}\,u^{\alpha}\,\bar{\delta}\left[\nabla_{\alpha}\nabla_{\beta}\,u^{\lambda}\right] - h^{\alpha}\,\bar{\delta}\left[\nabla_{\alpha}\nabla_{\beta}\,h^{\lambda}\right] \simeq 0$$

Multipliant la 1ere relation par h^{β}, la seconde par u^{β} et retranchant, il vient

$$(\beta^{2}u^{\alpha}u^{\beta} - h^{\alpha}h^{\beta})\,\bar{\delta}\left[\nabla_{\alpha}\nabla_{\beta}\,u^{\lambda}\right] \simeq 0$$

soit en explicitant:

$$(\beta^{2}u^{\alpha}u^{\beta} - h^{\alpha}h^{\beta})(l_{\alpha}\nabla_{\beta}\,\delta u^{\lambda} + l_{\beta}\nabla_{\alpha}\,\delta u^{\lambda} + l_{\alpha}l_{\beta}\,\bar{u}^{\lambda}) \simeq 0$$

Le coefficient de $\bar{u}^{\lambda}$ est nul et il reste:

$$\left\{\beta^{2}(u^{\alpha}\,l_{\alpha})u^{\beta} - (h^{\alpha}\,l_{\alpha})h^{\beta}\right\}\nabla_{\beta}\,\delta\,u^{\lambda} \simeq 0$$

soit d'après (15.1):

$$\beta(u^{\alpha}\,l_{\alpha})A^{\beta}\,\nabla_{\beta}\,\delta\,u^{\lambda} \simeq 0$$

Nous obtenons:

$$A^{\beta} \nabla_{\beta} \delta v^{\lambda} \simeq 0$$

Théorème. Les distributions δv^{λ}, δt^{λ} à supports sur une onde d'Alfven Σ d'espèce A se propagent le long des rayons associés selon les systèmes différentiels:

$$A^{\beta} \nabla_{\beta} \delta v^{\lambda} \simeq 0 \qquad A^{\beta} \nabla_{\beta} \delta t^{\lambda} \simeq 0$$

où $\simeq$ signifie modulo des termes linéaires par rapport aux δv^{μ} (resp. δt^{μ}).

Des résultats symétriques sont valables pour une onde d'Alfven d'espèce B.

c) Reprenons une onde d'espèce A et étudions l'action de la dérivation δ sur le vecteur

$$A^{\beta} = \beta u^{\beta} + h^{\beta}$$

D'après l'étude du a, on a:

$$\delta \beta = 0 \qquad l_{\alpha} \delta u^{\alpha} = 0 \qquad l_{\alpha} \delta h^{\alpha} = 0$$

Il en résulte:

$$\delta A^{\beta} = \beta \delta v^{\beta} + \delta t^{\beta}$$

soit d'après (11.9) ou le second membre est nul:

$$(u^{\alpha} l_{\alpha}) \delta A^{\beta} = (\beta u^{\alpha} l_{\alpha} + h^{\alpha} l_{\alpha}) \delta v^{\beta}$$

Il vient $\delta A^{\beta} = 0$ et le vecteur A^{β} lui-même, et sa direction en particulier, sont invariants par la dérivation δ.

IV. ONDES DE CHOC EN MAGNETOHYDRODYNAMIQUE

16. Le système fondamental des ondes de choc.

a) Dans un domaine Ω de V_4, soit encore Σ une hypersurface régulière d'équation $\varphi = 0$ (φ de classe C^4, avec $l = d\varphi$). L'hypersurface Σ est une onde de choc magnétohydrodynamique si u^{α}, h^{α} ou l'une au moins des variables thermodynamiques sont discontinus à la traversée de Σ, conformément aux hypothèses A_1 et A_2 du § 3. Nous établirons que sous les conditions de compressibilité H_1, Σ est nécessairement orientée dans le temps, donc de vitesse admissible au point de vue relativiste.

Soit Y un état du système fluide-champ, défini par les valeurs de $p, S, u^{\alpha}, h^{\alpha}$ en un point x de Σ. Un tel état est défini par la donnée de 8 quantités scalaires, c'est-à-dire dépend de 8 paramètres. Nous notons Y_0 l'état antérieur au choc, Y_1 l'état postérieur au choc et $[Q]$ la discontinuité, $Q_1 - Q_0$ d'une quantité à la traversée de Σ.

Le système fondamental (9.2), (9.3), (9.4) est supposé satisfait au sens des distributions:

$$\nabla_{\alpha}(r u^{\alpha})^D = 0 \qquad \nabla_{\alpha}(h^{\alpha} u^{\beta} - u^{\alpha} h^{\beta})^D = 0 \qquad \nabla_{\alpha}(T^{\alpha\beta})^D = 0$$

Ces équations impliquent:

$$(\nabla_\alpha(ru^\alpha))^D=0 \qquad (\nabla_\alpha(h^\alpha u^\beta - u^\alpha h^\beta))^D=0 \qquad (\nabla_\alpha T^{\alpha\beta})^D=0$$

Il résulte de l'équation (3.2) que l'on a:

$$(16.1) \qquad l_\alpha \bar{\delta}[ru^\alpha]=0 \qquad l_\alpha \bar{\delta}[h^\alpha u^\beta - u^\alpha h^\beta]=0 \qquad l_\alpha \bar{\delta}[T^{\alpha\beta}]=0$$

Posons

$$a(Y)=ru^\alpha l_\alpha \qquad V^\beta(Y)=(h^\alpha l_\alpha)u^\beta - \frac{a(Y)}{r}h^\beta$$

et:

$$W^\beta(Y)=(c^2\tau + \mu\frac{|h|^2}{r^2})\, a(Y)ru^\beta - ql^\beta - \mu(h^\alpha l_\alpha)h^\beta$$

Il est clair que le vecteur V^β est tangent à Σ. Le système (16.1) exprime que:

$$a(Y_1)=a(Y_o) \qquad V^\beta(Y_1)=V^\beta(Y_o) \qquad W^\beta(Y_1)=W^\beta(Y_o).$$

Le scalaire a et les vecteurs V^β et W^β définissent <u>les invariants du choc</u>.

b) Un choc est dit <u>tangentiel</u> si a=0. S'il en est ainsi:

$$u_o^\alpha l_\alpha = u_1^\alpha l_\alpha = 0$$

est de vitesse nulle par rapport au fluide dans les deux états et orientée dans le temps. De l'invariance de V^β et W^β il résulte:

(16.2) $$(h_1^\alpha l_\alpha)u_1^\beta = (h_o^\alpha l_\alpha)u_o^\beta$$

(16.3) $$(q_1-q_o)l^\beta + \mu\left\{(h_1^\alpha l_\alpha)h_1^\beta - (h_o^\alpha l_\alpha)h_o^\beta\right\} = 0$$

si $h_o^\alpha l_\alpha \neq 0$, u_1^β et u_o^β étant unitaires et colinéaires d'après (16.2), orientés vers le futur, on a $u_1^\beta = u_o^\beta$ et par suite $h_1^\alpha l_\alpha = h_o^\alpha l_\alpha$. De (16.3) il résulte:

$$l^\beta l_\beta (q_1-q_o) + \mu\left\{(h_1^\alpha l_\alpha)^2 - (h_o^\alpha l_\alpha)^2\right\} = 0$$

et par suite $[q]=0$. De (16.3) résulte ainsi $[h^\beta]=0$, $[p]=0$, la discontinuité de r restant indéterminée si $h_o^\alpha l_\alpha = 0$ il résulte de (16.2) $h_1^\alpha l_\alpha = 0$ et les équations de choc donnent seulement ici:

$$u_1^\alpha l_\alpha = u_o^\alpha l_\alpha = 0 \qquad h_1^\alpha l_\alpha = h_o^\alpha l_\alpha = 0 \qquad q_1 = q_o$$

les autres discontinuités restant indéterminées.

c) Nous supposons désormais $a \neq 0$ (choc non tangentiel) et introduisons le scalaire H, invariant par le choc défini par:

(16.4) $$H(Y) = \frac{1}{a^2} V^\beta V_\beta = \frac{(h^\alpha l_\alpha)^2}{a^2} - \frac{|h|^2}{r^2}$$

En substituant dans W^β à h^β son expression en termes de V^β et u^β:

$$h^\beta = \frac{h^\alpha l_\alpha}{a} r u^\beta - \frac{r}{a} V^\beta$$

il vient (nous supprimons la référence à Y lorsqu'elle est inutile):

$$W^{\beta} = (c^2\tau + \mu \frac{|h|^2}{r^2}) \, aru^{\beta} - ql^{\beta} - \mu \frac{(h^{\alpha} l_{\alpha})^2}{a} ru^{\beta} + \mu \frac{rh^{\alpha} l_{\alpha}}{a} V^{\beta}$$

soit:

$$W^{\beta} = a\alpha ru^{\beta} - ql^{\beta} + \mu \frac{r}{a} (h^{\alpha} l_{\alpha}) V^{\beta}$$

où l'on a introduit la variable importante:

$$\alpha = c^2\tau - \mu H = D(l)/a^2$$

Si nous décomposons W^{β} en sa partie tangentielle et sa partie normale à Σ, on obtient

$$W^{\beta} = X^{\beta} - (q - \frac{a^2\alpha}{l^{\alpha} l_{\alpha}}) \, l^{\beta}$$

où

$$X^{\beta} = a\alpha rv^{\beta} + \mu \frac{r}{a} (h^{\alpha} l_{\alpha}) \, V^{\beta} \tag{16.5}$$

est tangent à Σ. L'invariance de W^{β} est équivalente à l'ensemble de celle de X^{β} et de celle de la composante normale. Ainsi le scalaire:

$$e = q - \frac{c^2 a^2}{l^{\alpha} l_{\alpha}} \tau \tag{16.6}$$

est invariant.

Considérons en particulier le produit scalaire invariant au cours du. choc:

$$X^{\beta} V_{\beta} = W^{\beta} V_{\beta} = ar(h^{\alpha} l_{\alpha})(\alpha + \mu H)$$

D'après la définition de: α , on obtient $X^{\beta} V_{\beta} = c^2 ab$, où b est le scalaire invariant:

(16.7) $$b = f(h^{\alpha} l_{\alpha})$$

d) Considérons enfin le scalaire invariant:

$$K = \frac{1}{c^2 a^2} X^{\beta} X_{\beta}$$

donc nous allons donner deux expressions importantes:

Lemme 1. L'invariant K admet l'expression:

(16.8) $$K = c^2 f^2 - \frac{c^2 a^2}{l^{\alpha} l_{\alpha}} \tau^2 + 2\mu\tau\chi - \frac{\mu^2 H}{c^2} \chi$$

où l'on a posé:

(16.9) $$\chi = |h|^2 + \frac{a^2}{l^{\alpha} l_{\alpha}} H$$

Il admet aussi l'expression:

(16.10) $$K = c^2 f^2 + \mu |h|^2 \tau + \frac{\mu |h|^2}{c^2} \alpha - \frac{a^2}{c^2 l^{\alpha} l_{\alpha}} \alpha^2$$

En effet de:

$$r u^{\beta} = r v^{\beta} + \frac{a}{l^{\alpha} l_{\alpha}} l^{\beta}$$

on déduit

$$(r v^{\beta})(r v_{\beta}) = r^2 - \frac{a^2}{l^{\alpha} l_{\alpha}} \qquad r v^{\beta} V_{\beta} = r(h^{\alpha} l_{\alpha})$$

Il vient aussi:

$$K = (r^2 - \frac{a^2}{l^{\alpha} l_{\alpha}}) \frac{\alpha^2}{c^2} + 2 \frac{\mu \alpha}{c^2 a^2} r^2 (h^{\alpha} l_{\alpha})^2 + \mu^2 \frac{r^2 (h^{\alpha} l_{\alpha})^2}{c^2 a^2} H \tag{16.11}$$

En substituant à sa valeur, on a:

$$K = c^2 (r^2 - \frac{a^2}{l^{\alpha} l_{\alpha}})(\tau - \frac{\mu H}{c^2})^2 + 2\mu \frac{r^2 (h^{\alpha} l_{\alpha})^2}{c a^2} (\tau - \frac{\mu H}{c^2}) + \mu^2 \frac{r^2 (h^{\alpha} l_{\alpha})^2}{c^2 a^2} H$$

soit:

$$K = c^2 (r^2 - \frac{a^2}{l^{\alpha} l_{\alpha}}) \tau^2 + 2 \mu \tau \left\{ \frac{r^2 (h^{\alpha} l_{\alpha})^2}{a^2} - (r^2 - \frac{a^2}{l^{\alpha} l_{\alpha}}) H \right\} - \frac{\mu^2 H}{c^2} \left\{ \frac{r^2 (h^{\alpha} l_{\alpha})^2}{a^2} - (r^2 - \frac{a^2}{l^{\alpha} l_{\alpha}}) H \right\}$$

Or d'après la définition de H:

$$\bar{\bar{\chi}} = \frac{r^2 (h^\alpha l_\alpha)^2}{a^2} - (r^2 - \frac{a^2}{l^\alpha l_\alpha}) H = \frac{r^2 (h^\alpha l_\alpha)^2}{a^2} - r^2 (\frac{(h^\alpha l_\alpha)^2}{a^2} - \frac{|h|^2}{r^2}) +$$

$$+ \frac{a^2}{l^\alpha l_\alpha} H = |h|^2 + \frac{a^2}{l^\alpha l_\alpha} H$$

Il vient aussi:

$$K = c^2 f^2 - \frac{c^2 a^2}{l^\alpha l_\alpha} \tau^2 + 2\mu\tau\chi - \frac{\mu^2 H}{c^2} \chi$$

ce qui établit (16. 8). De cette relation on déduit:

$$K = c^2 f^2 - \frac{c^2 a^2}{l^\alpha l_\alpha} \tau^2 + 2\mu\tau \left(|h|^2 + \frac{a^2}{l^\alpha l_\alpha} H \right) - \frac{\mu^2 H}{c^2} \left(|h|^2 + \frac{a^2}{l^\alpha l_\alpha} H \right)$$

Ce qui peut s'écrire:

$$K = c^2 f^2 + \mu |h|^2 (2\tau - \frac{\mu H}{c^2}) - \frac{a^2}{c^2 l^\alpha l_\alpha} (c^4 \tau^2 - 2c^2 \tau \mu H + \mu^2 H^2)$$

En réintroduisant α , on obtient (16. 10), ce qui démontre le lemme.

Transformons enfin l'expression de HK, où K est donné par (16. 11). De

$$HK = H(r^2 - \frac{a^2}{l^\alpha l_\alpha}) \frac{\alpha^2}{c^2} + \frac{r^2 (h^\alpha l_\alpha)^2}{c^2 a^2} \mu H (2\alpha + \mu H)$$

il résulte en substituant à μH sa valeur $c^2 \tau - \alpha$:

Lichnerowicz

$$HK = H(r^2 - \frac{a^2}{l^\alpha l_\alpha})\frac{\alpha^2}{c^2} + \frac{r^2(h^\alpha l_\alpha)^2}{c^2 a^2}(c^2\tau + \alpha)(c^2\tau - \alpha)$$

soit:

$$HK = H(r^2 - (\frac{a^2}{l^\alpha l_\alpha})\frac{\alpha^2}{c^2} + \frac{r^2(h^\alpha l_\alpha)^2}{c^2 a^2}(c^4\tau^2 - \alpha^2)$$

Nous obtenons ainsi:

$$HK = \left\{ H(r^2 - \frac{a^2}{l^\alpha l_\alpha}) - \frac{r^2(h^\alpha l_\alpha)^2}{a^2} \right\} \frac{\alpha^2}{c^2} + c^2 \frac{r^2(h^\alpha l_\alpha)^2}{a^2}$$

soit:

(16.2) $$HK = c^2 \frac{b^2}{a^2} - \frac{1}{c^2}\chi\alpha^2$$

Ainsi $L = \chi\alpha^2$ est un invariant qui q, H, K étant fixés peut être substitué à b en convenant que le signe de $(h^\alpha l_\alpha)$ demeure inchangé au cours du choc.

e) En ce qui concerne χ, nous allons établir le lemme suivant qui nous sera utile à différentes reprises.

Lemme 2. 1) On a $(l^\alpha l_\alpha)\chi \leq 0$.

2) Si $l^\alpha l_\alpha$ est $\neq 0$, pour que $\chi = 0$, il faut et il suffit que l appartienne au 2-plan π défini par (u, h)

3) Si $l^\alpha l_\alpha$ est ≥ 0, on a $H \leq 0$ et par suite $\alpha > 0$.

En effet d'apres les définitions de χ et de H

$$(l^\alpha l_\alpha)\chi = |h|^2 l^\alpha l_\alpha + (h^\alpha l_\alpha)^2 - |h|^2 (u^\alpha l_\alpha)^2$$

Soit en introduisant h_n^2:

$$(l^\alpha l_\alpha)\chi = ((u^\alpha l_\alpha)^2 - l^\alpha l_\alpha)(h_n^2 - |h|^2)$$

ce qui, compte-tenu du lemme du § 12, démontre les 1) et 2).

L'inégalité $h_n^2 \leq |h|^2$ peut s'écrire:

$$(h^\alpha l_\alpha)^2 \leq |h|^2 ((u^\alpha l_\alpha)^2 - l^\alpha l_\alpha)$$

En divisant par $a^2 = (ru^\alpha l_\alpha)^2$, il vient:

$$H = \frac{(h^\alpha l_\alpha)^2}{a^2} - \frac{|h|^2}{r^2} \leq - \frac{|h|^2}{a^2} l^\alpha l_\alpha$$

Pour $l^\alpha l_\alpha \geq 0$, on a donc $H \leq 0$ et $\alpha = c^2 \tau - \mu H > 0$.

17. Analyse des équations de choc et chocs d'Alfven.

a) De l'étude du § 16, il résulte que les deux variables thermodynamiques et les trois scalaires $|h|^2, u^\alpha l_\alpha$, $h^\alpha l_\alpha$ vérifient les cinq relations scalaires fondamentales:

$$r_1 u_1^\alpha l_\alpha = r_o u_o^\alpha l_\alpha = a \tag{17.1}$$

$$\frac{(h_1^\alpha l_\alpha)^2}{a^2} - \frac{|h_1|^2}{r_1^2} = \frac{(h_o^\alpha l_\alpha)^2}{a^2} - \frac{|h_o|^2}{r_o^2} = H \tag{17.2}$$

(17.3) $$q_1 - \frac{c^2 a^2}{l^\alpha l_\alpha} \tau_1 = q_o - \frac{c^2 a^2}{l^\alpha l_\alpha} \tau_o = e$$

(17.4) $$c^2 f_1^2 - \frac{c^2 a^2}{l^\alpha l_\alpha} \tau_1^2 + 2\mu \tau_1 \chi_1 - \frac{\mu^2 H}{c^2} \chi_1 = c^2 f_o^2 - \frac{c^2 a^2}{l^\alpha l_\alpha} \tau_o^2 + 2\mu \chi_o \tau_o - \frac{\mu^2 H}{c^2} \chi_o = K$$

(17.5) $$\chi_1 \alpha_1^2 = \chi_o \alpha_o^2 = L$$

ainsi que la relation onon indépendante des précédentes:

(17.6) $$f_1 (h_1^\alpha l_\alpha) = f_o (h_o^\alpha l_\alpha) = b$$

Nous établions que, sous des conditions convenables, les système précédent définit d'une manière et d'une seule les valeurs après le choc des deux variables theormodynamiques et des trois scalaires $|h_1|^2, u_1^\alpha l_\alpha, h_1^\alpha l_\alpha$. Il est clair que si (17.3), (17.4), (17.5) définissent les valeurs de χ (ou de $|h|^2$) et des variables thermodynamiques, (17.1) fournit alors la composante normale de la vitesse et (17.2) ou (17.6) celle du champ magnétique.

D'autre part, d'après l'invariance de V^β et W^β, les composantes tangentielles de la vitesse et du champ magnétique vérifient:

(17.7) $$(h_1^\alpha l_\alpha) v_1^\beta - (u_1^\alpha l_\alpha) t_1^\beta = (h_o^\alpha l_\alpha) v_o^\beta - (u_o^\alpha l_\alpha) t_o^\beta$$

(17.8) $$(c^2 r_1 f_1 + \mu |h_1|^2)(u_1^\alpha l_\alpha) v_1^\beta - (h_1^\alpha l_\alpha) t_1^\beta = (c^2 r_o f_o + \mu |h_o|^2)(u_o^\alpha l_\alpha) v_o^\beta - (h_o^\alpha l_\alpha) t_o^\beta$$

Le déterminant des premiers membres de (17.7), (17.8) aux inconnues v_1^β, t_1^β s'écrit:

$$D_1(l)=(c^2 r_1 f_1+\mu|h_1|^2)(u_1^\alpha l_\alpha)^2-\mu(h_1^\alpha l_\alpha)^2=a^2\alpha_1$$

Si $\alpha_1 \neq 0$, (17.7), (17.8) détermine v_1^β, t_1^β en fonction des quantités déterminées par le système des 5 équations scalaires.

b) Supposons $\alpha_1=0$ en un point x de Σ ; l'hypersurface est alors onde d'Alfven en x pour l'état postérieur au choc et elle est orientée dans le temps ($l^\alpha l_\alpha < 0$). La relation (17.5) donne $\chi_o \alpha_o^2=0$ et ou bien $\alpha_o=0$ ou bien $\chi_o=0$.

Dans le cas $\alpha_o=\alpha_1=0$, Σ définit <u>un choc d'Alfven</u>. Les cas $\alpha_o \neq 0$, $\chi_o=0$, $\alpha_1=0$ et $\alpha_o=0$, $\chi_1=0$, $\alpha_1 \neq 0$ sont dits <u>des chocs singuliers</u>. Nous établirons qu'ils sont incompatibles avec les ondes d'Alfven infinitésimales et que, par suite, il n'ont pas de réalité physique. On a

<u>Théorème</u> : <u>1) Si dans un choc $\alpha_1=\alpha_o \neq 0$, le choc est nul.</u>

2) <u>Dans un choc d'Alfven ($\alpha_1=\alpha_o=0$) les variables thermodynamiques $|h|^2$, $(u^\alpha l_\alpha)$, $(h^\alpha l_\alpha)$ sont continus si l'hypothèse $\tau'_p < 0$ est satisfaite.</u>

1) Si $[\alpha]=0$, on a

$$[\tau]=0$$

Sous notre hypothèse $\chi_1=\chi_o$ et par suite:

$$[|h|^2]=0$$

D'après (17.3), on a $[q]=0$ et par suite

$$[p]=0$$

Dans K donné par (16.10), les trois derniers termes sont invariants au cours du choc. On a donc

$$[f]=0 \qquad\qquad [r]=0$$

et d'après (17.1) et (17.6)

$$[u^\alpha l_\alpha]=0 \qquad\qquad [h^\alpha l_\alpha]=0$$

Σ n'étant pas onde d'Alfven, il résulte de (17.7) et (17.8) que le choc est nul.

2) Pour un choc d'Alfven, on a toujours $[\tau]=0$ et il résulte des équations de choc, K étant donné par (16.10):

$$\left[p+\frac{1}{2}\mu|h|^2\right]=0 \qquad \left[c^2f^2+\mu|h|^2\tau\right]=0 \quad\Rightarrow\quad \left[c^2rf+\mu|h|^2\right]=0$$

De $[\mu|h|^2]=-[2p]$, on déduit:

$$[c^2rf-2p]=0 \tag{17.9}$$

Examinons l'indépendance des deux variables theormodynamiques:

$$\varphi=c^2rf-2p \qquad\qquad \tau$$

Par dérivation de φ en p il vient; d'après le § 6:

$$\varphi'_p = c^2 f r'_p - 1 = \gamma - 1 = -c^2 r^2 \tau'_p$$

De même en dérivant en S:

$$\varphi'_S = c^2 f r'_S + r\Theta = r(c^2 f \frac{r'_S}{r} + \Theta)$$

Or:

$$c^2 \tau'_S = \Theta V + c^2 f V'_S$$

et par suite:

$$c^2 r \tau'_S = \Theta - c^2 f \frac{r'_S}{r}$$

Il en résulte:

$$\varphi'_S = r(2\Theta - c^2 r \tau'_S)$$

Le jacobien de φ et τ par rapport à p et S a pour valeur:

$$\frac{d(\varphi, \tau)}{d(p, S)} = c^2 r^2 \tau'_p \tau'_S - r(2\Theta - c^2 r \tau'_S) \tau'_p$$

c'est-à-dire

$$\frac{d(\varphi, \tau)}{d(p, S)} = 2 - 2r \Theta \tau'_p$$

Le jacobien étant différent de zéro dans l'hypothèse τ'_p 0, les variables φ et τ sont indépendantes. Ces variables étant continues à la traversée de Σ, il en est de même pour toutes les variables theormodynamiques. En particulier $[r] = 0$, $[f] = 0$, $[p] = 0$. D'après (17.1) et (17.6), on a:

$$[u^\alpha l_\alpha] = 0 \qquad [h^\alpha l_\alpha] = 0$$

et $[|h|^2] = 0$, ce qui démontre le théorème.

c) Dans un choc d'Alfven, la vitesse tangentielle du fluide après le choc demeure indéterminée, la direction du champ magnétique tangentiel étant alors déterminée par example à l'aide du résultat qui suit.

Considérons un choc d'Alfven de type A; Σ est engendrée par les trajectoires du champ A^B et l'on a

$$A^\beta = \beta v^\beta + t^\beta$$

ainsi que:

$$A^\beta l_\beta = \beta u^\beta l_\beta + h^\beta l_\beta = 0$$

De (17.8) on déduit:

$$\beta^2(u^\alpha l_\alpha)\left[v^\beta\right] - (h^\alpha l_\alpha)\left[t^\beta\right] = 0$$

c'est-à-dire:

$$\beta(u^\alpha l_\alpha)\left[\beta v^\beta + t^\beta\right] = 0$$

Comme le choc est non tangentiel $\left[A^\beta\right] = 0$. Il vient:

Théorème. Dans un choc d'Alfven d'espèce A (resp. B), le vecteur A^β (resp. B^β) reste invariant dans le choc.

V. FONCTION D'HUGONIOT ET ORIENTATION DES ONDES DE CHOC.

18. Relation d'Hugoniot relativiste.

La relation (17.4) peut s'écrire:

$$c^2\left[f^2\right] - \frac{c^2 a^2}{l^\alpha l_\alpha}\left[\tau^2\right] + 2\mu\left[\chi\tau\right] - \frac{\mu^2 H}{c^2}\left[\chi\right] = 0$$

En tenant compte de $c^2 a^2/l^\alpha l_\alpha = q_1 - q_0 / \tau_1 - \tau_0$ tiré de (7.3), il vient:

$$c^2\left[f^2\right] - (\tau_0 + \tau_1)\left[q\right] + 2\mu\left[\chi\tau\right] - \frac{\mu^2 H}{c^2}\left[\chi\right] = 0$$

soit comme $\left[|h|^2\right] = \left[\chi\right]$

$$c^2\left[f^2\right] - (\tau_0 + \tau_1)\left[p\right] - \frac{1}{2}\mu(\tau_0 + \tau_1)\left[\chi\right] + 2\mu\left[\chi\tau\right] - \frac{\mu^2 H}{c^2}\left[\chi\right] = 0 \tag{18.1}$$

Nous nous proposons de substituer à (18.1) une relation plus maniable.

a) Examinons le terme $[\chi\tau]$ qui peut s'écrire:

$$[\chi\tau] = (\chi_1 - \chi_o)\ \tau_1 + \chi_o(\tau_1 - \tau_o)$$

ou

$$[\chi\tau] = \chi_1(\tau_1 - \tau_o) + (\chi_1 - \chi_o)\tau_o$$

Par addition membre à membre il vient:

$$(18.2) \qquad 2[\chi\tau] = (\tau_o + \tau_1)[\chi] + (\chi_o + \chi_1)[\tau]$$

En reportant dans (18.1), on obtient après simplifications:

$$c^2[f^2] - (\tau_o + \tau_1)[p] + \frac{1}{2}\mu(\tau_o + \tau_1 - \frac{2\mu H}{c^2})[\chi] + \mu(\chi_o + \chi_1)[\tau] = 0$$

soit:

$$(18.3) \quad c^4[f^2] - c^2(\tau_o + \tau_1)[p] + \frac{1}{2}\mu(\alpha_o + \alpha_1)[\chi] + \mu(\chi_o + \chi_1)[\alpha] = 0$$

b) Proposons-nous de transformer la somme des deux derniers termes. On a d'abord, comme dans (18.2):

$$\frac{1}{2}(\alpha_o + \alpha_1)[\chi] + \frac{1}{2}(\chi_o + \chi_1)[\alpha] = [\chi\alpha]$$

Or d'après (17.5), on peut écrire:

$$[\chi\alpha] = \chi_1 \alpha_1 - \chi_o \alpha_o = \frac{\chi_o \alpha_o^2}{\alpha_1} - \chi_o \alpha_o = -\frac{\chi_o \alpha_o}{\alpha_1} (\alpha_1 - \alpha_o)$$

La relation (18. 3) devient ainsi:

$$(18.4) \quad c^2(f_1^2 - f_o^2) - (\tau_o + \tau_1)(p_1 - p_o) + (\tau_1 - \tau_o)\, \frac{1}{2}\mu(\chi_o + \chi_1 - 2\frac{\chi_o \alpha_o}{\alpha_1}) = 0$$

Nous appelons relation d'Hugoniot cette relation que nous substituerons désormais à ((7. 4) dans le système des cinqu équations scalaires.

c) Un état initial Y_o du fluide étant donné, nous considérons jusqu'à nouvel ordre l'ensemble des états Y vérifiant les conditions:

$$(18.5) \qquad H(Y) = H(Y_o) = H \qquad\qquad L(Y) = L(Y_o) = L$$

de telle sorte que, pour $\tau \neq \frac{\mu H}{c^2}$, on a:

$$\chi = \frac{L}{(c^2\tau - \mu H)^2} = \frac{\chi_o \alpha_o^2}{(c^2\tau - \mu H)^2}$$

Il est commode de substituer à q la variable:

$$\bar{q} = p + \frac{1}{2}\mu\chi = q + \frac{1}{2}\mu \frac{a^2}{1^\alpha 1_\alpha} H$$

La relation (17. 3) peut ainsi s'écrire sous la forme:

$$(18.6) \qquad \bar{q}_1 - \bar{q}_0 = \frac{c^2 a^2}{1^{\alpha} 1_{\alpha}} (\tau_1 - \tau_0)$$

Sous les conditions (18.5) un état thermodynamique (τ, p) du fluide définit (pour $\tau \neq \frac{\mu H}{c^2}$) un point Z du plan ($\tau$, $\bar{q}$) et inversement. Entre les variables τ, S et $\bar{q}$, nous avons la relation fonctionnelle:

$$(18.7) \qquad q = \frac{1}{2} \frac{L}{(c^2\tau - \mu H)^2} + p(\tau, S)$$

où p(τ, S) est définie par inversion de l'équation d'état.

Nous sommes conduits à introduire la fonction d'Hugoniot $\mathcal{H}(Z_0, Z)$ de la magnétohydrodynamique, considérée comme une fonction de Z pour un point initial donné Z_0:

$$(18.8) \qquad \mathcal{H}(Z_0, Z) = c^2(f^2 - f_0^2) - (\tau + \tau_0)(p - p_0) + (\tau - \tau_0)\frac{1}{2}\mu\left(\chi + \chi_0 - 2\frac{\chi_0 \alpha_0}{\alpha}\right)$$

On a manifestement $\mathcal{H}(Z_0, Z_0) = 0$ et (18.4) peut s'écrire $\mathcal{H}(Z_0, Z_1) = 0$. C'est l'étude détaillée du comportement de la fonction $\mathcal{H}$ qui permet d'établir la plupart des théorèmes cherchés.

19. Différentielle de la fonction .

En différentiant (18.8) sous les conditions (18.5), il vient:

$$d\mathcal{H} = 2c^2 f df - (\tau + \tau_0)dp - (p - p_0)d\tau + \frac{1}{2}\mu\left(\chi + \chi_0 - 2\frac{\chi_0 \alpha_0}{\alpha}\right)d\tau +$$
$$+ (\tau - \tau_0)\frac{1}{2}\mu\left(d\chi + 2\frac{\chi_0 \alpha_0}{\alpha^2} d\alpha\right)$$

Or:

$$c^2 f df = f\,\Theta\, dS + \tau\, dp \qquad\qquad d\alpha = c^2 d\tau$$

On en déduit:

$$d\mathcal{H} = 2f\,\Theta\, dS + (\tau - \tau_0)dp - (p - p_0)d\tau + \frac{1}{2}\mu(\chi + \chi_0 - 2\,\frac{\chi_0\alpha_0}{\alpha})d\tau +$$

$$+(\tau - \tau\ )\,\frac{1}{2}\,\mu\,(d\chi + 2c^2\,\frac{\chi_0\alpha_0}{\alpha^2}\,d\tau)$$

En introduisant la variable $\bar{q}$ à la place de la pression p il vient:

$$d\mathcal{H} = 2f\,\Theta\, dS + (\tau - \tau_0)d\bar{q} - (\bar{q} - \bar{q}_0)d\tau + \mu(\chi - \frac{\chi_0\,\alpha_0}{\alpha})d\tau + \ (\ - \ _0$$

$$+\ \mu(\alpha - \alpha_0)\frac{\chi_0\alpha_0}{\alpha^2}\,d\tau$$

Soit après simplification:

$$c\,d\mathcal{H} = 2f\,\Theta\, dS + (\tau - \tau_0)d\bar{q} - (\bar{q} - \bar{q}_0)d\tau + \mu(\chi - \frac{\chi_0\,\alpha_0^2}{\alpha^2})d\tau$$

Ce qui s'écrit, compte-tenu de (17.5):

(19.1) $\qquad d\mathcal{H} = 2f\,\Theta\, dS + (\tau - \tau_0)d\bar{q} - (\bar{q} - \bar{q}_0)d\tau$

Dans le plan $(\tau, \bar{q})$ introduisons commme paramètre la pente m de la droite joignant Z_o à Z. On a:

(19. 2) $$\bar{q}-\bar{q}_o = m\,(\tau - \tau_o)$$

En différentiant il vient

$$d\bar{q} = m\,d\tau + (\tau - \tau_o)dm$$

et en multipliant par $(\tau - \tau_o)$:

$$(\tau - \tau_o)d\bar{q} - (\bar{q}-\bar{q}_o)d\tau = (\tau - \tau_o)^2 dm$$

La relation (19. 1) peut donc s'écrire:

(19. 3) $$d\mathcal{H} = 2f\,\Theta\,dS + (\tau - \tau_o)^2 dm$$

20. <u>Différentielle de S le long de la droite (Z_o, Z_1).</u>

Considérons en $x \in \Sigma$ un choc $Y_o \rightarrow Y_1$. L'état Y_1 postérieur au choc vérifie manifestement les conditions (18. 5). Nous nous proposons d'évaluer la différentielle de S pour une famille d'états telle que le point Z correspondant décrive la droite du plan $(\tau, \bar{q})$ joignant Z_o à Z_1. D'après (17. 3), cette droite a pour pente:

$$m = \frac{c^2 a^2}{l^\alpha l_\alpha}$$

a) De:

$$d\tau = \tau'_p dp + \tau'_S dS$$

on déduit:

$$\tau'_S\, dS = \frac{\gamma - 1}{c^2 r^2}\, dp + d\tau$$

Or, le long de la droite envisagée:

$$d\bar{q} = dp + \frac{1}{2}\mu\, d\chi = \frac{c^2 a^2}{1^{\alpha} 1_{\alpha}}\, d\tau$$

et en différentiant $\chi\alpha^2 = \text{const.}$:

$$\alpha\, d\chi + 2\chi c^2 d\tau = 0$$

On aen déduit:

$$\tau'_S\, \alpha\, dS = \left\{ \frac{\gamma - 1}{r^2} \left(\frac{a^2}{1^{\alpha} 1_{\alpha}}\, \alpha + \mu\chi \right) + \alpha \right\} d\tau$$

Ce qui peut s'écrire, compte-tenu des valeurs de α et χ :

$$(\tau'_S \alpha\, dS = \left\{ \frac{\gamma - 1}{r^2} \left(\frac{a^2}{1^{\alpha} 1_{\alpha}}\, c^2 \tau + \mu |h|^2 \right) + \alpha \right\} d\tau \tag{20.1}$$

b) Le second membre de (20. 1) peut s'évaluer aisément en fonction de P(l). La relation (17. 1) définissant P(l) peut se mettre sous la forme:

$$\frac{P(l)}{a^2} = c^2\tau \frac{a^2}{r^2} (\gamma-1)+(c^2\tau + \mu \frac{|h|^2}{r^2} \gamma)l^\alpha l_\alpha - \mu \frac{(h^\alpha l_\alpha)^2}{a^2} l^\alpha l_\alpha$$

ce qui peut s'écrire:

$$\frac{P(l)}{a^2} = \frac{\gamma-1}{r^2} (c^2 a^2 \tau + \mu |h|^2 l^\alpha l_\alpha)+(c^2\tau - \mu \frac{(h^\alpha l_\alpha)^2}{a^2} + \mu \frac{|h|^2}{r^2})l^\alpha l_\alpha$$

Il en résulte:

$$\frac{P(l)}{a^2} = \frac{\gamma-1}{r^2} (c^2 a^2 \tau + \mu |h|^2 l^\alpha l_\alpha)+ \alpha l^\alpha l_\alpha$$

On déduit ainsi de (20.1) la relation importante: le long de la droite (Z_o, Z_1):

(20.2) $$\tau'_S \, \alpha \, dS = \frac{P(l)}{a^2 l^\alpha l_\alpha} d\tau = \frac{P(l)}{c^2 a^4} d\bar{q}$$

21. Orientation dans le tempos des ondes de choc.

a) Considérons au point $x \in \Sigma$ un choc qui n'est ni nul, ni d'Alfven. Nous nous proposons de montrer que sous les hypothèses (H_1) de compressibilité ($\tau'_p < 0$, $\tau'_S > 0$) que nous postulons, l'onde de choc magnétohydrodynamique Σ est orientée dans le temps.

Considérons la famille des états Y vérifiant (18.5) et telle que le point Z correspondant décrive la droite (Z_o, Z_1). Il résulte de

(19.3) et (20.2) que l'on a dans ces conditions:

(21.1) $$d\mathcal{H} = 2f\,\Theta\, dS$$

et

(21.1) $$\tau'_S\,\alpha\, dS = \frac{P(l)}{c^2 a^4}\, d\bar{q}$$

b) Si $l^\alpha l_\alpha$ est > 0, y^Σ est > 1 et on déduit de (12.2), soit:

$$\frac{(1-y^\Sigma)^2\, P(l)}{(l^\beta l_\beta)^2} = \Pi(y^\Sigma)$$

que P(l) est > 0.
Si $l^\alpha l_\alpha = 0$, on a $P(l) = c^2 rf(\gamma - 1)(u^\alpha l_\alpha)^4 > 0$
Aussi si $l^\alpha l_\alpha \geqslant 0$, on a $P(l) > 0$ et d'après le lemme 2 du § 16, α est > 0. Il résulte de (21.2) que $dS/d\bar{q}$ est > 0 le long de la droite Z_o, Z_1.

D'autre part, le long de cette droite, $\mathcal{H}(Z_o, Z_o) = 0$, $\mathcal{H}(Z_o, Z_1) = 0$ et la fonction $\mathcal{H}$ est stationnaire en un point au moins du segment (Z_o, Z_1). D'après (21.1), il en est de même pour S, ce qui est en contradiction avec $dS/d\bar{q} > 0$.

C'est pourquoi $l^\alpha l_\alpha < 0$ et Σ est orienté dans le temps. Nous énonçons:

Théorème. Sous les hypothèses (H_1) $(\tau'_p < 0, \tau'_S > 0)$, toute onde de choc magnétohydrodynamique est nécessairement orientée dans le temps. Si v_o^Σ et v_1^Σ sont les vitesses de Σ par rapport au fluide avent et après mle choc, on a $v_o^{\Sigma} < c$, $v_1^{\Sigma} < c$.

D'après le lemme 2 du § 16, χ est $\geqslant 0$ et nous posons $\chi = k^2$; nous pouvons, pour un choid convenable des signes substituer à la seconde condition (18.5) la relation $k\alpha = k_o\alpha_o$. On a alors:

$$\chi + \chi_o - 2\frac{\chi_o\alpha_o}{\alpha} = k^2 + k_o^2 - 2kk_o = (k-k_o)^2$$

La fonction d'Hugoniot (18.8) peut s'écrire:

$$\mathcal{H}(Z_o, Z) = c^2(f^2 - f_o^2) - (\tau + \tau_o)(p - p_o) + (\tau - \tau_o)\,\frac{1}{2}\mu(k-k_o)^2 \tag{21.3}$$

et la relation d'Hugoniot:

$$c^2(f_1^2 - f_o^2) - (\tau_1 + \tau_o)(p_1 - p_o) + (\tau_1 - \tau_o)\,\frac{1}{2}\mu(k_1 - k_o)^2 = 0. \tag{21.4}$$

22. Approximation classique de la magnétohydrodynamique relativiste.

Nous nous proposons de déduire les équations de choc de la magnétohydrodynamique classique par approximation à partir de celles correspondant au cadre relativiste. Nous nous limitons aux chocs non tangentiels ($a \neq 0$). Nous posons dans la suite:

$$u^{\alpha} l_{\alpha} = \frac{w}{c}, \qquad j = rw, \qquad l^{\alpha} l_{\alpha} = -1$$

de telle sorte que $a = j/c$.

Cherchons les parties principales, relativemente à c^{-2}, des équations de choc (17.1), (17.2), (17.3), (17.5), (17.6). On a d'abord:

(22.1) $$[j] = [rw] = 0$$

L'invariance de $b = fh^{\alpha} l_{\alpha}$ donne l'équation:

$$\left(1+\frac{i_1}{c^2}\right) h_1^{\alpha} l_{\alpha} = \left(1+\frac{i_o}{c^2}\right) h_o^{\alpha} l_{\alpha}$$

On en déduit qu'à des termes en c^{-4} près:

(22.2) $$h_1^{\alpha} l_{\alpha} = \left(1+\frac{i_o - i_1}{c^2}\right) h_o^{\alpha} l_{\alpha}.$$

On retrouve qu'à l'approximation classique:

(22.3) $$[h^{\alpha} l_{\alpha}] = 0$$

De (17.3) il résulte:

$$[q + j^2 \tau] = 0$$

ce qui peut s'écrire à l'approximation classique:

(22/4) $$rw^2 + p + \frac{1}{2}\mu |h|^2 = 0$$

Compte-tenu de (22.2), l'invariance de

$$H = \frac{c^2}{j^2}(h^{\alpha} l_{\alpha})^2 - \frac{|h|^2}{r^2}$$

donne alors à des termes d'ordre supérieur près:

$$\frac{c^2}{j^2}\left(1+2\frac{i_o-i_1}{c^2}\right)(h_o^\alpha l_\alpha)^2-\frac{|h_1|^2}{r_1^2}=\frac{c^2}{j^2}(h_o^\alpha l_\alpha)^2-\frac{|h_o|^2}{r_o^2}$$

Il en résulte:

$$\left[2i\frac{(h_o^\alpha l_\alpha)^2}{j^2}+\frac{|h|^2}{r^2}\right]=0$$

soit, d'après les relations précédentes:

(22.5)
$$\left[i+\frac{1}{2}\frac{|h|^2}{(h^\alpha l_\alpha)^2}w^2\right]=0$$

La relation (22.5) peut encore s'écrire:

$$\left[i+\frac{1}{2}\frac{|t|^2+(h^\alpha l_\alpha)^2}{(h^\alpha l_\alpha)^2}w^2\right]=0$$

soit

(22.6)
$$\left[i+\frac{1}{2}w^2+\frac{1}{2}\frac{|t|^2}{(h^\alpha l_\alpha)^2}w^2\right]=0$$

Considérons enfin l'invariance de L . Tout d'abord:

$$c^{-2}\alpha=\frac{1}{r}\left(1+\frac{i}{c^2}\right)-\mu\left(\frac{(h^\alpha l_\alpha)^2}{j^2}-\frac{|h|^2}{c^2r^2}\right)$$

est équivalent à $\frac{1}{r} - \mu \frac{(h^{\alpha} l_{\alpha})^2}{j^2}$ à l'approximation classique. D'autre part:

$$k^2 = |t|^2 + \frac{j^2}{c^2} \frac{|h|^2}{r^2}$$

est équivalent à $|t|^2$. Ainsi:

$$\left[|t|^2 \left(\frac{1}{r} - \mu \frac{(h^{\alpha} l_{\alpha})^2}{j^2} \right)^2 \right] = 0 \tag{22.7}$$

Nous avons retrouvé en (22.1), (22.3) (22.4), (22.6), (22.7) les équations de choc de la magnétohydrodynamique classique, écrites dans un repère lié au choc.

23. Thermodynamique des chocs.

a) p et S étant prises comme variables theormodynamiques de base, f(p, S) vérifie d'après (5.3):

$$c^2 f'_p = V > 0 \qquad c^2 f'_S = \Theta > 0 \tag{23.1}$$

On en déduit par dérivation:

$$\frac{\partial}{\partial p} (c^2 f^2) = 2 \tau. \tag{23.2}$$

Les états Z_0 et Z_1 sont reliés par la relation d'Hugoniot (21.4) qui est symétrique en 0 et 1. Au cours d'un choc, on a bien entendu $S_0 \leqslant S_1$ en chaque point de Σ . Nous allons établir les résultats suivants, valables en chaque point de Σ .

Théorème 1. Pour un choc qui cn'est ni nul, ni d'Alfven, on a sous les hypothèses de compressibilité (H_1), (H_2):

$$S_o < S_1$$

En effet supposons qu'au point x de Σ , on ait $S_o = S_1$ et $p_o \neq p_1$. En modifiant au besoin le mnumérotage des états Z_o et Z_1, on peut supposer $p_o < p_1$. On a alors $\tau_o > \tau_1$ puisque $\tau'_p < 0$. De (23.2) on déduit

$$c^2 \left\{ f^2(p_1; S_o) - f^2(p_o, S_o) \right\} = 2 \int_{p_o}^{p_1} \tau(p, S_o) dp$$

Il en résulte d'après la condition de convevité (H_2):

$$c^2(f_1^2 - f_o^2) < (p_1 - p_o)(\tau(p_o, S_o) + \tau(p_1, S_o))$$

soit

$$c^2(f_1^2 - f_o^2) - (\tau_1 + \tau_o)(p_1 - p_o) \leqslant 0$$

On déduit lde la relation d'Hugoniot que $\tau_1 > \tau_o$ ce qui implique contradiction. On a donc $p_1 = p_o$ et le choc envisagé ne peut être que nul ou d'Alfven.

Théorème 2. Pour un choc qui n'est ni nul, ni d'Alfven, on a sous les hypothèses de compressibilité (H_1), (H_2):

$$p_1 > p_o \qquad\qquad f_1 > f_o \qquad\qquad \tau_1 < \tau_o$$

En particulier toute onde de choc est une onde de compression et $\underline{v}_1 \leq \underline{v}_o$.

Supposons en effet $p_1 \leq p_o$ au point x de Σ. De (23. 2), on déduit:

$$c^2\left\{ f^2(p_o, S_o) - f^2(p_1, S_o)\right\} = 2\int_{p_1}^{p_o} \tau(p, S_o)dp$$

D'après la condition de convexité (H_2), il en résulterait:

$$c^2\left\{ f^2(p_o, S_o) - f^2(p_1, S_o)\right\} \leq (p_o - p_1)\left\{\tau(p_o, S_o) + \tau(p_1, S_o)\right\}$$

Comme $S_o < S_1$, on aurait a fortiori puisque $f'_S > 0$, $\tau'_S > 0$

$$c^2\left\{ f^2(p_o, S_o) - f^2(p_1; S_1)\right\} < (p_o - p_1)\left\{\tau(p_o, S_o) + \tau(p_1, S_1)\right\}$$

soit:

$$c^2(f_1^2 - f_o^2) - (\tau_1 + \tau_o)(p_1 - p_o) > 0$$

La relation d'Hugoniot donne alors $\tau_1 < \tau_o$. D'après (H_1) cela est contradictoire avec $p_1 \leq p_o$, $S_1 > S_o$. On a donc $p_1 > p_o$ et, d'après (23. 1), $f_1 > f_o$.

Pour établir $\tau_1 < \tau_o$, on part de:

$$c^2\left\{f^2(p_1,S_1)-f^2(p_o,S_1)\right\}=2\int_{p_o}^{p_1}\tau(p,S_1)\,dp$$

Il en résulte puisque $\tau'_p < 0$:

$$c^2\left\{f^2(p_1,S_1)-f^2(p_o,S_1)\right\}>2\,\tau(p_1,S_1)(p_1-p_o)$$

ou a fortiori:

$$c^2(f_1^2-f_o^2)-2\tau_1(p_1-p_o)>0$$

De la relation d'Hugoniot il résulte alors:

$$(\tau_1-\tau_o)\left\{p_1-p_o+\frac{1}{2}\mu(k_1-k_o)^2\right\}<0$$

soit $\tau_1 < \tau_o$, ce qui démontre le théorème. On a par suite $\alpha_1 \leq \alpha_o$.

VI. ONDES DE CHOC ET ONDES D'ALFVEN.

24. Ondes de choc et ondes d'Alfven.

Considérons à la traversée de Σ un choc non tangentiel qui ne soit pas choc d'Alfven

a) Nous nous proposons d'établir le lemme suivant

Lemme. Il existe toujours au moins une direction n orthogonale à l, $\underline{u}_o, \underline{u}_1, \underline{h}_o, \underline{h}_1$.

Examinons les différents cas possibles :

Lichnerowicz

1) Si $\alpha_o \alpha_1 \neq 0$, il existe une direction n au moins orthogonale aux vecteurs 1;, V et X avec:

$$(24.1) \qquad X^{\beta} = a\alpha_o r_o v_o^{\beta} + \mu \frac{r_o}{a} h_o^{\alpha} l_{\alpha} V^{\beta} = a\alpha_1 r_1 v_1 + \mu \frac{r_1}{a} h_1^{\alpha} l_{\alpha} V^{\beta}$$

De (24.1) il résulte que n est orthogonal à v_o et v_1, donc à u_o et u_1. Comme:

$$(24.2) \qquad V^{\beta} = (h_o^{\alpha} l_{\alpha}) u_o^{\beta} - \frac{a}{r} h_o^{\beta} = (h_1^{\alpha} l_{\alpha}) u_1^{\beta} - \frac{a}{r_1} h_1^{\beta}$$

n est aussi orthogonal à h_o et h_1

2) <u>Si $\alpha_o \neq 0$, $\alpha_1 = 0$,</u> on a $\chi_o = 0$ et il résulte du lemme 2 du § 16 que l est dans le 2-plan (u_o, h_o). Le vecteur V est orthogonal à l dans ce 2-plan. Il existe une direction n au moins orthogonale à l, V et u_1. Cette direction est orthogonale au 2-plan (u_o, h_o) et étant orthogonale à V et u_1 est orthogonale à h_1.

3) <u>Si $\alpha_o = 0$ $\alpha_1 \neq 0$,</u> il suffit d'échanger le rôle des indices 0 et 1.

b) En $x \in \Sigma$, introduisons une perturbation infinitésimale de l'état antérieur au choc. Il en résulte une perturbation infinitésimale de l'état postérieur au choc reliée à la précédente par les relations obtenues en différentiant les équations fondamentales de choc. Il vient ainsi:

$$(24.3)\quad \begin{cases} \left[\delta r u^{\alpha} + r\delta u^{\alpha}\right] l_{\alpha} = 0 \\ \left[\delta(h^{\alpha} l_{\alpha})u^{\beta} + (h^{\alpha} l_{\alpha})\delta u^{\beta} - \delta(\frac{a}{r})h^{\beta} - \frac{a}{r}\delta h^{\beta}\right] = 0 \\ \left[\delta(\beta^2 \frac{a}{r})u^{\beta} + \beta^2 \frac{a}{r}\delta u^{\beta} - \mu\,\delta q l^{\beta} - \delta(h^{\alpha} l_{\alpha})h^{\beta} - h^{\alpha} l_{\alpha}\,\delta h^{\beta}\right] = 0 \end{cases}$$

Adoptons en x un repère orthonormée $\{e_{(\alpha)}\}$ tel que $e_{(1)}$ soit colinéaire à l et $e_{(3)}$ à la direction n. Dans ce repère il vient:

$$u^3_o = 0 \qquad h^3_o = 0 \qquad u^3_1 = 0 \qquad h^3_1 = 0$$

Le système différentiel (24. 3) se partage en deux systèmes dont le premier contient exclusivement les perturbations δu^3, δh^3, soit:

$$(24.4)\qquad (h^{\alpha}_1 l_{\alpha})\,\delta u^3_1 - \frac{a}{r}\delta h^3_1 = (h^{\alpha}_o l_{\alpha})\delta u^3_o - \frac{a}{r_o}\delta h^3_o$$

$$(24.5)\qquad \beta^2_1 \frac{a}{r_1}\,\delta u^3_1 - (h^{\alpha}_1 l_{\alpha})\,\delta h^3_1 = \beta^2_o \frac{a}{r_o}\,\delta u^3_o - (h^{\alpha}_o l_{\alpha})\,\delta h^3_o$$

Nous supposons que <u>seuls δu^3_o, δh^3_o sont ≠0 avant le choc</u>. Les variables thermodynamiques n'ayant pas été perturbées, il en résulte que, dans les états respectivement antérieur ou postérieur au choc Σ , de telles perturbations correspondent à des chocs d'Alfven infinitésimaux, c'est-à-dire à des ondes d'Alfven.

Considérons, dans l'état antérieur à Σ , une onde d'Alfven de type A. Le vecteur A^{α}_o étant invariant à la traversée de cette onde infinitésimale, une telle onde porte en x une perturbation (δu^3_{oA}, δh^3_{oA})

telle que:

(24.6) $$\beta_o \, \delta u^3_{oA} + \delta h^3_{oA} = 0$$

De même une onde d'Alfven de type B porte en x une perturbation $(\delta u^3_{oB}, \delta h^3_{oB})$ telle que:

(24.7) $$\beta_o \, \delta u^3_{oB} - \delta h^3_{oB} = 0$$

La superposition en x d'une onde de type A et d'une onde de type B fournit une perturbation $(\delta u^3_o, \delta h^3_o)$ arbitraire avec:

$$\delta u^3_o = \delta u^3_{oA} + \delta u^3_{oB} \qquad \delta h^3_o = \delta h^3_{oA} + \delta h^3_{oB}$$

c) Les vecteurs A^α_o et B^α_o vérifient en $x \in \Sigma$:

(24.8) $$A^\alpha_o l_\alpha = \beta_o \frac{a}{r_o} + h^\alpha_o l_\alpha \qquad B^\alpha_o l_\alpha = \beta_o \frac{a}{r_o} - h^\alpha_o l_\alpha$$

On en déduit:

$$(A^\alpha_o l_\alpha)(B^\alpha_o l_\alpha) = \beta^2_o \frac{a^2}{r^2_o} - (h^\alpha_o l_\alpha)^2 = \frac{a^2}{\mu} \alpha_o$$

Convenons d'orienter l de l'état antérieur vers l'état postérieur au choc Σ. On a alors $a<0$. Supposons, pour fixer les idées, $b>0$ et par suite $h^\alpha_o l_\alpha > 0$ (resp. $h^\alpha_1 l_\alpha > 0$). D'après (24.8), le vecteur B_o (resp. B_1) est orienté par rapport à Σ du même côté que l. Quant au vecteur A_o (resp. A_1), son orientation par rapport à Σ est celle de 1 ou l'orientation opposée selon que α_o (resp. α_1) est positif ou négatif. Pour α nul, A est tangent à Σ. Si b était < 0, les rôles des vecteurs A et B seraient simplement inversés.

25. Compatibilité d'une onde de choc avec les ondes d'Alfven.

a) Nous examinons d'abord les cas où le choc Σ envisagé, non choc d'Alfven, est tel que $\alpha_o \alpha_1 = 0$. Supposons en premier lieu

$$\alpha_1 = 0 \qquad\qquad \alpha_o > 0$$

Dans l'état Y_o;, les ondes d'Alfven de type A et B qui abortissent en $x \in \Sigma$ peuvent créer en ce point une perturbation (δu^3_o, δh^3_o) arbitraire et, dans (24.4), (24.5), on a puisque $\alpha_1 = 0$

$$\beta_1^2 \frac{a^2}{r_1^2} - (h_1^\alpha l_\alpha)^2 = 0$$

Pour que ces relations (24.4), (24.5) admettent une solution quels que soient δu^3_o, δh^3_o, il faut et il suffit que l'on ait identiquement par rapport à ces variables:

$$\beta_1^2 \frac{a}{r_1} (h_o^\alpha l_\alpha \delta u^3_o - \frac{a}{r_o} \delta h^3_o) - h_1^\alpha l_\alpha (\beta_o^2 \frac{a}{r_o} \delta u^3_o - h_o^\alpha l_\alpha \delta h^3_o) = 0$$

soit:

$$\beta_1^2 \frac{a}{r_1} (h_o^\alpha l_\alpha) - \beta_o^2 \frac{a}{r_o} (h_1^\alpha l_\alpha) = 0 \tag{25.1}$$

(25.2) $$\beta_1^2 \frac{a}{r_1} \frac{a}{r_o} - (h_1^\alpha l_\alpha)(h_o^\alpha l_\alpha)=0$$

Si nous mettons (25.1) sous la forme:

$$\frac{\beta_1^2 \frac{a}{r_1}}{\beta_o^2 \frac{a}{r_o}} = \frac{h_1^\alpha l_\alpha}{h_o^\alpha l_\alpha} = \lambda$$

il vient d'après (25.2):

$$\lambda \alpha_o = 0$$

soit $\lambda=0$ ce qui est impossible. Il y a incompatibilité de l'onde de choc Σ avec les ondes d'Alfven

b) Supposons maintenant:

$$\alpha_1 < 0 \qquad \alpha_o = 0$$

ainsi $\mu H = c^2 \tau_o$ est $>$ 0. On a:

(25.3) $$\beta_o \frac{a}{r_o} + h_o^\alpha l_\alpha = 0$$

Avant le choc, une onde de type B et une onde de type A tangents à Σ peuvent abortir en $x \in \Sigma$ créant une perturbation arbitraire (δu_o^3, δh_o^3). Mais peuvent s'éloigner de x d'une part l'onde de type A pour l'état Y_o portant une perturbation (δu_{oA}^3, δh_{oA}^3) vérifiant:

(25.4) $$\beta_o \, \delta u^3_{oA} + \delta h^3_{oA} = 0$$

d'autre part une onde d'Alfven de type B pour l'état Y_1 portant une perturbation (δu^3_{1B}, δh^3_{1B}) vérifiant

(25.5) $$\beta_1 \, \delta u^3_{1B} - \delta h^3_{1B} = 0$$

Les relations (24.4), (24.5) relient cette dernière perturbation à une perturbation ($\bar{\delta} u^3_o$, $\bar{\delta} h^3_o$) antérieure au choc avec

$$(h_1^\alpha l_\alpha - \beta_1 \frac{a}{r_1}) \, \delta u^3_{1B} = (h_o^\alpha l_\alpha) \, \bar{\delta} u^3_o - \frac{a}{r_o} \, \bar{\delta} h^3_o$$

$$\beta_1 \, (\beta_1 \frac{a}{r_1} - h_1^\alpha l_\alpha) \, \delta u^3_{1B} = \beta_o^2 \, \frac{a}{r_o} \, \bar{\delta} u^3_o - (h_o^\alpha l_\alpha) \, \bar{\delta} h^3_o$$

Ainsi pour qu'une perturbation ($\bar{\delta} u^3_o$, $\bar{\delta} h^3_o$) puisse être transmise à travers le choc Σ et s'éloigner, il faut et il suffit que:

(25.6) $$(\beta_1 h_o^\alpha l_\alpha + \beta_o^2 \, \frac{a}{r_o}) \, \bar{\delta} u^3_o - (\beta_1 \, \frac{a}{r_o} + h_o^\alpha l_\alpha) \, \bar{\delta} h^3_o = 0$$

Nous établirons dans un instant que:

$$\beta_1 \, \frac{a}{r_o} + h_o^\alpha l_\alpha \neq 0$$

L'onde de choc Σ sera compatible avec la perturbation arbitraire (δu^3_o, δh^3_o) s'il existe toujpurs une décomposition

$$\begin{cases} \bar{\delta} u_o^3 + \delta u_{oA}^3 = \delta u_o^3 \\ \bar{\delta} h_o^3 + \delta h_{oA}^3 = \delta h_o^3 \end{cases}$$

avec:

$$\bar{\delta} h_o^3 = \Pi \, \bar{\delta} u_o^3 \qquad\qquad \delta h_{oA}^3 = - \beta_o \delta u_{oA}^3$$

où l'on a posé:

$$\Pi = \frac{\beta_1 h_o^\alpha 1_\alpha + \beta_o^2 \frac{a}{r_o}}{\beta_1 \frac{a}{r_o} + h_o^\alpha 1_\alpha}$$

On est ainsi amené à étudier le système linéaire:

$$\begin{cases} \bar{\delta} u_o^3 + \delta u_{oA}^3 = \delta u_o^3 \\ \Pi \, \bar{\delta} u_o^3 - \beta_o \delta u_{oA}^3 = \delta h_o^3 \end{cases}$$

Le déterminant de (25.7) est donné par:

$$\Pi + \beta_o = \frac{\beta_1 h_o^\alpha 1_\alpha + \beta_o^2 \frac{a}{r_o} + \beta_o (\beta_1 \frac{a}{r_o} + h_o^\alpha 1_\alpha)}{\beta_1 \frac{a}{r_o} + h_o^\alpha 1_\alpha} = \frac{o \quad \#+ \quad 1}{\quad}$$

$$= \frac{\beta_o + \beta_1}{\beta_1 \frac{a}{r_o} + h_o^\alpha 1_\alpha} \left(\beta_o \frac{a}{r_o} + h_o^\alpha 1_\alpha \right) = 0$$

Ainsi (25.7) n'admet pas de solution pour toutes valeurs des seconds membres et Σ n'est pas compatible avec les ondes d'Alfven.

Il nous reste à montrer que $\beta_1 \frac{a}{r_o} + h_o^\alpha l_\alpha \neq 0$. Si non, on aurait $\beta_1 = \beta_o$ ou

$$c^2 r_1 f_1 + \mu |h_1|^2 = c^2 r_o f_o + \mu |h_o|^2$$

ce qui peut s'écrire:

$$r_1^2 \left(c^2 \tau_1 + \mu \frac{|h_1|}{r_1^2}\right) = r_o^2 \left(c^2 \tau_o + \mu \frac{|h_o|^2}{r_o^2}\right)$$

ou

$$r_1^2 \left(\alpha_1 + \mu \frac{(h_1^\alpha l_\alpha)^2}{a^2}\right) = r_o^2 \mu \frac{(h_o^\alpha l_\alpha)^2}{a^2} = r_o^2 \mu \frac{(h_1^\alpha l_\alpha)^2}{a^2} \frac{f_1^2}{f_o^2}$$

Il en résulte:

$$\frac{1}{\tau_1^2} \alpha_1 = \mu \frac{(h_1^2 l_\alpha)^2}{a^2} \left(\frac{1}{\tau_o^2} - \frac{1}{\tau_1^2} \right)$$

soit en divisant par $\alpha_1 / c^2 = \tau_1 - \tau_o$

$$\frac{(h_1^\alpha l_\alpha)^2}{a^2} (\tau_o + \tau_1) = \frac{c^2 \tau_o^2}{} = H \tau_o \left(\frac{(h_1^\alpha l_\alpha)^2}{a^2} - \frac{|h_1|^2}{r_1^2} \right) \tau_o$$

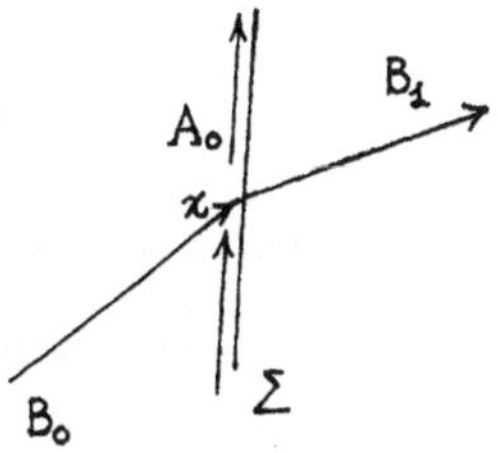

Après simplification il vient:

$$\frac{|h_1|^2}{r_1^2}\tau_o + \frac{(h_1^{\alpha} l_{\alpha})^2}{a^2}\tau_1 = 0$$

Cette relation implique $h_1 = 0$ donc $H=0$, ce qui est absurde.

Nous pouvons énoncer pour un choc non tangentiel.

Théorème 1. Si Σ est une onde de choc telle que $\alpha_o \alpha_1 = 0$, elle est incompatible avec les ondes d'Alfven à moins qu'elle ne corresponde à un choc d'Alfven ($\alpha_o = \alpha_1 = 0$) .

Les chocs singuliers n'ont pas de réalité physique.

c) Supposons maintenant que pour le choc Σ envisagé, on ait

$$0 < \alpha_1 < \alpha_o$$

Les ondes d'Alfven de type A et B dans l'état Y_o créent en x une perturbation (δu_o^3, δh_o^3) arbitraires; α_1 étant $\neq 0$, le système (24. 4), (24. 5) admet une solution unique définissant une perturbation de l'état Y_1 pouvant s'éloigner de x selon les ondes d'Alfven A_1 et B_1 correspondant à cet état. Il y a compatibilité de l'onde de choc avec les ondes d'Alfven

d) Supposons que l'on ait

$$\alpha_1 < 0 < \alpha_o$$

Les ondes d'Alfven de type A et B dans l'état Y_o créent toujours en x une perturbation arbitraire (δu_o^3 δh_o^3)

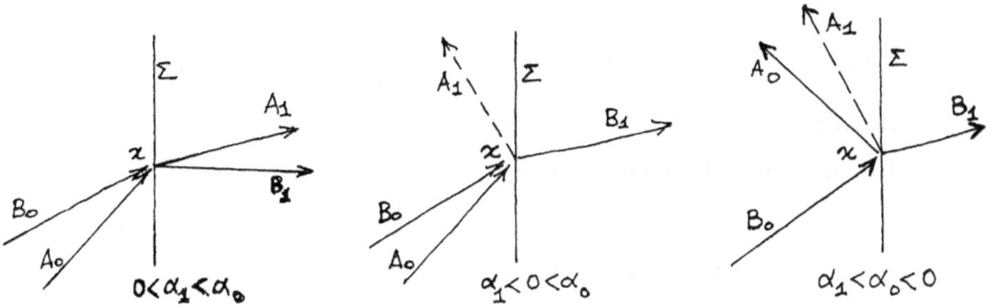

Mais seule s'éloigne de x, dans l'état Y_1, une onde d'Alfven de tipe B portant une perturbation vérifiant

$$\beta_1 \delta u^3_{1B} - \delta h^3_{1B} = 0$$

De (24.4) et (24.5) résulte:

$$\begin{cases} h^\alpha_o l_\alpha \delta u^3_o - \dfrac{a}{r_o} \delta h^3_o = -(\beta_1 \dfrac{a}{r_1} - h^\alpha_1 l_\alpha) \delta u^3_{1B} \\ \beta^2_o \dfrac{a}{r_o} \delta u^3_o - h^\alpha_o l_\alpha \delta h^3_o = \beta_1 (\beta_1 \dfrac{a}{r_1} - h^\alpha_1 l_\alpha) \delta u^3_{1B} \end{cases}$$

Il est nécessaire que $(\delta u^3_o, \delta h^3_o)$ vérifie la rellation:

(25.3) $$\left(\beta^2_o \frac{a}{r_o} + \beta_1 h^\alpha_o l_\alpha\right) \delta u^3_o - \left(\beta_1 \frac{a}{r_o} + h^\alpha_o l_\alpha\right) \delta h^3_o = 0$$

Cette relation ne peut être une identité en $(\delta u^3_o, \delta h^3_o)$, sinon l'on aurait

$$\beta_1 = \beta_o \qquad \beta_o \frac{a}{r_o} + h^\alpha_o l_\alpha = 0$$

et par suite $\alpha_o=0$. Il ya a ainsi incompatibilité de l'onde de choc Σ avec les ondes d'Alfven

e) Supposons enfin que l'on ait:

$$\alpha_1 < \alpha_o < 0$$

Avant le choc, seule une onde d'Alfven de type B aboutit en x créant une perturbation (δu^3_{oB}, δh^3_{oB}) vérifiant

$$\beta_o \ \delta u^3_{oB} - \delta h^3_{oB} = 0$$

Mais peuvent s'éloigner de x une onde d'Alfven de type A pour l'état Y_o et une onde d'Alfven de type B pour l'état Y_1 qui avant le choc correspond à une perturbation ($\bar{\delta} u^3_o$, $\bar{\delta} h^3_o$). Nous posons:

$$\delta u^3_{oB} = \bar{\delta} u^3_o + \delta u^3_{oA} \qquad \delta h^3_{oB} = \bar{\delta} h^3_o + \delta h^3_{oA}$$

avec, d'après (25.3):

$$\bar{\delta} h^3_o = \Pi \bar{\delta} u^3_o \qquad \delta h^3_{oA} = -\beta_o \delta u^3_{oA}$$

où:

$$\Pi = \frac{\beta_o^2 \frac{a}{r_o} + \beta_1 h_o^\alpha l_\alpha}{\beta_1 \frac{a}{r_o} + h_o^\alpha l_\alpha} \qquad \left(\text{pour } \beta_1 \frac{a}{r_o} + h_o^\alpha l_\alpha = 0\right)$$

Lichnerowiczz

Noas sommes amenés comme au b à resoudre les équations:

$$\bar{\delta} u_o^3 + \delta u_{oA}^3 = \delta u_{oB}^3$$

$$\Pi \bar{\delta} u_o^3 - \beta_o \delta u_{uoA}^3 = \beta_o \delta u_{oB}^3$$

où le déterminant

$$\Pi + \beta_o = \frac{\beta_o + \beta_1}{\beta_1 \frac{a}{r_o} + h_o^\alpha l_\alpha} \quad (\beta_o \frac{a}{r_o} + h_o^\alpha l_\alpha) \neq 0$$

Si $\beta_1 \frac{a}{r_o} + h_o^\alpha l_\alpha = 0$ on obtient aussi trivialement une solution.

Il y a compatibilité de l'onde de choc Σ avec les ondes d'Alfven. Nous énonçons

Théorème 2. Pour qu'un choc non d'Alfven soit compatible avec les ondes d'Alfven, il faut et il suffit que $\alpha_o \alpha_1 \geq 0$

f) Les chocs envisagés peuvent donc être décomposés en chocs lents tels que

$$\alpha_1 < \alpha_o < 0$$

et chocs rapides tels que:

$$0 < \alpha_1 < \alpha_o$$

Pourtant de la relation fondamentale:

$$k_1^2 \alpha_1^2 = k_o^2 \alpha_o^2$$

Pour un choc lent $\alpha_1^2 > \alpha_o^2$, donc $k_1^2 < k_o^2$ et par suite $|h_1|^2 < |h_o|^2$. Pour un choc rapide, les inégalités sont inversées. On a donc :

Proposition. Dans un choc lent, la grandeur de champ magnétique diminue; dans un choc rapide elle augmente.

VII. VITESSES DES ONDES DE CHOC ET THEOREMES FONDAMENTAUX.

26. Entropie et courbes isentropiques.

a) Un état initial Y_o étant donné, nous considérons la suite des états Y vérifiant

$$(26.1) \qquad H(Y)=H(Y_o)=H \qquad k\alpha = k_o \alpha_o \quad (\text{avec } \alpha\alpha_o > 0)$$

et étudions les points représentatifs correspondants Z dans le plan $(\tau, \bar{q})$; si H est > 0 la droite $\tau = \mu H/c^2$ est interdite. Dans le plan $(\tau, \bar{q})$, la courbe isentropique $\mathcal{S}$ correspondant à la valeur S de l'eutropie est définie par:

$$(26.2) \qquad \bar{q} = p(\tau, S) + \frac{1}{2} \frac{\mu k_o^2 \alpha_o^2}{(c^2\tau - \mu H)^2} = \varphi(\tau, S)$$

En dérivant (26.2) le long de $\mathcal{S}$, il vient:

$$\left(\frac{d\bar{q}}{d\tau}\right)_{\mathcal{S}} = p'_{\tau} - \frac{c^2 \mu k_o^2 \alpha_o^2}{\alpha^3}$$

et en dérivant une seconde fois:

$$\left(\frac{d^2\bar{q}}{d\tau^2}\right)_{\mathcal{S}} = p''_{\tau^2} + 3c^4 \mu \frac{k_o^2 \alpha_o^2}{\alpha^4}$$

On en déduit d'après (8.5)

(26.3) $$\left(\frac{d^2\bar{q}}{d\tau^2}\right)_{\mathcal{S}} = - \frac{1}{\tau'^3_p} M$$

où l'on a posé:

(26.4) $$M = \tau''_{p^2} - 3c^4 \mu \frac{k^2}{\alpha^2} \tau'^3_p$$

qui est <u>strictement positif</u> en vertu des hypothèses (H_1), (H_2). Ainsi <u>les courbes isentropiques $\mathcal{S}$ du plan $(\tau, \bar{q})$ sont convexes</u>

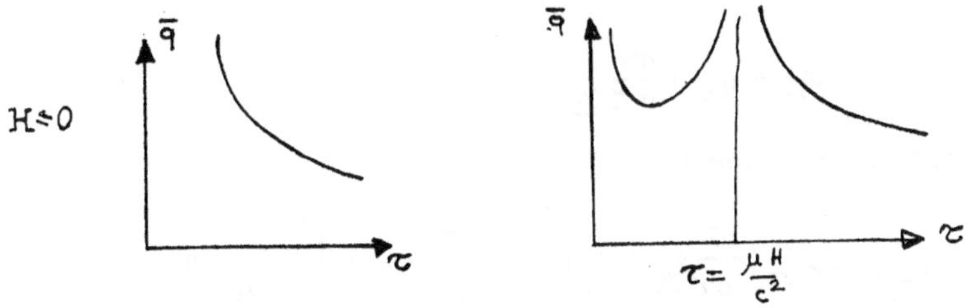

Notons que la fonction $\varphi(\tau, S)$ définie par (26.2) vérifie

(26.5) $$\varphi'_S = - \frac{\tau'_S}{\tau'_p} \qquad \varphi''_{\tau^2} = - \frac{1}{\tau'^3_p} M$$

Lichnerowicz

b) Soit Δ une droite de pente m du plan $(\tau, \bar{q})$ issue du point Z_0. On a, le long de Δ

$$\bar{q}-\bar{q}_0 = m(\tau - \tau_0) \qquad d\bar{q}=m\,d\tau$$

Lemme. En tout point $Z_{\mathcal{S}}$ de Δ où S est stationnaire, on a:

$$\left(\frac{d^2S}{d\tau^2}\right)_{\mathcal{S}} = -\left(\frac{1}{\tau_p'^2\,\tau_S'}M\right)_{\mathcal{S}}$$

Aussi $\left(\frac{d^2S}{d\tau^2}\right)_{\mathcal{S}}$ est <0, sous les hypothèses (H_1), (H_2).

Partons de la relation fonctionnelle:

$$\bar{q} = \varphi(\tau, S)$$

et dérivons la le long de Δ par rapport au paramètre τ. Il vient:

$$\varphi'_\tau + \varphi'_S \left(\frac{dS}{d\tau}\right)_\Delta = m$$

et en dérivant une seconde fois:

$$\varphi''_{\tau^2} + 2\varphi''_{S\tau}\left(\frac{dS}{d\tau}\right)_\Delta + \varphi''_{S^2}\left(\frac{dS}{d\tau}\right)^2_\Delta + \varphi'_S\left(\frac{d^2S}{d\tau^2}\right)_\Delta = 0$$

En un point $Z_{\mathcal{S}}$ de Δ où $(dS/d\tau)_{\mathcal{S}}$ s'annule, on a:

$$(\varphi''_{\tau^2})_{\mathcal{S}} + (\varphi'_S)_{\mathcal{S}}\left(\frac{d^2S}{d\tau^2}\right)_{\mathcal{S}} = 0$$

On en déduit d'après (26.5)

$$\left(\frac{d^2S}{d\tau^2}\right)_{\mathcal{S}} = -\left(\frac{\tau'_p}{\tau'_S}\right)\frac{1}{\tau'^3_p} M)_{\mathcal{S}} = -\left(\frac{1}{\tau'^2_p\,\tau'_S} M\right)_{\mathcal{S}}$$

ce qui démontre le lemme.

c) Considérons dans le plan $(\tau, \bar{q})$, la courbe d'Hugoniot $\mathcal{H}$ définie par l'équation $\mathcal{H}(Z_o, Z)=0$. On déduit de (19.1) que le long de $\mathcal{H}$:

(26.6) $$2f\,\Theta\,\left(\frac{dS}{d\tau}\right)_{\mathcal{H}} + (\tau - \tau_o)\left(\frac{d\bar{q}}{d\tau}\right)_{\mathcal{H}} - (\bar{q} - \bar{q}_o) = 0$$

Aussi au point Z_o, on a $(dS/d\tau)_{\mathcal{H}} = 0$. En dérivant (26.6), on obtient:

(26.7) $$2f\,\Theta\,\left(\frac{d^2S}{d\tau^2}\right)_{\mathcal{H}} + 2\left(\frac{d(f\,\Theta)}{d\tau}\right)_{\mathcal{H}}\left(\frac{dS}{d\tau}\right)_{\mathcal{H}} + (\tau - \tau_o)\left(\frac{d^2\bar{q}}{d\tau^2}\right)_{\mathcal{H}} = 0$$

et en Z_o, on a aussi $(d^2S/d\tau^2)_{\mathcal{H}} = 0$. Aussi la courbe d'Hugoniot et l'isentropique $S=S_o$ ont un contact du second ordre en Z_o. En ce point, nous avons donc:

(26.8) $$\left(\frac{d^2\bar{q}}{d\tau^2}\right)_{\mathcal{H}} = \left(\frac{d^2\bar{q}}{d\tau^2}\right)_{\mathcal{S}} = -\left(\frac{1}{\tau'^3_p} M\right)_{Z_o}$$

En dérivant (26.7), on obtient en Z_o:

$$2f\,\Theta\,\left(\frac{d^3S}{d\tau^3}\right)_{\mathcal{H}} + \left(\frac{d^2\bar{q}}{d\tau^2}\right)_{\mathcal{H}} = 0$$

Il résulte ainsi de (26.8):

(26.9) $$\left(\frac{d^3S}{d\tau^3}\right)_{\mathcal{H}} = \left(\frac{1}{2f\,\Theta}\,\frac{1}{{\tau'}_p^3}\,M\right)_{Z_0}$$

D'après le contact entre $\mathcal{H}$ et l'isentropique $\mathcal{S}$ $(S=S_0)$, on a en Z_0:

$$\left(\frac{d\tau}{dp}\right)_{\mathcal{H}} = \left(\frac{d\tau}{dp}\right)_{\mathcal{S}} = (\tau'_p)_{Z_0}$$

et il vient en ce point:

$$\left(\frac{d^3S}{dp^3}\right)_{\mathcal{H}} = \left(\frac{d^3S}{d\tau^3}\right)_{\mathcal{H}} \left(\frac{d\tau}{dp}\right)^3_{\mathcal{H}} = \left(\frac{d^3S}{d\tau^3}\right)_{\mathcal{H}} ({\tau'}_p^3)_{Z_0}$$

On obtient ainsi en Z_0:

(26.10) $$\left(\frac{d^3S}{dp^3}\right)_{\mathcal{H}} = \left(\frac{M}{2f\,\Theta}\right)_{Z_0}$$

et dans le voisinage de Z_0, nous avons le long de $\mathcal{H}$:

(26.11) $$S-S_0 = \left(\frac{M}{2f\,\Theta}\right)_{Z_0} (p-p_0)^3 + \dots$$

où le coefficient de $(p-p_0)^3$ est positif. Nous énonçons:

Théorème. La courbe d'Hugoniot et l'isentropique $S=S_0$ du plan $(\tau, \bar{q})$ ont au point initial Z_0 un contact du second ordre. Pour un choc faible, l'accroissement d'entropie est du troisième ordre par rapport à la puissance du choc mesurée par l'accroissement de pression

27. Conditions nécessaires sur les vitesses des ondes de choc.

a) Considérons un choc qui, au point x de Σ , fait passer de Z_o à Z_1. Nous avons vu que le long de la droite Δ joignant Z_o à Z_1, la fonction $\mathcal{H}(Z_o, Z)$ est stationnaire en un point au moins Z_s situé entre Z_o et Z_1 et que d'après (21.1), il en est de même pour S. Du lemme du § 26, il résulte que le point Z_s de Δ où S est stationnaire est unique et correspond à un maximum strict de S sur Δ La variable τ décroissant le long de Δ orientée de Z_o vers Z_1, on a:

$$\frac{dS}{d\tau} < 0 \quad \text{pour } \tau > \tau_s \qquad\qquad \frac{dS}{d\tau} > 0 \text{ pour } \tau < \tau_s$$

et aussi:

$$\left(\frac{dS}{d\tau}\right)_o < 0 \qquad\qquad \left(\frac{dS}{d\tau}\right)_1 > 0$$

On en déduit d'après (20.2):

$$(27.1) \qquad \alpha_o P(1)_o > 0 \qquad\qquad \alpha_1 P(1)_1 < 0$$

b) La relation (12.5) peut s'écrire:

$$(27.2) \qquad \frac{a^2}{(c^2 rf + \mu |h|^2)((u^\alpha l_\alpha)^2 - l^\alpha l_\alpha)} \alpha = \frac{(v^\Sigma)^2}{c^2} - \frac{(v^A)^2}{c^2}$$

Lichnerowicz

Considérons un choc rapide $(0 < \alpha_1 < \alpha_o)$. D'après (27.2), on a:

$$v_o^{\Sigma} > v_o^{A} \qquad v_1^{\Sigma} > v_1^{A}$$

De (27.1) il résulte que $P(1)_o$ est > 0 et $P(1)_1 < 0$; d'après (12.2), il en est de même pour $\Pi \left(\frac{(v^{\Sigma})^2}{c^2}\right)$. On en déduit

$$v_o^{\Sigma} > v_o^{MR} \qquad v_1^{\Sigma} > v_1^{MR}$$

Considérons maintenant un choc lent $(\alpha_1 < \alpha_o < 0)$. D'après (27.2):

$$v_o^{\Sigma} < v_o^{A} \qquad v_1^{\Sigma} < v_1^{A}$$

De (27.1) il résulte que $P(1)_o$ est < 0 et $P(1)_1 > 0$..On en déduit de même

$$v_o^{\Sigma} > v_o^{ML} \qquad v_1^{\Sigma} < v_1^{ML}$$

Nous obtenons:

Théorème. Sous les hypothèses de compressibilité (H_1), (H_2), les vitesses v_o^{Σ} et v_1^{Σ} par rapport au fluide d'une onde de choc, respectivement avant ou après le choc, vérifient vnécessairement les inégalités suivantes:

1) pour un choc rapide

$$v_o^{ML} < v_o^{A} < v_o^{MR} < vv_o^{\Sigma} \qquad v_1^{ML} < v_1^{A} < v_1^{\Sigma} < v_1^{MR}$$

2) pour un choc lent:

$$v_o^{ML} < v_o^{\Sigma} < v_o^{A} < v_o^{MR} \qquad v_1^{\Sigma} < v_1^{ML} < v_1^{A} < v_1^{MR}$$

où interviennent les vitesses magnétosoniques lentes et rapides et les vitesses d'Alfven avant et après le choc.

(27.1) résume la situation.

28. Théorème d'existence et d'unicité.

a) En inversant la fonction $\tau = \tau(p, S)$, on obtient une fonction $S = S(p, \tau)$ exprimant l'entropie en fonction des variables p et τ. En raisonnant comme au § 8, on obtient:

$$\tau'_p = -\frac{S'_p}{S'_\tau} \qquad \tau'_S = \frac{1}{S'_\tau}$$

Les conditions de compressibilité (H_1) se traduisent par les inégalités $S'_\tau > 0$, $S'_p > 0$.

Par une nouvelle dérivation, on obtient:

$$\tau''_{p^2} = -\frac{S''_{p^2}(S'_\tau)^2 - 2S''_{p\tau}S'_p S'_\tau + S''_{\tau^2}(S'_p)^2}{(S'_\tau)^3} > 0$$

Nous supposons ici que la fonction $S(p, \tau)$ satisfait les hypothèses (H_1) et (H_2) pour $\tau \leq \tau_o$ et paur des valeurs arbitrairement grandes de p. Nous nous proposons de montrer que si $v_o^A < v_o^{MR} < v_o^{\Sigma}$ (ou $v_o^{ML} < v_o^{\Sigma} < v_o^{A}$) c'est-à-dire $\alpha_o P(1)_o > 0$, les équations générales de choc admettent une solution unique non triviale avec $\alpha_o \alpha_1 > 0$.

b) Considérons dans le plan $(\tau, \bar{q})$ la branche de l'isentropique $\mathcal{S}$ correspondant à $S=S_o$ qui est telle que $\tau \leqslant \tau_o$ avec $\alpha\alpha_o > 0$. Le long de $\mathcal{S}$, nous avons d'après (19.1)

$$\left(\frac{d\mathcal{H}}{d\tau}\right)_{\mathcal{S}} = (\tau - \tau_o)\left(\frac{d\bar{q}}{d\tau}\right)_{\mathcal{S}} - (\bar{q} - \bar{q}_o)$$

En dérivant une seconde fois, il vient:

$$\left(\frac{d^2\mathcal{H}}{d\tau^2}\right)_{\mathcal{S}} = (\tau - \tau_o)\left(\frac{d^2\bar{q}}{d\tau^2}\right)_{\mathcal{S}}$$

soit d'après (26.3):

$$\left(\frac{d^2\mathcal{H}}{d\tau^2}\right)_{\mathcal{S}} = -(\tau - \tau_o)\frac{1}{\tau'^3_p} M < 0$$

Aussi sur la branche considérée, $(d\mathcal{H}/d\tau)_{\mathcal{S}}$ est >0 et <u>$\mathcal{H}(Z_o, Z)$ est <0</u>.

c) Soit Δ une droite de pente m issue de Z_o. Il résulte de (19.3) que le long de Δ :

$$\left(\frac{d\mathcal{H}}{d\tau}\right)_{\Delta} = 2f\,\Theta\left(\frac{dS}{d\tau}\right)_{\Delta} \tag{28.1}$$

Soit m_o la pente de l'isentropique $\mathcal{S}$ en Z_o. De l'hypothèse globale faite, il résulte que le long de la branche envisagée de $\mathcal{S}$, $\bar{q}$ peut prendre des valeurs arbitrairement grandes. Or la courbe $\mathcal{S}$ est convexe. On en déduit que si <u>$m<m_o$</u>; la droite Δ rencontre la branche considérée de $\mathcal{S}$ en un point unique $Z_A \neq Z_o$ et pour ce point:

$$\mathcal{H}(Z_o, Z_A) < 0$$

Mais le long de Δ , $S(Z_o)=S(Z_A)=S_o$ et $S(\tau)$ est nécessairement stationnaire entre Z_o et Z_A en un point $Z_{\mathcal{S}}$ nécessairement unique d'après le lemme et qui est un maximum strict pour S sur Δ

Nous avons donc $(dS/d\tau)_\Delta < 0$ en Z_o et d'après (28.1), $(d\mathcal{H}/d\tau)_\Delta < 0$ en Z_o. Quand Z décrit Δ de Z_o vers Z_A, $\mathcal{H}(Z_o, Z)$ commence par être positif et il existe donc au moins un point Z_1 sur Δ entre Z_o et Z_A pour lequel $\mathcal{H}(Z_o, Z_1)=0$.

Nous savons que sur Δ, S croit de Z_o à $Z_{\mathcal{S}}$, passe par un maximum en $Z_{\mathcal{S}}$ et décroit constamment ensuite. D'après (28.1), $\mathcal{H}$ a le même comportement et le point Z_1 est nécessairement unique.

Aussi <u>à tout nombre $m < m_o$, correspond un point unique Z_1 de la courbe d'Hugoniot de Z_o avec $\tau_1 < \tau_o$ (et $\alpha_1 \alpha_o > 0$) tel que $m = \bar{q}_1 - \bar{q}_o / \tau_1 - \tau_o$</u>.

Nous voyons que m fournit une paramétrisation simple de la branche considérée de la courbe d'Hugoniot. Nous notons qu'il reésulte de (19.3) que le long de $\mathcal{H}$:

$$2f\, \Theta\, dS + (\tau - \tau_o)^2 dm = 0$$

et que par suite <u>S croit quand m décroit.</u>

Pour $m = \frac{c^2 a^2}{1^{\alpha}\, 1^{\alpha}} < 0$, Z_1 satisfait (17.3) et $\mathcal{H}(Z_o, Z_1)=0$; Z_1 étant connu, τ_1 et p_1 le soit et f_1 est donné par exemple par la relation d'Hugoniot. De (17.5), on déduit la valeur de $|h_1|^2$, de (17.1) celle de $u^{\alpha}_{1} l_{\alpha}$ et de (17.2) la valeur de $h^{\alpha}_{1} l_{\alpha}$ (qui a le signe de $h^{\alpha}_{o} l_{\alpha}$). Les équations (17.1), (17.2), (17.3), (17.4), (17.5) ont aussi une solution unique telle que $\alpha_1 \alpha_o > 0$

d) Composons la tcondition

$$c^2 a^2 / l_{\alpha} l^{\alpha} < m_o$$

et la condition nécessaire trouvée au § 27 :

$$\alpha_o \, P(1)_o) > 0$$

D'après les résultats du § 320, on a:

$$\frac{P(1)}{a^2 (l^{\alpha} l_{\alpha})} = \frac{\gamma - 1}{r^2} \left(\frac{a^2}{l^{\alpha} l_{\alpha}} \alpha + \mu k^2 \right) + \alpha \tag{28.2}$$

Or

$$c^2 \tau'_p = - \frac{\gamma - 1}{r^2}$$

En divisant (28.2) par $\tau'_p \alpha$, il vient:

$$\frac{1}{a^2 \tau'_p (l^{\alpha} l_{\alpha})} \frac{P(1)}{\alpha} = - \frac{c^2 a^2}{l^{\alpha} l_{\alpha}} + \frac{1}{\tau'_p} - \frac{c^2 \mu k^2}{\alpha}$$

Or d'après le § 26 a):

$$\left(\frac{d\bar{q}}{d\tau}\right)_{\mathcal{J}} = \frac{1}{\tau'_p} - \frac{c^2\mu\, k_o^2\alpha_o^2}{\alpha^3} = \frac{1}{\tau'_p} - \frac{c^2\mu k^2}{\alpha}$$

On obtient aussi:

$$(28.3) \qquad \frac{1}{a^2\,\tau'_p\,(l^\alpha l_\alpha)}\;\frac{P(l)}{\alpha} = \left(\frac{d\bar{q}}{d\tau}\right)_{\mathcal{J}} - \frac{c^2a^2}{l^\alpha l_\alpha}$$

Au point Z_o il vient:

$$\frac{1}{a^2(\tau'_p)_o(l^\alpha l_\alpha)}\;\frac{P(l)_o}{\alpha_o} = m_o - \frac{c^2a^2}{l^\alpha l_\alpha}$$

On voit aussi que $c^2a^2/l^\alpha l_\alpha < m_o$ est équivalent à $\alpha_o P(l)_o > 0$. Nous énonçons

Théorème. Si la fonction $S(p,\tau)$ satisfait les hypothèses (H_1), (H_2) pour $\tau \leq \tau_o$ et pour des valeurs arbitrairement grandes de p, à tout état Y_o vérifiant $\alpha_o P(l)_o > 0$ correspond une solution unique non triviale des équations de choc telle que $\alpha_1\alpha_o \geq 0$.

Les inégalités figurant dans cet énoncé s'interprètent immédiatement, comme au § 27, en termes de vitesses de l'onde de choc et des vitesses magnétosoniques et d'Alfven.

Lichnerowicz

VIII. RETOUR AUX RAYONS MAGNETOSONIQUES.

29. Etude de la direction N^{β}.

a) A partir des équations fondamental de la magnétohydrodynamique, on déduit que pour une onde infinitésimale $\dot{\Sigma}$ non engendrée par des lignes de courant ($u^{\alpha} l_{\alpha} \neq 0$), on a d'abord:

$$\delta(ru^{\alpha} l_{\alpha})=0 \qquad \delta(h^{\alpha} l_{\alpha} u^{\beta} - u^{\alpha} l_{\alpha} h^{\beta})=0 \qquad \delta(T^{\alpha\beta} l_{\alpha})=0$$

Aussi on a l'invariance par δ du scalaire et des deux vecteurs:

$$a = ru^{\alpha} l_{\alpha} \neq 0 \qquad V^{\beta} \qquad X^{\beta}$$

Les invariants scalaires de choc mis en évidence aux §§ 16 et 17 sont donc aussi des invariants pour l'opérateur δ de discontinuité infinitésimale.

Nous supposons l'onde envisagé non pas d'Alfven ($\alpha \neq 0$), mais magnétosonique de telle sorte que

$$P(l)=0$$

D'après les calculs du § 20, ceci peut s'écrire:

$$\frac{\gamma - 1}{r^2}(c^2 a^2 \tau + \mu |h|^2 l^{\alpha} l_{\alpha}) + \alpha\, l^{\alpha} l_{\alpha} = 0 \quad (c^2 a^2 \tau + \mu |h|^2 l^{\alpha} l_{\alpha} \neq 0) \tag{29.1}$$

b) Transformons de même l'expression du vecteur N^{β} qui donne la direction du rayon (voir (14.3))

$$N^{\beta} = 2c^2\tau(\gamma-1)\frac{a^2}{r^2}\, ar\, v^{\beta} + \left(c^2\tau + \mu\frac{|h|^2}{r^2}\gamma\right)(l^{\alpha} l_{\alpha})ar\, v^{\beta} -$$

$$-\mu(h^{\alpha} l_{\alpha})(l^{\rho} l_{\rho})\, t^{\beta}$$

Compte-tenu de:

$$t^{\beta} = \frac{r}{a}(h^{\alpha} l_{\alpha})\, v^{\beta} - \frac{r}{a}\, V^{\beta}$$

il vient:

$$N^{\beta} = \left\{2c^2\tau\frac{a^2}{r^2}(\gamma-1) + \left(c^2\tau + \mu\frac{|h|^2}{r^2}\gamma\right) l^{\alpha} l_{\alpha} - \mu\frac{(h^{\alpha} l_{\alpha})^2}{a^2}\, l^{\rho} l_{\rho}\right\} ar\, v^{\beta} +$$

$$+\mu\frac{r}{a}(h^{\alpha} l_{\alpha}) l^{\rho} l_{\rho}\, V^{\beta}$$

Or d'après (29. 1) (c'est-à-dire P(l)=0)

$$c^2\tau\frac{a^2}{r^2}(\gamma-1) + \left(c^2\tau + \mu\frac{|h|^2}{r^2}\gamma\right) l^{\rho} l_{\rho} - \mu\frac{(h^{\alpha} l_{\alpha})^2}{a^2}\, l^{\rho} l_{\rho} = 0$$

et l'on a:

$$N^{\beta} = c^2\tau\frac{a^2}{r^2}(\gamma-1)ar\, v^{\beta} + \mu\frac{r}{a}(h^{\alpha} l_{\alpha}) l^{\rho} l_{\rho}\, V^{\beta}$$

c'est-à-dire en tirant $\gamma - 1/r^2$ de (29. 1):

$$N^{\beta} = -\frac{c^2a^2\tau\, l^{\rho} l_{\rho}}{c^2a^2\tau + \mu|h|^2 l^{\rho} l_{\rho}}\, a\alpha r\, v^{\beta} + \mu \frac{r}{a}(h^{\alpha} l_{\alpha}) l^{\rho} l_{\rho} V^{\beta}$$

On déduit de (16/5):

$$N^{\beta} = -\frac{c^2a^2\tau\, l^{\rho} l_{\rho}}{c^2a^2\tau + \mu|h|^2 l^{\rho} l_{\rho}} (X^{\beta} - \mu \frac{r}{a}(h^{\alpha} l_{\alpha}) V^{\beta}) + \mu \frac{r}{a}(h^{\alpha} l_{\alpha}) l^{\rho} l_{\rho} V^{\beta}$$

La direction de N^{β} est donc celle du vecteur proportionnel:

$$-c^2a^2\tau X^{\beta} + \mu \frac{r}{a}(h^{\alpha} l_{\alpha})(2c^2a^2\tau + \mu|h|^2 l^{\rho} l_{\rho}) V^{\beta}$$

En utilisant $b = f h^{\alpha} l_{\alpha}$ et en divisant par τ, on obtient le vecteur colinéaire à N^{β}:

(29. 2) $$M^{\beta} = -c^2a^2X^{\beta} + 2\mu \frac{b}{a} Q\, V^{\beta} \quad (X^{\beta}, V^{\beta} \text{ non colinéaires})$$

où l'on a posé:

(29. 3) $$Q = \frac{c^2a^2}{\tau} + \frac{1}{2}\frac{\mu|h|^2}{\tau^2} l^{\rho} l_{\rho}$$

30. Action de δ sur la direction du rayon.

a) Pour que la direction de N^{β} soit invariante par δ, il faut et il suffit que:

$$\delta M^{\beta} = 2\mu \frac{b}{a} \delta Q V^{\beta} = 0$$

Il en est donc en particulier ainsi pour b=0, c'est-à-dire si le champ magnétique est tangentiel

b) Cherchons à évaluer δQ. Il vient:

$$\tau^2 \delta Q = -c^2 a^2 \delta\tau + \frac{1}{2} l^\rho l_\rho \mu |h|^2 - l^\rho l_\rho \mu |h|^2 \frac{\delta\tau}{\tau} \tag{30.1}$$

D'autre part, d'après l'invariance de l par δ, on a:

$$\delta \left(p + \frac{1}{2}\mu |h|^2 \right) - \frac{c^2 a^2}{l^\rho l_\rho} \delta\tau = 0$$

c'est-à-dire:

$$-c^2 a^2 \delta\tau + \frac{1}{2} l^\rho l_\rho \mu \delta |h|^2 = -l^\rho l_\rho \delta p$$

En reportant dans (30.1), on en déduit, compte-tenu de $\delta S=0$

$$\tau^2 \delta Q = \frac{-l^\rho l_\rho}{\tau} \left\{ \tau + \mu |h|^2 \tau'_p \right\} \delta p$$

soit, comme:

$$c^2 \tau'_p = -\frac{\gamma - 1}{r^2}$$

on a:

$$\tau^2 \delta Q = -\frac{l^\rho l_\rho}{c^2_{rf}} \left\{ c^2_{rf} - \mu |h|^2 (\gamma - 1) \right\} \delta p \tag{30.2}$$

Pour que δQ soit nul, il faut et il suffit que

$$\gamma = \frac{c^2 rf + \mu |h|^2}{\mu |h|^2}$$

c'est-à-dire: que:

(30. 3) $$\frac{v^2}{\mathbf{v}^2} = \frac{\mu |h|^2}{c^2 rf + \mu |h|^2}$$

Nous avons ainsi établi que, contrairement aux résultats concernant les rayons associés aux ondes soniques et aux ondes d'Alfven (en magnétohydrodynamique) , la direction des rayons associés aux ondes magnétosoniques n'est pas en général invariante par l'opérateur δ de discontinuité infinitésimale.

Il y a invariance seulement si le champ magnétique est tangentiel ou si la relation (30. 3) est satisfaite. Mais celle-ci n'est autre que (13. 1). Cette relation correspond donc au cas singulier où le cône P(l)=0 admet deux génératrices doubles.

A l'approximation classique la relation (30. 3) s'écrit:

(30. 4) $$\mu |h|^2 = r\, v^2 .$$

BIBLIOGRAPHIE

[1] F. Hoffman et E. Teller, Phys. Rev. t. 80, p. 692, (1950).

[2] A. H. Taub, Arch. Rat. Mech. Anal. t. 3, p. 312, (1959).

[3] Y. Choquet-Bruhat, Astron. Acta t. 6, p. 354, (1960)

[4] W. Israel, Proc. Roy. Soc. A. 259, p. 129, (1960).

[5] Pham-Mau-quan, Ann. Inst. H. Poincaré t. 2, p. 151, (1965).

[6] A. Lichnerowicz, Relativistic Hydrodynamics and Magnetohydrodynamics, W. A. Benjamin New York (1967).

[7] A. Lichnerowicz, Ann. Inst. Poincaré t. 5, p. 37, (1966).

[8] A. Lichnerowicz, Ann. Inst. Poincaré t. 7, p. 271, (1967).

[9] A. Lichnerowicz, Comm. Math. Phys. t. 12, p. 145, (1969).

[10] A. Lichnerowicz, Comptes rendus Acad. Sc. Paris t. 268, p. 256, (1969).

CENTRO INTERNAZIONALE MATEMATICO ESTIVO

(C. I. M. E.)

VARIATIONAL PRINCIPLES IN GENERAL RELATIVITY

A. H. TAUB

Corso tenuto a Bressanone dal 7 al 16 Giugno 1970

LECTURE I

VARIATIONAL PRINCIPLES IN GENERAL RELATIVITY

1. Introduction

In these lectures we shall derive the Einstein field and the equations of motion for uncharged and charged self-gravitating fluids from variational principles. We shall also see how singular hyper-surfaces (shock waves) and the equations governing their behavior may be treated by means of these principles. In addition we shall show how the "second variation" problem is related to the discussion of the stability of the solutions of the Einstein field equations.

Before taking up these problems we shall discuss some general properties of variational principles and show how a form of the principle of equivalence may be used to formulate in general relativity a field theory described in special relativity by a variational principle involving a Lagrangian function which is a scalar function in Minkowski space-time and which in turn depends on tensor fields and the first derivatives.

Given a four dimensional space-time with a metric tensor $g_{\mu\nu}$ and a tensor field $\phi^{\mu_1 \dots \mu_u}_{\nu_1 \dots \nu_m}$ (written as ϕ^A) defined over the space-time. Let f be a scalar function formed from the metric tensor of the tensor field ϕ^A and the derivatives of these tensor fields.

$$I(g, \phi^A) = \int_V f \sqrt{-g}\, d^4 x$$

where V is a fixed but arbitrary four volume in the space-time, is a functional of the metric tensor and the tensor field ϕ^A.

For given $g_{\mu\nu}$, I may be evaluated in any coordinate system in the space-time by using the assumed transformation properties of the function f . Thus under the

$$x^\mu = x^\mu (x^{*\mu})$$

$$f(x) = f(x(x^*)) = f^*(x^*)$$

and we may write

$$I(g, \phi^A) = \int_V f(x) \sqrt{-g}\, d^4 x$$

$$= \int_{V^*} f(x(x^*)) \sqrt{-g}\, (x(x^*))\, J\, d^4 x^*$$

$$= \int_{V^*} f^*(x^*) \sqrt{-g^*}\, d^4 x^* \qquad (1.1)$$

where J is the Jacobian of the transformation of coordinates, that is,

$$J = \det \left\| \frac{\partial x^\mu}{\partial x^{*\nu}} \right\| ,$$

we of course assume that $J \neq 0$,
and V^* is the same four volume as V but now expressed in the $x^{*\mu}$ coordinate system.

If the tensor fields $g_{\mu\nu}(x)$ and $\phi^A(x)$ are embedded in families of tensor fields $g_{\mu\nu}(x;e)$, $\phi^A(x;e)$ such that $g_{\mu\nu}(x) = g_{\mu\nu}(x;o)$ and $\phi^A(x) = \phi^A(x;o)$ the functional

$I(g, \phi^A)$ becomes a function of the parameter e. Thus we have

$$I(e) = I(g_{\mu\nu}(x;e) ; \phi^A (x ; e)).$$

An example of the embedding referred to above is given by

$$g_{\mu\nu}(x\ e) = g_{\mu\nu}(x) + e\, h_{\mu\nu}(x), \quad \phi^A(x;e) = \phi^A(x) + e\, \psi^A(x)$$

Where $h_{\mu\nu}$ is an arbitrary symmetric second order tensor field and ψ^A is a tensor field with the same transformation properties as ϕ^A.

A variational principle is said to apply if

$$\left(\frac{dI}{de}\right)_{e=0} = I'(0) = 0$$

for arbitrary

$$\left(\frac{dg_{\mu\nu}}{de}\right)_{e=0} = g'_{\mu\nu}(0)$$

or

$$\left(\frac{d\phi A}{de}\right)_{e=0} = \phi'^A(0)$$

or both.

2. Embeddings Induced by Coordinate Transformations.

Let $g^*_{\mu\nu}(x^*)$ and $\phi^{*A}(x^*)$ be the components of fixed tensor fields $g_{\mu\nu}$ and ϕ^A in the $x^{*\mu}$ coordinate system and let the equations of transformation to another coordinate system x^μ depend on the parameter e, that is, let

$$x^\mu = x^\mu(x^{*\nu};e). \tag{2.1}$$

Then since

$$g_{\mu\nu}(x) = g^*_{\sigma\tau}\ (x^*(x))\frac{\partial x^{*\sigma}}{\partial x^\mu}\ \frac{\partial x^{*\tau}}{\partial x^\nu} = g_{\mu\nu}(;e)$$

and

$$\phi^A(x) \sim \phi^{\mu_1\ \cdots\ \mu_n}_{\nu_1\ \cdots\ \nu_m}\ (x)$$

$$= \phi^{*\sigma_1\ \cdots\ \sigma_n}_{\tau_1\ \cdots\ \tau_m}\ \frac{\partial x^{\mu_1}}{\partial x^{*\sigma_1}} \cdots \frac{\partial x^{\mu_n}}{\partial x^{*\sigma_n}}\ \frac{\partial x^{*\tau_1}}{\partial x^{\nu_1}} \cdots \frac{\partial x^{*\tau_m}}{\partial x^{\nu_m}}$$

$$\sim \phi^A\ (x;e)$$

the $g_{\mu\nu}$ and ϕ^A also depend on the parameter e. Such dependence will be said to be induced by the coordinate transformation (2.1).

It may be readily verified that if

$$x^\mu(x^{*\nu}_1 0) = x^{*\mu}$$

$$\left(\frac{dx^\mu}{de}\right)_{e=0} = +\ \xi^\mu\ (x)$$

then

$$g'_{\mu\nu}\ (0) = -\ \xi_{\mu;\nu} - \xi_{\nu;\mu} \qquad (2.2)$$

where the covariant derivative is taken with respect to the metric

$$g^*_{\mu\nu}\ (x^*) = g_{\mu\nu}\ (x^*;0),$$

and

$$\phi'^{A}(0) = - \underset{\xi}{\pounds} \phi^{A}$$

$$\sim -\phi^{\sigma_1 \cdots \sigma_n}{}_{\tau_1 \cdots \tau_m;\rho}\xi^{\rho} + \sum_i \phi^{\sigma_1 \cdots \sigma_{i-1}\rho\sigma_{i+1} \cdots \sigma_n}{}_{\tau_1 \cdots \tau_m}\xi^{\sigma_i}{}_{;\rho}$$

$$-\sum_j \phi^{\sigma_1 \cdots \sigma_n}{}_{\tau_1 \cdots \tau_{j-1}\rho\tau_{j+1} \cdots \tau_m}\xi^{\rho}{}_{;\tau_j} \ . \qquad (2.3)$$

These expressions for $g'_{\mu\nu}(0)$ and $\phi'^{A}(0)$ when the e dependence is induced by a coordinate transformation will be used below.

When the volume over which $I(e)$ is defined is independent of e, that is the limits of integration in the right hand side of the first of equations (1.1) are independent of e, then the limits of integration in the other integrals occurring in that equation depend on e. Thus if $f(x)$ is a scalar function depending on the $g_{\mu\nu}(x;e)$ whose dependence on e is due only to the fact that this dependence has been induced by a coordinate transformation then the expression for I,

$$I(e) = \int_V f \sqrt{g} \, d^4 x$$

depends on e only because the integrand depends on e. Whereas the equivalent formula

$$I(e) = \int_{V^*} f^* \sqrt{g^*} \, d^4 x^*$$

depends on e only because the limits of integration which determine the volume over which the integral is to be carried out depends on e. The integrand is by assumption independent of e. We may write the last equation for I(e) as

$$I(e) = \int_V f^{*}(x^{*}(x)) \sqrt{g^{*}(x^{*}(x))}\, J^{-1}\, d^4 x$$

and now the integrand depends on e through the dependence of the functions $x^{*}(x)$ on e. It follows then that

$$I'(0) = - \int_V (f \xi^{\sigma})_{;\sigma} \sqrt{-g}\, d^4 x \qquad (2.4)$$

3. The Principle of Equivalence

If in the special theory of relativity a field theory can be described by a variational principle then there exists a scalar Lagrangian function $\mathcal{L}$ which depends on the dependent variables of the theory and their derivatives. We shall assume that the dependent variables are given by fields which behave as tensors under Lorentz transformations. The case of spinor fields can be treated in a fashion similar to that described below. Thus we are considering the case where the function $\mathcal{L}$ is a scalar function of a tensor field and its derivatives under Lorentz transformations. We postulate that the dependent variables of the theory are tensors under general coordinate transformations in Minkowski space-time and write $\mathcal{L}$ in such a coordinate system as a function of the tensor field ϕ^A, its covariant derivatives and the metric tensor evaluated in the general coordinate system.

We now form the integral

$$I = \int_V (R - k\mathcal{L}) \sqrt{-g}\, d^4 x. \qquad (3.1)$$

In this integral, the $g_{\mu\nu}$ are no longer assumed to be the metric of a flat space-time and hence one can compute a combination of it, its first, and second derivatives which is a scalar curvature of a space-time with metric tensor $g_{\mu\nu}$. The volume of integration entering on the right-hand side of equation (3.6) is an arbitrary one, and K is a constant which may be related to the Einstein graviataional constant.

Next we form the function $I(e)$ by embedding the tensor fields $g_{\mu\nu}$ and ϕ^A in I into the families $g_{\mu\nu}(x;e)$ and $\phi^A(x;e)$ as discussed earlier and study the conditions for

$$I'(0) = 0.$$

We shall denote

$$\delta_g I = I'(0)$$

when

$$\phi'^A(0) = 0$$

and

$$\delta_\phi I = I'(0)$$

when

$$g'_{\mu\nu}(0) = 0.$$

In general we shall have

$$I'(0) = \delta g\, I + \delta\phi\, I$$

The general relativistic formulation of the field equations determining the field tensor ϕ^A will be given by

$$\delta_\phi I = 0 \tag{3.2}$$

and the Einstein field equations for the gravitational field created by the sources dependent on the tensor field ϕ^A will be taken to be

$$\delta_g I = 0 \tag{3.3}$$

It is evident that the equations obtained from equations (3.2) are those that would hold in a general coordinate system in Minkowski space-time. Hence in view of the principle of equivalence which states that some non-galilean coordinate systems in Minkowski space-time are locally equivalent to the equivalent to the presence of gravitational fields, equations (3.2) should represent the equation determining the tensor field ϕ^A in general relativity.

In the subsequent discussion we shall see that equations (3.3) lead to the Einstein field equations in the form

$$G^{\mu\nu} \equiv R^{\mu\nu} - \frac{1}{2} g^{\mu\nu} R = -KT^{\mu\nu} \tag{3.4}$$

and $T^{\mu\nu}$ will be a symmetric tensor determined from the tensor field ϕ^A.

4. Notation

In order to avoid an excessive use of indices, we employ the following notation

$$\phi^A \sim \phi^{\sigma_1 \ldots \sigma_n}{}_{\tau_1 \ldots \tau_n} . \tag{4.1}$$

The symbol $\sim$ is to be read as "stands for."

$$\phi^A{}_{;\mu} \sim \phi^{\sigma_1 \cdots \sigma_n}{}_{\tau_1 \cdots \tau_m ; \mu}$$

$$= \phi^{\sigma_1 \cdots \sigma_n}{}_{\tau_1 \cdots \tau_m , \mu} + \sum_i \phi^{\sigma_1 \cdots \sigma_{i-1} \rho \sigma_{i+1} \cdots \sigma_n}{}_{\tau_1 \cdots \tau_m} \Gamma^{\sigma_i}_{\rho\mu}$$

$$- \sum \phi^{\sigma_1 \cdots \sigma_n}{}_{\tau_1 \cdots \tau_{j-1} \rho \tau_{j+1} \cdots \tau_m} \Gamma^{\rho}_{\tau_j \mu} \tag{4.2}$$

where the comma denotes the ordinary derivative and

$$\Gamma^{\rho}_{\sigma\tau} = g^{\rho\lambda} \frac{1}{2} (g_{\sigma\lambda,\tau} + g_{\lambda\tau,\sigma} - g_{\sigma\tau,\lambda}). \tag{4.3}$$

We also write ψ_A, and ψ^{μ}_A where

$$\psi_A \sim \psi^{\tau_1 \cdots \tau_m}{}_{\sigma_1 \cdots \sigma_n} ,$$
$$\psi^{\mu}_A \sim \psi^{\tau_1 \cdots \tau_m}{}_{\sigma_1 \cdots \sigma_n}{}^{\mu} \tag{4.4}$$

and

$$\psi_A \phi^A \sim \psi^{\tau_1 \cdots \tau_m}{}_{\sigma_1 \cdots \sigma_n} \phi^{\sigma_1 \cdots \sigma_n}{}_{\tau_1 \cdots \tau_m} \tag{4.5}$$

$$\psi^{\mu}_A \phi^A \sim \psi^{\tau_1 \cdots \tau_m}{}_{\sigma_1 \cdots \sigma_n}{}^{\mu} \phi^{\sigma_1 \cdots \sigma_n}{}_{\tau_1 \cdots \tau_m} \tag{4.6}$$

That is, the quantity given in (4.5) is a scalar and that in (4.6) is a vector.

We consider $\mathcal{L}$ as a function of three sets of variables: $g_{\mu\nu}, \phi^A$ and $\phi^A{}_{;\mu}$ and define

$$\theta^{\mu\nu} = \frac{\partial \mathcal{L}}{\partial g^{\mu\nu}} + \frac{1}{2} g^{\mu\nu} \mathcal{L} \tag{4.7}$$

$$q_A = \frac{\partial \mathcal{L}}{\partial \phi^A} \tag{4.8}$$

$$p^{\mu}_A = \frac{\partial \mathcal{L}}{\partial \phi^A{}_{;\mu}} . \tag{4.9}$$

In each case the remaining two sets of variables are kept constant in the partial differentiations.

We also define

$$\begin{aligned} p^{\rho\lambda\mu} = {} & \sum_i p^{\tau_1\ldots\tau_m}{}_{\sigma_1\ldots\sigma_n}{}^{\mu}\, \phi^{\sigma_1\ldots\sigma_{i-1}\lambda\sigma_{i+1}\ldots\sigma_n}{}_{\tau_1\ldots\tau_m}\, g^{\sigma_i\rho} \\ & -\sum_j p^{\tau_1\ldots\tau_{j-1}\lambda\tau_{j+1}\ldots\tau_m}{}_{\sigma_1\ldots\sigma_n}{}^{\mu}\, \phi^{\sigma_1\ldots\sigma_n}{}_{\tau_1\ldots\tau_{j-1}\sigma\tau_{j+1}\ldots\tau_m}\, g^{\sigma\rho} \end{aligned} \tag{4.10}$$

It follows from equations (4.3) that

$$\left(\frac{d\Gamma^{\rho}_{\sigma\tau}}{de}\right)_{e=o} = \Gamma'^{\rho}_{\sigma\tau} = g^{\rho\lambda}\, \frac{1}{2}\, (g'_{\sigma\lambda;\tau} + g'_{\lambda\tau;\sigma} - g'_{\sigma\tau;\lambda}) \tag{4.11}$$

and from (4.2) that

$$\begin{aligned} \left(\frac{d\phi^A{}_{;\mu}}{de}\right)_{e=o} - \left[\left(\frac{d\phi^A}{de}\right)_{e=o}\right]_{;\mu} = {} & (\phi^A{}_{;\mu})' - (\phi^{A'})_{;\mu} \\ \sim {} & \sum_i \phi^{\sigma_1\ldots\sigma_{i-1}\rho\sigma_{i+1}\ldots\sigma_n}{}_{\tau_1\ldots\tau_m}\, \Gamma'^{\sigma_i}_{\rho\mu} \\ & -\sum_j \phi^{\sigma_1\ldots\sigma_m}{}_{\tau_1\ldots\tau_{j-1}\rho\tau_{j+1}\ldots\tau_m}\, \Gamma'^{\rho}_{\tau_j\mu} \end{aligned}$$

where the variations are produced by varying the ϕ^A and the $g_{\mu\nu}$.

Hence

$$p_A^\mu (\phi^A_{;\mu})' - p_A^\mu (\phi^{A'})_{;\mu} = \frac{1}{2} P^{\lambda\rho\mu} (g'_{\lambda\rho;\mu} + g'_{\lambda\mu;\rho} - g'_{\mu\rho;\lambda})$$
$$= M^{\lambda\rho\mu} g'_{\lambda\rho;\mu} \tag{4.12}$$

where

$$M^{\lambda\rho\mu} = \frac{1}{4} (P^{\lambda\rho\mu} + P^{\rho\lambda\mu} + P^{\lambda\mu\rho} + P^{\rho\mu\lambda} - P^{\mu\rho\lambda} - P^{\mu\lambda\rho}) = M^{\rho\lambda\mu} \tag{4.13}$$

It follows from this equation that

$$2M^{\lambda\rho\mu} - P^{\lambda\rho\mu} = \frac{1}{2} (P^{\rho\lambda\mu} - P^{\lambda\rho\mu} + P^{\lambda\mu\rho} + P^{\rho\mu\lambda} - P^{\mu\rho\lambda} - P^{\mu\lambda\rho})$$
$$\equiv N^{\lambda\rho\mu} = -N^{\lambda\mu\rho} \tag{4.14}$$

It is a consequence of the definition of the Ricci tensor,

$$R_{\mu\nu} = \Gamma^\sigma_{\mu\sigma,\nu} - \Gamma^\sigma_{\mu\nu,\sigma} + \Gamma^\rho_{\mu\sigma}\Gamma^\sigma_{\rho\nu} - \Gamma^\rho_{\mu\nu}\Gamma^\sigma_{\rho\sigma} \tag{4.15}$$

and equation (4.11) that

$$\left(\frac{dR_{\mu\nu}}{de}\right)_{e=o} = R'_{\mu\nu} = \Gamma'^\sigma_{\mu\sigma;\nu} - \Gamma'^\sigma_{\mu\nu;\sigma}$$

and that

$$g^{\mu\nu}R'_{\mu\nu} = [(g^{\mu\nu}g^{\rho\sigma} - g^{\mu\rho}g^{\nu\sigma})g'_{\mu\nu;\rho}]_{;\sigma} \tag{4.16}$$

5. The Euler Equations

The relation between the energy-momentum tensor for the ϕ-field which appears in the Einstein field equations and the variation of $\mathcal{L}$ with respect to the $g_{\mu\nu}$ was first pointed out by Hilbert [1] as is noted in Pauli's classical discussion of the theory of relativity[2] where additional references may be found. In this section we shall derive the Einstein field equations and the equations that must be satisfied by the ϕ-field.

It may be verified that

$$
\begin{aligned}
I'(0) &= \delta_g I + \delta_\phi I \\
&= - \int_V \Big[(G^{\mu\nu} + K\theta^{\mu\nu}) \, g'_{\mu\nu} + K(q_A \phi^{A'} + p^\mu_A \, (\phi^A_{;\mu})') \\
&\quad - \{(g^{\rho\sigma} g^{\mu\nu} - g^{\rho\mu} g^{\nu\sigma}) \, g'_{\mu\nu;\rho}\}_{;\sigma} \Big] \sqrt{-g} \, d^4 x
\end{aligned} \tag{5.1}
$$

where $G^{\mu\nu}$ is defined in equations (3.4), q_A in (4.8) and p^μ_A in equation (4.9).

It follows from equation (4.12) that equation (5.1) may be written as

$$
\begin{aligned}
-I'(0) = \int_V \Big[&(G^{\mu\nu} + K\theta^{\mu\nu}) \, g'_{\mu\nu} + K(q_A \phi^{A'} + p^\mu_A \, (\phi^{A'})_{;\mu} \\
&+ KM^{\mu\nu\sigma} \, g'_{\mu\nu;\sigma}) - \{(g^{\rho\sigma} g^{\mu\nu} - g^{\rho\mu} g^{\nu\sigma}) \, g'_{\mu\nu;\rho}\}_{;\sigma} \Big] \sqrt{-g} \, d^4 x
\end{aligned}
$$

On integrating by parts this in turn may be written as

$$-I'(0) = \int_V \{[G^{\mu\nu} + KT^{\mu\nu}]\, g'_{\mu\nu} + K(q_A - p^{\mu}_{A;\mu})\, \phi^{A'}\} \sqrt{-g}\, d^4 x$$

$$+ \int_V [KM^{\mu\nu\sigma}\, g'_{\mu\nu} + Kp^{\sigma}_{A}\, \phi^{A'} - (g^{\rho\sigma} g^{\mu\nu} - g^{\rho\mu} g^{\nu\sigma}) g'_{\mu\nu;\rho}]_{;\sigma} \sqrt{-g}\, d^4 x \tag{5.2}$$

where the symmetric tensor

$$T^{\mu\nu} = \theta^{\mu\nu} - M^{\mu\nu\sigma}_{;\sigma} = T^{\nu\mu} \tag{5.3}$$

and $M^{\mu\nu\sigma}$ is defined by equations (4.10) and (4.13).

By requiring $I'(0)$ to vanish for arbitrary variations which vanish on the boundary of the region V we obtain the Euler equations

$$G^{\mu\nu} + KT^{\mu\nu} = 0 \tag{5.4}$$

and

$$F_A \equiv q_A - p^{\mu}_{A;\mu} = 0 \tag{5.5}$$

Equations (5.4) are the Einstein field equations with a matter tensor given by equation (5.3). Because this tensor arises from the variation of I with respect to the gravitational field $g_{\mu\nu}$ we call $T^{\mu\nu}$ the gravitational matter tensor. Note that even in the Minkowski space-time $T^{\mu\nu}$ is different from $\theta^{\mu\nu}$ in a general coordinate system.

Equations (5.5) are the equations for the field ϕ^A. They may be obtained from the special relativity equations in a galilean coordinate system by replacing every ordinary derivative by a covariant one. They obviously reduce to the equations of special relativity in case the tensor $g_{\mu\nu}$ is the metric tensor of Minkowski space-time.

6 Conservation Laws

In this section we shall use a technique similar to that of E. Noether [3] to relate the tensor $T^{\mu\nu}$ occurring above with a not-necessarily-symmetric tensor which will be called the inertial stress energy tensor and to derive various conservation laws. The inertial stress energy tensor of the field ϕ^A is defined by the equation

$$t^{\sigma}{}_{\rho} = p_A{}^{\sigma} \phi^A{}_{;\rho} - \delta^{\sigma}_{\rho} \mathcal{L} \tag{6.1}$$

and it is evident that in general

$$t_{\sigma\rho} = g_{\sigma\nu} t^{\nu}_{\rho} \neq t_{\sigma\rho}$$

Further we see that as a consequence of the equations satisfied by the ϕ field, equations (5.5), we have

$$\begin{aligned} t^{\sigma}{}_{\rho;\sigma} &= (p_{A;\sigma}^{\sigma} - q_A)\phi^A_{;\rho} + p^{\sigma}_A (\phi^A_{;\rho\sigma} - \phi^A_{;\sigma\rho}) \\ &= p^{\sigma}_A (\phi^A_{;\rho\sigma} - \phi^A_{;\sigma\rho}) \end{aligned}$$

when $\mathcal{L}$ does not depend explicitly on the coordinates. In special relativity we have $t^{\sigma}_{\rho;\sigma} = 0$. In general $t^{\sigma}_{\rho;\sigma}$ can at most depend linearly on p^{σ}_A, ϕ^A and the curvature tensor in view of the Ricci identity. We shall evaluate this dependence in equation (6.7).

The computations carried out in the sequel make use of the fact that when $g'_{\mu\nu}$ and ϕ'^{A} are given by equations (2.2) and 2.3), I'(0) must be given by equation (2.4) with $f = R - K\mathcal{L}$. That is, we must have

$$I'(0) = -\int_V ((R - K\mathcal{L})\xi^{\sigma})_{;\sigma}\sqrt{-g}\, d^4 x$$

when equations (2.2) and (2.3) are substituted into equation (5.2).

When the former equations are substituted into the latter one we use the identity

$$2G^{\mu\nu}\xi_{\mu;\nu} - (g^{\rho\sigma}g^{\mu\nu} - g^{\rho\mu}g^{\sigma\nu})(\xi_{\mu;\nu} + \xi_{\nu;\mu})_{;\rho\sigma} = -(R\xi^{\sigma})_{;\sigma} ,$$

equation (26.1) and the definition

$$\iota^{\mu\nu} = -\sum_i F^{\tau_1\ldots\tau_m}{}_{\sigma_1\ldots\sigma_{i-1}\sigma\sigma_{i+1}\ldots\sigma_n} g^{\sigma\mu}\phi^{\sigma_1\ldots\sigma_{i-1}\nu\sigma_{i+1}\ldots\sigma_\nu}{}_{\tau_1\ldots\tau_m}$$

$$+\sum_j F^{\tau_1\ldots\tau_{j-1}\nu\tau_{j+1}\ldots\tau_m}{}_{\sigma_1\ldots\sigma_n}\phi^{\sigma_1\ldots\sigma_n}{}_{\tau_1\ldots\tau_{j-1}\rho\tau_{j+1}\ldots\tau_m} g^{\rho\mu}$$

to obtain

$$I'(0) = -\int_V [(R - K\mathcal{L})\xi^{\sigma}]_{;\sigma}\sqrt{-g}\, d^4 x$$

$$+ K\int_V [(2T^{\mu\nu} + \iota^{\mu\nu})\xi_{\mu;\nu} + F_A\phi^A{}_{;\rho}\xi^{\rho}]\sqrt{-g}\, d^4 x$$

$$+ K\int_V (N^{\mu\nu\sigma}\xi_{\mu;\nu} + t_\rho{}^{\sigma}\xi^{\rho})_{;\sigma}\sqrt{-g}\, d^4 x \qquad (6.2)$$

where $N^{\mu\nu\sigma}$ is defined by equation (4.14).

Note that

$$L^{\mu\nu} = 0$$

when equations (5.5) are satisfied.

Since the tensor $N^{\mu\nu\sigma}$ is antisymmetric in ν and σ we have

$$(\xi_\mu N^{\mu\nu\sigma})_{;\nu\sigma} = 0$$

That is,

$$(\xi_{\mu;\nu} N^{\mu\nu\sigma} + \xi_\mu N^{\mu\nu\sigma}{}_{;\nu})_{;\sigma} = 0$$

or

$$(\xi_{\mu;\nu} N^{\mu\nu\sigma})_{;\sigma} = (\xi_\mu N^{\mu\sigma\nu}{}_{;\nu})_{;\sigma}$$

In view of this equation and equation (2.4) we may write equation (6.1) as

$$\int_V [(2T^{\mu\nu} + L^{\mu\nu})\, \xi_{\mu;\nu} + F_A \phi^A{}_{;\rho} \xi^\rho] \sqrt{-g}\, d^4 x$$

$$+ \int_V [(t_\rho{}^\sigma + N_\rho{}^{\sigma\nu}{}_{;\nu})\, \xi^\rho]_{;\sigma} \sqrt{-g}\, d^4 x = 0$$

On integrating this equation by parts we obtain

$$\int_V (F_A \phi^A{}_{;\rho} - 2T_\rho{}^\sigma{}_{;\sigma} - L_\rho{}^\sigma{}_{;\sigma})\, \xi^\rho \sqrt{-g}\, d^4 x$$

$$+ \int_V [(2T_\rho{}^\sigma + L_\rho{}^\sigma + t_\rho{}^\sigma + N_\rho{}^{\sigma\nu}{}_{;\nu}) \xi^\rho]_{;\sigma} \sqrt{-g} d^4 x = 0$$

When the tensor field ϕ^A is a solution of the Euler equations (5.5) the above equations become

$$2\int_V T^{\mu\nu}\xi_{\mu;\nu}\sqrt{-g}\, d^4x + \int_V [(t_\rho{}^\sigma + N_\rho{}^{\sigma\nu}{}_{;\nu})\xi^\rho]_{;\sigma}\sqrt{-g}\, d^4x = 0 \tag{6.3}$$

and

$$-2\int_V T^{\mu\nu}{}_{;\nu}\, \xi_\mu \sqrt{-g}\, d^4x + \int_V [(2T_\rho{}^\sigma + t_\rho{}^\sigma + N_\rho{}^{\sigma\nu}{}_{;\nu})\xi^\rho]_{;\sigma}\sqrt{-g}\, d^4x = 0 \tag{6.4}$$

Both equations must hold for arbitrary volumes and arbitrary vectors ξ_μ.

If ξ_μ is a Killing vector, that is, satisfies

$$\xi_{\mu;\nu} + \xi_{\nu;\mu} = 0$$

then equation (6.2) inplies for such vectors that

$$[(t_\rho{}^\sigma + N_\rho{}^{\sigma\nu}{}_{;\nu})\xi^\rho]_{;\sigma} = [t_\rho{}^\sigma\xi^\rho + N^{\rho\nu\sigma}\xi_{\rho;\nu}]_{;\sigma} = 0 \tag{6.5}$$

Killing vectors need not exist in a space-time satisfying the Einstein field equations. As is well-known there are ten linearly independent Killing vectors in Minkowski space-time--the generators of the inhomogeneous Lorentz group. Equations (6.5) for these Killing vectors are the conservation laws of energy momentum and angular momentum discussed by Belinfante and by Rosenfeld. Since the tensor $t^{\rho\sigma}$ satisfies these conservation laws, in special relativity, we may consider it as the inertial energy tensor.

Since equation (6.4) must hold for arbitrary vectors and arbitrary volumes, we must have

$$T^{\mu\nu}{}_{;\nu} = 0 \tag{6.6}$$

and

$$2T^{\rho\sigma} + t^{\rho\sigma} + N^{\rho\sigma\nu}{}_{;\nu} = 0 \tag{6.7}$$

Equation (6.6) is a consequence of the Einstein field equations. We have now shown that it follows from the invariance properties of $\mathcal{L}$ and the ϕ^A field equations, equations (5.1). Thus equations (6.6) and (6.8) hold in special relativity.

Equations (6.7) relate two energy tensors, the gravitational one $T^{\mu\nu}$ and the inertial one $t^{\mu\nu}$.

It follows from equations (6.6) and (6.7) that

$$\begin{aligned} t^{\rho\sigma}{}_{;\sigma} &= -N^{\rho\sigma\nu}{}_{;\nu\alpha} = -\frac{1}{2}(N^{\rho\sigma\nu}{}_{;\nu\sigma} - N^{\rho\sigma\nu}{}_{;\sigma\nu}) \\ &= \frac{1}{2} N^{\tau\sigma\nu} R^{\rho}{}_{\tau\nu\sigma} \end{aligned} \tag{6.8}$$

This equation, which may be derived directly from the definition of $t^{\rho\sigma}$ and equations (5.5), reduces to

$$t^{\rho\sigma}{}_{;\sigma} = 0$$

in the case of special relativity.

It should be noted that as a consequence of equation (6.6), equation (6.3) may be written as

$$\int_V [(2T_\rho{}^\sigma + t_\rho{}^\sigma + N_\rho{}^{\sigma\nu}{}_{;\nu})\, \xi^\rho]_{;\sigma} \sqrt{-g}\, d^4x = 0$$

which in turn may be written as an integral over the hypersurface S boundary the volume V, namely

$$\int_S [2T_\rho{}^\sigma + t_\rho{}^\sigma + N_\rho{}^{\sigma\nu}{}_{;\nu}]\, \xi^\rho\, n_\sigma dS = 0 \qquad (6.9)$$

where $n_\sigma dS$ is the element of volume in S, and ξ^ρ is an arbitrary vector. Equation (6.9) may be used to relate the time rate of change of the three-dimensional volume integrals of T^4_ρ and t^4_ρ by choosing the hypersurface surface S to consist of the hyperplanes t = constant and t + dt = constant.

7. Generalizations

The results obtained above may be readily generalized to the case where there are a number of ϕ-fields present. In such a case for each such field there will be an associated $T^{\mu\nu}$ and a corresponding $t^{\mu\nu}$. The right-hand side of the Einstein field equations will contain the sum of the $T^{\mu\nu}$ and this tensor will be related to the sum of the $t^{\mu\nu}$ by equations analogous to equations (6.6).

In case the ϕ-field is a spinor field a similar discussion to that given above can be made. The special relativistic Lagrangian must first be generalized by replacing ordinary derivatives of the spinor field by covariant ones. The variations in the metric tensor may be performed by varying the generalized Dirac matrices which satisfy the relation

$$\gamma_\mu \gamma_\nu + \gamma_\nu \gamma_\mu = 2g_{\mu\nu}.$$

It should be pointed out that if we define the scalar

$$\mathcal{H} = p_A^\sigma \phi^A_{;\sigma} - \mathcal{L} = \mathcal{H}(p_A, \phi^A)$$

then equations (5.5) become

$$p_{A;\sigma}^{\sigma} = -\frac{\partial \mathcal{H}}{\partial \phi^A} \tag{7.1}$$

and

$$\phi^A_{;\sigma} = \frac{\partial \mathcal{H}}{\partial p_A^{\sigma}} \tag{7.2}$$

These two equations are similar to the Hamiltonian equations for particles.

Note, however, that $\mathcal{H}$ is not t_ρ^{ρ} nor t_4^{4}. Thus if in virtue of equations (7.1) we call $\mathcal{H}$ the Hamiltonian it is related to t_ρ^{σ} through the equations

$$\mathcal{H} = t_\rho^{\rho} + 3\mathcal{L} \tag{7.3}$$

Thus the connection of the Hamiltonian with the stress energy tensor involves the Lagrangian function.

References

1. D. Hilbert, Nachr. Ges. Wiss. Göttingen, p. 395 (1915).

2. W. Pauli, Theory of Relativity, Pergamon Press, London, p. 158, (1958).

3. E. Noether, Göttingen Nachtrichten, p. 235, (1918).

4. L. Rosenfeld, Acad. Roy. Belgique, 18, p. 6, (1940).

5. F. J. Belinfante, Physics 7, p. 887 (1939).

LECTURE II

A VARIATIONAL PRINCIPLE FOR PERFECT FLUIDS

8. Co-Moving Coordinates

The general discussion of variational principles given above may be applied to the derivation of the equation of motion of a self-gravitating perfect fluid and the Einstein field equations for the case when such a fluid is the source of the gravitational field. In order to make such an application appropriate field variables ϕ^A must be chosen and a Lagrangian function $\mathcal{L}$ must be specified. We shall use as field variables the rest density of the fluid ρ, a variable related to the rest temperature Θ of the fluid and a set of functions which characterize a three-parameter congruence of curves, the world lines (particle paths) of elements of the fluid. We shall not vary these quantities arbitrarily but shall restrict the variations of ρ, the particle paths and the metric so that the mass of the fluid is conserved under the variation.

The use of co-moving coordinates will enable us to represent the congruence of particle paths and their variation in terms of the metric and its variation. Thus, some of the field variables ϕ^A are absorbed into the tensor $g_{\mu\nu}$. The use of this special coordinate system greatly simplifies the calculations. The resulting simplifications are one implication of the fact that as a field theory relativistic hydrodynamics is special in that the equations of motion of the fluid are a consequence of the Einstein field equations and are not independent of them. That is, in the case of a perfect fluid, we are dealing with a situation where equations (5.5) are implied by equations (5.4).

We shall be dealing with a one-parameter family of space-times with metrics $g_{\mu\nu}(x;e)$. In each space-time of the family there is a congruence of curves determined by the solutions to the ordinary differential equations

$$\frac{dx^{*\mu}}{ds} = U^{*\mu}(x^*;e) \tag{8.1}$$

where the $x^{*\mu}$ are the labels assigned to events in the space-time in an arbitrary coordinate system in which the metric has components $g^*_{\mu\nu}(x^*;e)$, and the $U^{*\mu}$ are the components of the velocity four vector of the fluid in this coordinate system. They satisfy

$$g^*_{\mu\nu}U^{*\mu}U^{*\nu} = 1 \tag{8.2}$$

We may write the solutions of Eqs. (8.1) as

$$x^{*\mu} = x^{*\mu}(\xi^i,s;e) \qquad (i=1,2,3) \tag{8.3}$$

where

$$x^{*\mu}_o = x^{*\mu}(\xi^i,0;e)$$

are required to be the parametric equations of a hypersurface $\Sigma(e)$. The four variables ξ^i,s which we shall denote as x^μ, form a comoving coordinate system in each of the space-times. Eqs. (8.3) which may be written more generally as

$$x^{*\mu} = x^{*\mu}(x;e) \tag{8.4}$$

with

$$x^i = \xi^i,$$
$$x^4 = x^4(\xi^i, s; e) \qquad (8.5)$$

may be regarded as the transformation between the $\overset{*}{x}$ coordinate system and a general comoving one which uses the x^μ as labels for events. Eq. (8.1) is then to be understood as

$$\frac{\partial x^{*\mu}}{\partial s} = u^{*\mu}(x^*(x;e);e) \qquad (8.6)$$

where in the partial differentiation the x^i are kept constant for when these variables are fixed a particular world-line is selected.

In the general comoving coordinate system we have the components of the four velocity vector given by

$$u^\mu(x) = u^{*\sigma}(x^*) \frac{\partial x^\mu}{\partial x^{*\sigma}} = \frac{\partial x^\mu}{\partial s}$$

as follows from Eqs. (8.6). Hence

$$u^\mu = \frac{\delta^\mu_4}{\sqrt{g_{44}}} \qquad (8.7)$$

where $g_{\mu\nu}$ are the compoenents of the metric tensor in the comoving coordinate system. Eqs. (8.7) are a consequence of Eqs. (8.5) and (8.2).

We shall be using a comoving coordinate system in each space-time of the one parameter family of space-times with which we shall be concerned. In a particular one of these the metric tensor in the comoving coordinate system will be written as $g_{\mu\nu}(x;e)$. The tensor

$$g'_{\mu\nu}(x;e) = \frac{\partial g_{\mu\nu}}{\partial e}, \qquad (8.8)$$

with x^μ kept constant will measure the change in the metric tensor evaluated in the comoving coordinate system at an event labelled by the coordinates x^μ, produced by a change in the parameter e. Similar statements will apply to other tensor fields which depend on e. In particular we shall have

$$u'^\mu = -\frac{1}{2}u^\mu \frac{g'_{44}}{g_{44}} = -\frac{1}{2}u^\mu g'_{\sigma\tau}u^\sigma u^\tau. \qquad (8.9)$$

That is, the transformations given by Eqs. (8.4) for various values of e, produce comoving coordinates in each of the space-times associated with that value of e.

We shall use the notation

$$\dot{V}^{*\mu}(x^*;e) = \frac{\partial V^{*\mu}}{\partial e} \qquad (8.10)$$

with $x^{*\mu}$ kept constant, where the $V^{*\mu}$ are the components of a vector field in a general coordinate system using the labels $x^{*\mu}$. It is of interest to determine the relation between V'^μ and $\dot{V}^\mu$. To do this we define

$$\xi^{*\mu} = \frac{\partial x^{*\mu}}{\partial e} \qquad (8.11)$$

where $x^{*\mu}$ is given as a function of x and e by equations (8.4) and x^μ is kept constant under the differentiation . Since

$$\frac{\partial x^{*\mu}}{\partial x^\nu}\frac{\partial x^\nu}{\partial x^{*\rho}} = \delta^\mu_\rho$$

must hold for all values of e, it follows from the differentiatior of this equation with respect to e keeping x^μ fixed that

$$\frac{\partial}{\partial e}\left(\frac{\partial x^\nu}{\partial x^{*\rho}}\right) = -\frac{\partial x^\nu}{\partial x^{*\mu}}\frac{\partial \xi^{*\mu}}{\partial x^{*\rho}} \tag{8.12}$$

that

$$\frac{\partial}{\partial e}\left(\frac{\partial x^{*\nu}}{\partial x^\rho}\right) = \frac{\partial}{\partial x^\rho}\xi^{*\nu}. \tag{8.13}$$

From the transformation law of vectors we have

$$V^\mu(x;e) = V^{*\nu}(x^*(x;e);e)\frac{\partial x^\mu}{\partial x^{*\nu}}.$$

On differentiating this equation with respect to e, keeping x^μ fixed we find

$$V' = \left(\frac{\partial V^{*\nu}}{\partial x^{*\rho}}\xi^{*\rho} + \dot{V}^{*\nu}\right)\frac{\partial x^\mu}{\partial x^{*\nu}} + V^{*\mu}\frac{\partial}{\partial e}\left(\frac{\partial x^\mu}{\partial x^{*\nu}}\right).$$

In virtue of Eq. (8.12) we may write this as

$$V'^\mu = (\dot{V}^{*\nu} + \mathcal{L}_{\xi^*}V^{*\nu})\frac{\partial x^\mu}{\partial x^{*\nu}} \tag{8.14}$$

where

$$\mathcal{L}_{\xi^*}V^{*\nu} = V^{*\nu}{}_{;\rho}\xi^{*\rho} - \xi^{*\nu}{}_{;\rho}V^{*\rho} \tag{8.15}$$

and is of course the Lie derivative of the vector $V^{*\nu}$ with respect to $\xi^{*\mu}$. It may be shown by similar arguments that for any tensor the operation of taking the prime derivative of the tensor compenents differs from the transform of taking the dot derivative by the appropriate Lie derivative of the tensor.

In particular for a scalar we have

$$f'(x;e) = \dot{f}^* + \frac{\partial f^*}{\partial x^{\rho *}} \xi^{*\rho} \tag{8.16}$$

where

$$f^*(x^*;e) = f(x(x^*;e);e).$$

For the metric tensor we have

$$g'_{\sigma\tau} = (\dot{g}^*_{\mu\nu} + \xi^*_{\mu;\nu} + \xi^*_{\nu;\mu}) \frac{\partial x^{*\mu}}{\partial x^\sigma} \frac{\partial x^{*\nu}}{\partial x^\tau}, \tag{8.17}$$

It follows from equation (8.9) that

$$(g_{\mu\nu}U^\mu U^\nu)' = g'_{\mu\nu}U^\mu U^\nu + 2g_{\mu\nu}U'^\mu U^\nu = 0$$

as is to be expected. If we apply equation (8.14) to the four-velocity field, we find that

$$\begin{aligned} \dot{U}^{*\nu} = & -\tfrac{1}{2} U^{*\nu}\dot{g}^*_{\sigma\tau}U^{*\sigma}U^{*\tau} + U^{*\rho}(g^{*\nu\sigma} - U^{*\nu}U^{*\sigma})\xi^*_{\sigma;\rho} \\ & - (g^{*\nu\sigma} - U^{*\sigma}U^{*\nu})U^*_{\sigma;\rho}\xi^{*\rho} \end{aligned} \tag{8.18}$$

Hence

$$(g^*_{\mu\nu}\dot{U}^{*\mu}U^{*\nu}) = 0 \tag{8.19}$$

as a consequence of equation (8.18) and the fact that equation (8.2) holds.

9. The Variations of the Field Variables

The particle paths which are a three parameter congruence of curves, are described by equations (8.4). In these equations the x^i $(i = 1,2,3)$ label a particular particle and x^4 measures a "co-moving time" which is allowed to differ from the proper time along the world line of a particle (cf. equation (8.5)). The vector field $\xi^{*\mu}$ defined in terms of the former equations describes the variations in this congruence of curves. That is, if $x^{*\mu}(x;o)$ is the description in the starred coordinates of the position of and time when particle x^i is located at the comoving time x^4, as this particle moves along an unperturbed path, then $x^{*\mu}(x;0) + e\xi^{*\mu}(x;0)$ describes the corresponding event as the particle moves along a perturbed world-line.

The quantities $g'_{\sigma\tau}$ measure a variation in the metric tensor in the sense that $g_{\sigma\tau}(x;0) + eg'_{\sigma\tau}(x;0)$ describe two almost equal space-times. In each of these space-times, the particle paths, whether perturbed or not, are described by the curves x^i = constant. Because we are allowing general space-time hypersurfaces to be described by the equations x^4 = constant, a variation in $g_{\sigma\tau}$ produces a variation in U^ν; the unit tangent vector to the particle paths.

Equations (8.17) and (8.18) describe the variations produced at the event labeled by $x^{*\mu}$ of the metric tensor and the four-velocity vector field when these tensors are varied in the comoving coordinate system and then evaluated in the $x^{*\mu}$ coordinate system. It should be noted that although four arbitrary functions, the $\xi^{*\mu}$, enter into equations (8.18), the manner in which they enter is restricted so that equation (8.19) holds.

We now turn to a discussion of the variations which we shall allow in the rest density ρ and the rest temperature Θ.

If $\phi(x;e)$ is the rest density of the fluid and U^μ is its four-velocity, then the conservation of matter is expressed by the equation

$$(\rho U^\mu)_{;\mu} = \frac{1}{\sqrt{-g}} (\sqrt{-g}\ \rho\ U^\mu)_{,\mu} = 0 \qquad (9.1)$$

In a comóving coordinate system when equation (817) holds we may integrate this equation to obtain

$$\rho\sqrt{-g} = \sqrt{g_{44}}\ M_o(x^1,x^2,x^3{}_;e) \qquad (9.2)$$

The function M_o which is independent of x^4 measures the amount of fluid in the hypersurface x^4 = constant between the world-lines labeled by x^1 and $x^1 + dx^1$, x^2 and $x^2 + dx^2$, x^3 and $x^3 + dx^3$. The requirement

$$M_o' = 0$$

is then the statement that for each of the space-times $g_{\mu\nu}(x;e)$ and for every perturbation of the world-lines of the particles, the amount of matter in the region described above, is constant.

It follows from the results of the preceeding section and equation (9.2) that the requirement $M' = 0$ is equivalent to

$$\frac{\rho'}{\rho} = -\frac{1}{2}(g^{\sigma\tau} - U^\sigma U^\tau)\ g'_{\sigma\tau} \qquad (9.3)$$

In the $x^{*\mu}$ coordinate system, we have

$$\frac{\dot{\overset{*}{\rho}}}{\rho^*} + \frac{1}{2} g^{*\sigma\tau}\dot{\overset{*}{g}}_{\sigma\tau} + \frac{1}{\rho^*}(\overset{*}{\rho}\xi^{\sigma*})_{;\sigma} - \frac{1}{2}U^{*\sigma}U^{*\tau}(\dot{\overset{*}{g}}_{\sigma\tau} + \overset{*}{\xi}_{\sigma;\tau} + \overset{*}{\xi}_{\tau;\sigma}) = 0. \qquad (9.4)$$

The rest temperature Θ and the rest specific entropy S of a fluid are defined by the equation

$$\Theta dS = d\varepsilon - \frac{p}{\rho^2} d\rho \tag{9.5}$$

where p is the pressure and $\varepsilon(p,\rho)$ is the rest specific internal energy given by the caloric equation of state of the fluid. In this equation $\Theta(p,\rho)$ is determined as an integrating factor of the right-hand side of the equation and S is then determined up to a constant of integration.

It follows from equation (9.5) that if p and ρ are functions of a parameter e, then

$$\Theta S' = \varepsilon' - \frac{p}{\rho^2}\rho' \tag{9.6}$$

Instead of considering the function Θ as a field variable we introduce another function α defined by the equation

$$\Theta = U^{\mu}\alpha_{,\mu}. \tag{9.7}$$

In a comoving coordinate system

$$\Theta = \frac{1}{\sqrt{g_{44}}} \frac{\partial \alpha}{\partial x^4}$$

and

$$\Theta' = -\frac{1}{2} \Theta g'_{\sigma\tau} U^{\sigma} U^{\tau} + U^{*}\alpha'_{,\mu} \tag{9.8}$$

10. The Lagrangean For A Perfect Fluid

We shall define this function as

$$\mathcal{L} = \rho(c^2 + \varepsilon - \Theta S) \tag{10.1}$$

where ρ, ε, Θ and S are defined as above. Then the variational principle discussed in Lecture I may be written as $I'(0) = 0$ where

$$I = I_g - 2KI_F \tag{10.2}$$

with

$$I_g = \int_V R\sqrt{-g}\, d^4x \tag{10.3}$$

$$I_f = \int \rho(c^2 + \varepsilon - \Theta S)\sqrt{-g}\, d^4x \tag{10.4}$$

It then follows from the results of the preceeding section that the equation

$$I'_F = \int \{(\rho\sqrt{-g})'(c^2 + \varepsilon - \Theta S) + (\rho\sqrt{-g})(\varepsilon' - \Theta S' - S\Theta')\}\, d^4x$$

is equivalent to

$$2I'_F = \int [T_F^{\mu\nu} g'_{\mu\nu} - 2\rho S U^\mu \alpha'_{,\mu}]\sqrt{-g}\, d^4x\ , \tag{10.5}$$

where

$$T_F^{\mu\nu} = \rho(c^2 + \varepsilon + \frac{p}{\rho})U^\mu U^\nu - pg^{\mu\nu} \tag{10.6}$$

and use has been made of equations (9.3), (9.6) and (9.8). We may integrate by parts the last term in equation (10.5) and obtain

$$2I'_F = \int [T_F^{\mu\nu} g'_{\mu\nu} + 2(\rho S U^\mu)_{;\mu}\alpha']\sqrt{-g}\, d^4x - \int 2(\alpha'\rho S U^\mu)_{;\mu}\sqrt{-g}\, d^4x \tag{10.7}$$

We now turn to the calculation of I_g'. We have

$$\sqrt{-g}\, R = \sqrt{-g}\, g^{\mu\nu}R_{\mu\nu}$$

and

$$R_{\mu\nu} = \Gamma^{\sigma}_{\mu\sigma,\nu} - \Gamma^{\sigma}_{\mu\nu,\sigma} + \Gamma^{\rho}_{\mu\sigma}\Gamma^{\sigma}_{\rho\nu} - \Gamma^{\rho}_{\mu\nu}\Gamma^{\sigma}_{\rho\sigma} \tag{10.8}$$

where

$$\Gamma^{\rho}_{\mu\nu} = \frac{1}{2}g^{\rho\lambda}(g_{\lambda\mu,\nu} + g_{\lambda\nu,\mu} - g_{\mu\nu,\lambda}). \tag{10.9}$$

Hence

$$\Gamma'^{\rho}_{\mu\nu} = \frac{1}{2}g^{\rho\lambda}(g'_{\lambda\mu;\nu} + g'_{\lambda\mu;\nu} - g'_{\mu\nu,\lambda}) \tag{10.10}$$

where the covariant derivative is taken with respect to the metric $g_{\mu\nu}$. In deriving this result we use the fact that

$$g'^{\mu\nu} = -g'_{\sigma\tau}g^{\sigma\mu}g^{\tau\nu}, \tag{10.11}$$

as follows from the equations

$$g^{\mu\nu}g_{\nu\sigma} = \delta^{\mu}{}_{\sigma}.$$

It is a consequence of the above that

$$g^{\mu\nu}R'_{\mu\nu} = g^{\mu\nu}(\Gamma'^{\sigma}_{\mu\sigma;\nu} - \Gamma'^{\sigma}_{\mu\nu;\sigma}) = [(g^{\mu\nu}g^{\rho\sigma} - g^{\mu\rho}g^{\nu\sigma})g'_{\mu\nu;\rho}]_{;\sigma} \tag{10.12}$$

Therefore

$$\begin{aligned} I' = & -\int\{(G^{\mu\nu} + KT_F^{\mu\nu})g'_{\mu\nu} + 2K(\rho SU^{\mu})_{;\mu}\alpha'\}\sqrt{-g}\, d^4x \\ & + \int\sqrt{-g}\{(g^{\mu\nu}g^{\rho\tau} - g^{\mu\rho}g^{\nu\sigma})g'_{\mu\nu;\rho} + 2\alpha'\rho SU^{\sigma}\}_{;\sigma}\, d^4x \end{aligned} \tag{10.13}$$

where

$$G^{\mu\nu} = R^{\mu\nu} - \frac{1}{2}g^{\mu\nu}R.$$

The Euler equations are obtained by requiring $I'(0) = 0$ for arbitrary $g'_{\mu\nu}(0)$ and $\alpha'(0)$ which vanish on the boundary of the region of integration. Since the second integral in equation (10.13) has an integrand which is a divergence of a vector field, it vanishes for such variations and the Euler equations become

$$G^{\mu\nu} + KT_F^{\mu\nu} = 0 \tag{10.14}$$

and

$$(\rho S U^{\mu})_{;\mu} = 0 \tag{10.15}$$

where $T_F{}^{\mu\nu}$ is given by equations (10.6).

In view of the definition of $T_F^{\mu\nu}$, and the definition of specific entropy, equations (10.14) and (10.15) are equivalent to (10.14) and

$$(\rho U^{\mu})_{;\mu} = 0 \tag{10.16}$$

This is so, because equations (10.14), together with the Bianchi identity

$$G^{\mu\nu}{}_{;\nu} = 0$$

imply that

$$T_{F;\nu}{}^{\mu\nu} = [\rho(c^2 + \varepsilon + \tfrac{p}{\rho})U^{\mu}U^{\nu} - pg^{\mu\nu}]_{;\nu} = 0 \tag{10.17}$$

Hence

$$U_{\mu}T^{\mu\nu}{}_{;\nu} = (c^2 + \varepsilon + \tfrac{p}{\rho})(\rho U^{\nu})_{;\nu} + \rho(\varepsilon_{,\nu} + p(\tfrac{1}{\rho})_{,\nu})U^{\nu} = 0$$

In view of equation (9.5) we may write this equation as

$$(c^2 + \varepsilon + \tfrac{p}{\rho})(\rho U^{\nu})_{;\nu} + \rho U^{\nu}\,\Theta S_{;\nu} = 0$$

or as

$$(c^2 + \varepsilon - \Theta S + \tfrac{p}{\rho})(\rho U^{\nu})_{;\nu} + \Theta(\rho U^{\nu} S)_{;\nu} = 0$$

Hence equation (10.15) implies equation (10.16) and conversely.

Equations (10.14) and (10.16) are the Einstein field equations for a self-gravitating fluid when matter is conserved. Equations (10.17) are the equations of motion of the fluid. Equations (10.16) and (10.17) reduce to the special relativistic equations of conservation of matter, energy and momentum when the space-time is Minkowski space. They may therefore be properly called the generalization of these equations to general relativity. The special relativistic equations reduce to the Newtonian ones in case the fluid particle velocities are small compared to the velocity of light.

It should be noted that through the use of the comoving coordinate system in which both the unperturbed and perturbed particle paths are represented by the curves x^i = constant. The variation of the particle paths does not appear in the evaluation of I'. Thus the field variables corresponding to the particle paths do not appear in this integral and there are no field equations for these variables. The equations describing the particle paths in a general coordinate system are of course derivable from equation (10.16) and (10.17). The latter ones may be written as equations (10.15) or as

$$\rho S_{,\mu} U^{\mu} = 0 \tag{10.18}$$

and

$$\rho(c^2 + i)U^{\nu}U^{\mu}_{;\nu} = p_{,\nu}(g^{\nu\mu} - U^{\mu}U^{\nu}) \tag{10.19}$$

where

$$i = \varepsilon + \frac{p}{\rho} \tag{10.20}$$

LECTURE III

SINGULAR HYPERSURFACES

11. The Existence of Shock Waves

It has been shown [1, 2] that solutions of equations (10.14) for $g_{\mu\nu}$, p, ρ, and u^{μ} may be found which reduce to solutions of problems in fluid flow in the special theory of relativity when $K = 0$. If we further take the limit of these solutions when $c \to \infty$, we obtain solutions to problems in classical hydrodynamics. Further, the approximation method which was used is such that any special relativistic solution of a plane-symmetric problem in hydrodynamics could be obtained by such a reduction process.

In particular those solutions of the special relativistic equations describing the motion of perfect fluids which are physically unacceptable in certain regions of Minkowski (flat) space-time have their counterpart among the solutions of equations (10.14). Such solutions are associated with the formation of shock waves in special relativity and in classical theory [3]. For this reason it was suggested that the theory of a perfect fluid in general relativity must allow for the existence of shock waves. That is, we must contemplate the existence of three-dimensional hypersurfaces in space-time across which there may be discontinuities in the stress energy tensor, the $g_{\mu\nu}$, and their derivatives.

If such hypersurfaces are to be considered, then we must consider equations (10.14) as holding on each side of such a hypersurface, and we must supplement these equations by conditions which relate the values of the $g_{\mu\nu}$, the derivatives of these quantities, and the stress energy tensor on both sides of such a hypersurface. The relations that must hold between the components of the stress energy tensor across a hypersurface of discontinuities must be the generalization of the Rankine-Hugoniot equations of classical and special relativistic hydrodynamics [3].

It is the purpose of this lecture to derive and discuss a set of conditions of the type described above for a space-time in which the matter present is a perfect fluid. The method used is based on the existence of the variational principle discussed previously.

If the field equations and the equations of motion, the conservation equations, can be derived from a variational principle, then we may generalize the variational principle by allowing the region of integration involved to include regions of space-time which contain hypersurfaces across which the matter distribution and the metric tensor and its derivatives are discontinuous. We may even vary these singular hypersurfaces. In the classical theory such a generalization is known to give the classical Rankine-Hugoniot equations [5].

It will be shown below that such a generalization of the variational principle leading to the field equations and the equations of motion for a space-time containing a perfect fluid leads to a general relativistic generalization of the Rankine-Hugoniot equations which reduces to the appropriate equations in the special relativistic and classical limits. We shall also obtain conditions that must be satisfied by the

$g_{\mu\nu}$ and their derivatives across hypersurfaces of discontinuities. Such conditions have been discussed by S O'Brien and J. L. Synge [6] and by A Lichnerowicz [7]. The results given below are obtained in a general coordinate system. When the coordinate system is chosen to be that used by O'Brien and Synge, the equations obtained reduce to those given by them.

A transformation of coordinates with a discontinuous second derivative may be show [8] to reduce these conditions to the requirement of Lichnerowicz, namely that $g_{\mu\nu}$ and the derivatives of the $g_{\mu\nu}$ be continuous in the new coordinate system.

The equations governing the behavior of thin shells of material discussed by W. Israel [9] and A. Papapetrou and A. Hamoni [10] may also be derived from a variational principle. These equation differ from those describing shocks because in the former situation, the integrand occurring in the integral defining the variational principle is such that its evaluation involves an integral over the singular hypersurface. In the case of shocks no such hypersurface integral occurs directly in the definition of the variational principle.

12. The Generalization of the Integral I

The variational principle we shall now consider will involve the integral I defined by equation (10.2). However, we shall assume that there exists a three-dimensional hypersurface Σ which divides the region of integration V into two four-dimensional regions V_1 and V_2 in each of which the integrand exists and is integrable. There may or may not be a contribution to I involving an integral over Σ. Thus

$$I = I_1 + I_2 + S \equiv \int_{V_1 + V_2} (R - 2K\mathcal{L})\sqrt{-g}\, d^4x + \S \tag{12.1}$$

where

$$I_A = \int_{V_A} (R - 2K\mathcal{L})\sqrt{-g}\, d^4x, \qquad (A = 1, 2) \tag{12.2}$$

$$S = \int_{\Sigma} f\, d\Sigma, \tag{12.3}$$

$d\Sigma$ is the invariant three dimensional volume measure on the hypersurface Σ and f is a function of the $g_{\mu\nu}$ and the field variables of the fluid evaluated on Σ. In case $f = 0$, we will see that the singular hypersurface is a shock wave or a boundary between two regions of space-time across which no matter flows (slip-surface). In case $f \neq 0$ the singular hypersurface will be shown to be a thin shell.

In a general coordinate system with labels $x^{*\mu}$, a family of hypersurfaces $\Sigma(e)$ may be defined by the parametric equations

$$x^{*\mu} = x^{*\mu}(\alpha, \beta, \gamma; e) \tag{12.4}$$

where the parameters α, β, γ define a point on the hypersurface and the parameter e labels particular hypersurface. The vector field

$$\Xi^{*\mu} = \frac{\partial x^{*\mu}}{\partial e}$$

where α, β and γ are kept constant in the differentiation is then defined on the hypersurface $\Sigma(e)$.

We observe that in the comoving coordinate system we have

$$\Xi^\mu \equiv \frac{\partial x^\mu}{\partial e} = \frac{\partial x^\mu}{\partial x^{*\rho}} (\Xi^{*\rho} - \xi^{*\rho})$$

where $\frac{\partial x^\mu}{\partial e}$ is evaluated for fixed α, β, γ and e by use of equations (12.4) and the inverse of equations (8.4).

The coordinate system used to write equation (12.1) may be an arbitrary one in space-time. In case it is the comoving one the variation in I due to a variation in the singular hypersurface is given by

$$\delta_\Sigma I = \int_\Sigma [\sqrt{-g}(R - 2K\mathcal{L})] \Xi^\mu \lambda_\mu \delta\sigma + \int_\Sigma \frac{\partial f}{\partial x^\mu} \Xi^\mu d\Sigma \tag{12.5}$$

where

$$\lambda\mu\delta\sigma = \varepsilon_{\mu\nu\sigma\tau} \frac{\partial x^\nu}{\partial\alpha} \frac{\partial x^\sigma}{\partial\beta} \frac{\partial x^\tau}{\partial\gamma} d\alpha d\beta d\gamma \tag{12.6}$$

and we have used the notation

$$[h] = \lim_{\varepsilon\to 0} \{h(x^\mu - \varepsilon\Xi^\mu) - h(x^\mu + \varepsilon\Xi^\mu)\} = h_1 - h_2 \tag{12.7}$$

Equation (10.13) which holds in each of the regions V_1 and V_2 may be used in the evaluation of I' where I is given by equation (12.1). Under a variation of the variables $g_{\mu\nu}$, the fluid field variables and the singular hypersurface we then have

$$I' = -\int_{V_1+V_2} \{(G^{\mu\nu} + KT_F^{\mu\nu})g'_{\mu\nu} + 2k(\rho SU^{\mu})_{;\mu}\alpha'\}\sqrt{-g}\, d^4x$$

$$+\int [\{\sqrt{-g}(g^{\mu\nu}g^{\rho\sigma} - g^{\rho\mu}g^{\nu\sigma})g'_{\mu\nu;\rho} + 2K\rho\alpha' SU^{\sigma}\sqrt{-g}\}\lambda\sigma]d\sigma \tag{12.8}$$

$$+\int_{\Sigma}\frac{\partial f}{\partial x^{\mu}}\Xi^{\mu}d\Sigma + \int_{\Sigma}[\sqrt{-g}(R - 2K\mathcal{L})\Xi^{\sigma}\lambda\sigma]d\sigma + \int_{\Sigma}\tau^{\mu\nu}g'_{\mu\nu}d\Sigma$$

In this equation the last term arises from the variation of S in equation (12.1) with respect to the field variables. The third and fourth terms come from the variation in I due to the variation of the hypersurface Σ and the first two terms come from the application of equation (10.13). The divergences occurring in that equation have been integrated under the assumption that $g'_{\mu\nu}$ vanishes on all boundaries of V_1 and V_2 other than the hypersurface Σ.

13. Shock Waves

We shall first discuss the conditions

$$I'(0) = 0$$

in the case $f = \tau^{\mu\nu} = 0$. That is the case for which

$$I = I_1 + I_2 = \int_{V_1+V_2} (R - 2K\mathcal{L})\sqrt{-g}\, d^4x.$$

In equation (12.8) we may set

$$g'_{\mu\nu} = \dot{g}_{\mu\nu} + \xi_{\mu;\nu} + \xi_{\nu;\mu}$$

where $\dot{g}_{\mu\nu}$ represents the variation in $g_{\mu\nu}$ due to the variation of the metric tensor at a fixed $x^{*\mu}$ and the second terms represent the variation in $g_{\mu\nu}$ due to a variation of the world-lines of the fluid particles. After an integration by parts, equation (12.8) may be written as

$$I' = -\int_{V_1+V_2} \{(G^{\mu\nu} + KT_F^{\mu\nu})\dot{g}_{\mu\nu} - 2KT_{F;\nu}^{\mu\nu}\xi_\mu + 2K(\rho SU^\mu)_{;\mu}\alpha'\}\sqrt{-g}\, d^4x$$
$$+ \int_\Sigma [\sqrt{-g}(g^{\mu\nu}g^{\rho\sigma} - g^{\rho\mu}g^{\nu\sigma})\dot{g}_{\mu\nu;\rho}\lambda\sigma]d\sigma - \int_\Sigma [\sqrt{-g}T_F^{\sigma\mu}\xi_\mu\lambda\sigma]d\sigma$$
$$+ \int_\Sigma [\sqrt{-g}\{R(\Xi^\sigma + \xi^\sigma) - 2K\mathcal{L}\Xi^\sigma + 2K\rho\alpha' SU^\sigma\}\lambda\sigma]d\sigma \qquad (13.1)$$

We now require that $I'(0) = 0$ for $\dot{g}_{\mu\nu}$, ξ_μ and α' which vanish on the hypersurface Σ and are otherwise arbitrary. It then follows that in region V_1 and in region V_2 equations (10.14) and (10.15) hold. It is a consequence of the first set of these that

$$T_{F;\nu}^{\mu\nu} = 0 \qquad (13.2)$$

in regions V_1 and V_2. Thus no additional conditions are obtained by setting to zero the coefficient of ξ_μ in the first integral on the right-hand side of equation (13.1).

We restrict the variations of α' on the hypersurface Σ so that

$$[\sqrt{-g}\{R(\Xi^\sigma + \xi^\sigma) - 2K\mathcal{L}\Xi^\sigma\}\lambda\sigma] = -[2K\rho\sqrt{-g}SU^\sigma\lambda_\sigma\alpha'] \qquad (13.3)$$

In view of the fact that equations (10.14) hold we may write this equation as

$$[\sqrt{-g}\{\tfrac{\rho}{2}(c^2 + \varepsilon - 3p)(\Xi^\sigma + \xi^\sigma) - \rho(c^2 + \varepsilon - \Theta S)\Xi^\sigma\}\lambda\sigma]$$

$$= \quad -[\sqrt{-g}\rho S U^\sigma \lambda\sigma\alpha'] \qquad (13.4)$$

This equation, which relates the variations of the hypersurface, the variations of the particle paths and the variations of the temperature is the general relativistic analogue of the relation which was previously discussed in the classical theory [5]. It arises because in the theory of shock waves, one does not have a conservation of entropy flow across shocks and hence one does not have the generalisation of equations (10.15) holding. Instead one requires a generalisation of the equation describing the conservation of mass.

If equation (13.3) holds, it follows from equation (13.1) and the requirement that $I'(0) = 0$ for arbitrary ξ_μ that we must have

$$[\sqrt{-g}T_F{}^{\mu\nu}\lambda_\nu] \quad = \quad 0 \qquad (13.5)$$

These equations give the conditions that must be satisfied by the discontinuities in the stress-energy tensor of the fluid across the hypersurface Σ. They are a natural generalization of equations (13.2) and may be derived from them by rewriting the latter equations as

$$(f_\mu T_F{}^{\mu\nu})_{;\nu} \quad = \quad T_F{}^{\mu\nu} f_{\mu;\nu}$$

for arbitrary vector fields f_μ and integrating these equations over an appropriate volume in space-time containing the hypersurface Σ. These equations have been discussed in some detail in reference [3].

14. The Conditions on the Metric Tensor Across a Shock

We now turn to a discussion of the second integral in equation (13.1). In order to evaluate this integral we introduce coordinates adapted to the hypersurface Σ. We shall treat the case for which

$$n_\mu n_\nu g^{\mu\nu} < 0 \tag{14.1}$$

and indicate the changes that must be made if this condition does not hold. We shall assume that in the region V_1 (and in the region V_2) the hypersurface Σ is defined by the equation

$$\begin{aligned} x^1 &= x_o^{\ 1} \\ x^i &= x^i(\alpha^j) \qquad \alpha, i, j = 1, 2, 3 \end{aligned} \tag{14.2}$$

where $x_o^{\ 1}$ is a constant.

The induced metric on the hypersurface Σ is then

$$ds^2 = g_{\mu\nu} x^\mu{}_{|i} x^\nu{}_{|j} d\alpha^i d\alpha^j = \gamma_{ij} d\alpha^i d\alpha^j \tag{14.3}$$

where we may choose the α^i so that

$$x^\mu{}_{|i} = \frac{\partial x^\mu}{\partial \alpha^i} = \delta^\mu_i \tag{14.4}$$

are three vectors tangent to the hypersurface. The unit normal to the hypersurface is

$$n_\mu = N\delta^1_\mu \tag{14.5}$$

where

$$N^2 g^{11} = -1 \tag{14.6}$$

and

$$n_\mu x^\mu{}_{|i} = 0 \tag{14.7}$$

If we define γ^{ij} so that

$$\gamma^{ij}\gamma_{jk} = \delta^i_k \tag{14.8}$$

we then have

$$g^{\mu\nu} = -n^\mu n^\nu + \gamma^{ij} x^\mu{}_{|i} x^\nu{}_{|j} \tag{14.9}$$

where

$$n^\mu = g^{\mu\nu} n_\nu = N g^{\mu 1} \tag{14.10}$$

Equations (14.6) and (14.7) may be written as

$$g_{\mu\nu} x^\mu{}_{|i} n^\nu = 0$$

and

$$g_{\mu\nu} n^\mu n^\nu = -1$$

These equations determine the components of n^μ to be

$$n^1 = \frac{1}{N} = -N g^{11}$$

$$n^i = -\frac{1}{N}\gamma^{ij} g_{1j} = N g^{1i}$$

$$N^2 = g^2{}_{11}(-1 + \gamma^{ij} g_{4i} g_{4j})^2$$

It follows from equation (14.9) that

$$\sqrt{-g} = N\sqrt{\gamma} \tag{14.11}$$

where γ is the determinant of the γ_{ij}.

We note that under the transformation of coordinates given by the equations

$$\bar{x}^\mu = x^\mu + (x^1 - x_o{}^1)\phi^\mu(x^i)$$

we have on the hypersurface Σ where $x^1 = x_o{}^1$,

$$\bar{n}^1 = n^1(1 + \phi^1)$$

$$\bar{n}^j = n^j + n^1\phi^j$$

as follows from the equations

$$\bar{n}^\mu = n^\sigma \frac{\partial \bar{x}^\mu}{\partial x^\sigma}.$$

Thus if $n^1 \neq 0$ we may always choose the coordinate system in V_1 so that

$$n^1 - 1 = n^j = 0$$

That is, so that

$$g_{11} = 1.$$

$$g_{1i} = 0.$$

Such coordinates are of course Gaussian coordinates based on the hypersurface Σ. We shall restrict ourselves to such coordinates in the remainder of this section.

In Gaussian coordinates, we have

$$N_\mu = \delta^1_\mu$$

and hence

$$n_{\mu;\nu} - n_{\nu;\mu} = 0 \tag{14.12}$$

Therefore

$$n^{\mu}n_{\mu;\nu} = n^{\nu}n_{\mu;\nu} = 0 \tag{14.13}$$

The second fundamental form of the hypersurface Σ is defined as

$$\Omega_{ij} = n_{\mu;\nu}x^{\mu}{}_{|i}x^{\nu}{}_{|j} = \Omega_{ji} \tag{14.14}$$

The last of the above equations holds in view of equations (14.12). It follows from the definition of Ω_{ij} and n_{μ} that in the Gaussian coordinate system

$$\Omega_{ij} = -n_{\rho}\Gamma^{\rho}_{ij} =. -\Gamma^{1}_{ij} = \tfrac{1}{2}\gamma_{ij,1}. \tag{14.15}$$

Since the four vectors n^{μ}, $x^{\mu}{}_{|i}$ are linearly independent we may write

$$n^{\mu}{}_{;\nu} = n_{\nu}(n^{\mu}{}_{;\rho}n^{\rho}) + \Omega^{ij}x^{\mu}{}_{|i}x^{\rho}{}_{|j}g_{\rho\nu}.$$

In view of equation (14.13) this becomes

$$n^{\mu}{}_{;\nu} = \Omega^{ij}x^{\nu}{}_{|i}x^{\rho}{}_{|j}g_{\rho\nu} \tag{14.16}$$

where

$$\Omega^{ij} = \gamma^{ik}\gamma^{je}\Omega_{ke}.$$

Hence

$$n^{\mu}{}_{;\mu} = \Omega = \gamma^{ij}\Omega_{ij}$$

We now apply the above results to the evaluation of the integral

$$J = \int_{V_1} \sqrt{-g}(g^{\mu\sigma}g^{\rho\sigma} - g^{\rho\mu}g^{\nu\sigma})\dot{g}_{\mu\nu;\rho\sigma}d^4x$$

where $\dot{g}_{\mu\nu}$ is an arbitrary tensor defined in V_1 and such that it vanishes on all boundaries of V_1 except the hypersurface Σ. We may evaluate J in an arbitrary coordinate system in V_1 ; in particular in the Gaussian coordinate system based on Σ. Then we have

$$J = \int_\Sigma (g^{\mu\nu}g^{\rho\sigma} - g^{\rho\mu}g^{\nu\sigma})\dot{g}_{\mu\nu;\rho}\sqrt{-g}\lambda_\sigma d\sigma$$

$$= \int_\Sigma (g^{\mu\nu}g^{\rho\sigma} - g^{\rho\mu}g^{\nu\sigma})\dot{g}_{\mu\nu;\rho}n_\sigma d\Sigma$$

where equation (12.6) defines $\lambda_\sigma d\sigma$ and n_σ is the unit normal to the hypersurface Σ and $d\Sigma$ is the invariant volume element of this hypersurface, that is, we may evaluate J in the Gaussian coordinate system based on Σ where we have

$$n_\sigma(g^{\mu\nu}g^{\rho\sigma} - g^{\rho\mu}g^{\nu\sigma})\dot{g}_{\mu\nu;\rho} = n_\sigma(g^{\mu\nu}g^{\rho\sigma} - g^{\rho\mu}g^{\nu\sigma})(\dot{g}_{\mu\nu,\rho} - \dot{g}_{\lambda\nu}\Gamma^\lambda_{\mu\rho})$$

$$= (n^\rho x^\mu_{|i}x^\nu_{|j} - n^\nu x^\mu_{|i}x^\rho_{|j})\gamma^{ij}(\dot{g}_{\mu\nu,\rho} - \dot{g}_{\lambda\nu}\Gamma^\lambda_{\mu\rho})$$

$$= \gamma^{ij}\dot{g}_{ij,1} + \gamma^{ij}P_{i|j} + \dot{g}_{11}\Omega - \dot{g}_{ij}\Omega^{ij}$$

where

$$P_i = \dot{g}_{1i},$$

$$P_{i|j} = P_{i,j} - P_k\{^k_{ij}\} ,$$

and the $\{^k_{ij}\}$ are the Christoffel symbols formed from the γ_{ij}.

Hence we have

$$J = \int_\Sigma (\gamma^{ij}\dot{g}_{ij,1} - \Omega g^{\mu\nu}\dot{g}_{\mu\nu} + (\gamma^{ij}\Omega - \Omega^{ij})\dot{g}_{ij})\sqrt{\gamma}d\Sigma$$

Thus the second term in equation (13.1) may be written as

$$\int_{\Sigma}[\sqrt{-g}(g^{\mu\nu}g^{\rho\sigma} - g^{\rho\mu}g^{\nu\sigma})\dot{g}_{\mu\nu;\rho}\lambda_{\sigma}]d\sigma$$

$$= \int_{\Sigma}[\sqrt{\gamma}(\gamma^{ij}\dot{g}_{ij,1} - \Omega g^{\mu\nu}\dot{g}_{\mu\nu} + (\gamma^{ij}\Omega - \Omega^{ij}\dot{g}_{ij})]d\Sigma \tag{14.17}$$

The vanishing of I' given by equation (13.1) for $\dot{g}_{\mu\nu}$ such that

$$(\sqrt{-g}\, g^{\mu\nu}g_{\mu\nu})_{+} = (\sqrt{-g}\, g^{\mu\nu}\dot{g}_{\mu\nu})_{-} = 0 \tag{14.18}$$

but otherwise arbitrary on Σ and for arbitrary $\dot{g}_{ij,1}$ then implies that

$$[\sqrt{\gamma}\gamma^{ij}] = 0 \tag{14.19}$$

and that

$$[\sqrt{\gamma}(\gamma^{ij}\Omega - \Omega^{ij})] = 0 \tag{14.20}$$

Equations (14.19) imply that the induced metric on the hypersurface Σ takes on the same values when computed from the metric in V_1 as when computed from the metric in V_2. Since we may use Gaussian coordinates based on the hypersurface Σ in each of these regions we have that in these coordinates the metric tensor $g_{\mu\nu}$ is continuous. Equations (14.20) then imply that the second fundamental form of the hypersurface Σ takes on the same values when this hypersurface is regarded as one in V_1 or one in V_2. This result obtains without the restriction imposed on $\dot{g}_{\mu\nu}$ by equation (14.18). However that equation is the statement that the variations in volume element, as measured by the variations in the $\sqrt{-g}$, vanish on the hypersurface Σ.

15. Thin Shells

The equations that determine the behavior of thin shells may be derived by considering the conditions for which $I' = 0$ where I' is given by equations (12.8) with $f \neq 0$. We shall discuss the case for which $\rho = 0$ inside V_1 and V_w but not one the hypersurface Σ. In that case we have $\mathcal{L} = 0$ inside V_1 and V_2 and equation (12.8) becomes

$$I' = -\int_{V_1+V_2} G^{\mu\nu} g'_{\mu\nu}\sqrt{-g}\, d^4x + \int_\Sigma [\sqrt{-g}(g^{\mu\nu}g^{\rho\sigma} - g^{\rho\mu}g^{\nu\sigma})g'_{\mu\nu;\rho\tau}]d\sigma$$

$$+ \int_\Sigma \tau^{\mu\nu} g'_{\mu\nu} d\Sigma$$

when we set $\Xi^\sigma = 0$ and $f = 0$.

It then follows by using the argument given above that in the regions V_1 and V_2 we must have

$$G^{\mu\nu} = 0 \tag{15.1}$$

Then on setting

$$g'_{\mu\nu} = \dot{g}_{\mu\nu} + \xi_{\mu;\nu} + \xi_{\nu;\mu}$$

we find that

$$I' = \int_\Sigma [\sqrt{-g}(g^{\mu\nu}g^{\rho\sigma} - g^{\rho\mu}g^{\nu\sigma})\dot{g}_{\mu\nu;\rho}\lambda_\sigma]d\sigma + \int_\Sigma \tau^{\mu\nu}\dot{g}_{\mu\nu}\sqrt{\gamma}\, d\Sigma$$

$$+ \int_\Sigma \tau^{\mu\nu}\xi_{\mu;\nu}\sqrt{\gamma}\, d\Sigma \tag{15.2}$$

The requirement that $I' = 0$ for arbitrary vectors ξ_μ which vanish on the boundaries of V_1 and V_2 other than Σ requires that we consider the last term in this expression.

We shall evaluate it in the Gaussian coordinate system used in the previous section. In such a coordinate system we have as the non-vanishing Christoffel symbols

$$\Gamma^{1}_{ij} = -\Omega_{ij} \qquad \Gamma^{i}_{j1} = \Omega^{i}_{j}$$

$$\Gamma^{k}_{ij} = \{^{k}_{ij}\}.$$

Hence we have

$$\xi_{1;1} = \xi_{1,1} \ , \qquad \xi_{1;i} = \xi_{1,i} - \xi_{k}\Omega^{k}{}_{i}$$

$$\xi_{i;1} = \xi_{\nu,1} - \xi_{k}\Omega^{k}{}_{i} \qquad \xi_{i;j} = \xi_{i,j} - \xi_{k}\{^{k}_{ij}\} - \xi_{1}\Omega_{ij}$$

The last of these equations may be written as

$$\xi_{i;j} = \xi_{i|j} - \xi_{1}\Omega_{ij}$$

when we have used the symbol $|$ to denote the covariant derivative with respect to the three-dimensional tensor γ_{ij}

We may then write

$$\int_{\Sigma} \tau^{\mu\nu}\xi_{\mu;\nu}d\Sigma = \int(\tau^{11}\xi_{1,1} + \tau^{1i}(\xi_{1,i} + \xi_{i,1} - 2\xi_{k}\Omega^{k}{}_{i}) + \tau^{ij}(\xi_{i|j} - \xi_{1}\Omega_{ij}))d\Sigma$$

The vanishing of this integral for arbitrary ξ_{μ} and $\xi_{\mu,1}$ then implies that

$$\begin{aligned} \tau^{11} &= 0 \\ \tau^{1i} &= 0 \\ \tau^{ij}\Omega_{ij} &= 0 \\ \tau^{ij}{}_{|j} &= 0 \end{aligned} \qquad (15.3)$$

These are the equations which govern the behavior of the thin shell described by the tensor $\tau^{\mu\nu}$ which vanishes off the hypersurface Σ. We may use the results of the preceeding section to determine the equations which determine $\tau^{\mu\nu}$ in terms of the geometry of the hypersurface Σ. If we use equation (14.17) we may write equation (15.2) as

$$I' = \int_\Sigma \{\tau^{\mu\nu}\dot{g}_{\mu\nu}\sqrt{\gamma} + [\sqrt{\gamma}(\gamma^{ij}\dot{g}_{ij,1} - \Omega g^{\mu\nu}\dot{g}_{\mu\nu} + (\gamma^{ij}\Omega - \Omega^{ij})\dot{g}_{ij}]\}d\Sigma$$

If we impose equations (14.18) on the $\dot{g}_{\mu\nu}$ we find that $I' = 0$ for arbitrary $\dot{g}_{\mu\nu}$ and $\dot{g}_{\mu\nu,1}$ subject to these conditions if equations (14.19) hold and

$$\tau^{ij} = [\Omega^{ij} - \gamma^{ij}\Omega] \qquad (15.4$$

Equations (15.3) and (15.4) are the equations given by Israel [9] and Papaetrou and Hamoui [10] for the theory of thin shells.

Taub

References

1. A. H. Taub, Isentropic hydrodynamics in plane symmetric space-time, Physical Rev., vol. 103 (1956), pp. 454-467.

2. _________, Approximate solutions of the Einstein equations for isentropic motions of plane-symmetric distributions of perfect fluids, Physical Rev. Vol 107 (1957) pp 884-900

3. _________, Relativistic Rankine-Hugoniot equations, Physical Rev., vol. 74 (1948), pp. 328-334.

4. _________, General relativistic variational principle for perfect fluids, Physical Rev., vol. 94 (1954), pp. 1468-1470.

5. _________, On Hamilton's principle for perfect compressible fluids, Proceedings of First Symposium of Applied Mathematics, American Mathematical Society, New York, 1949, pp. 148-157.

6. Stephen O'Brien and John L. Synge, Jump conditions at discontinuities in general relativity, Communications of the Dublin Institute for Advanced Studies, Ser. A, no. 9 (1953).

7. A. Lichnerowicz, Théories relativistes de la gravitation et de l'electromagnetisme, Masson et cie, Paris, 1955.

8. A. H. Taub, Singular hypersurfaces in General Relativity, Ill. Jour. Math. 1, (1957) pp. 371-388.

9. W. Israel, Singular Hypersurfaces and Thin Shells in General Relativity, Il Nuovo Cimento, 44 (1966) pp. 1-14.

10. A Papapetrou and A. Hamoui, Couches Simples de Materie en Relativité Generale, Ann. Inst. Henri Poincaré 9 (1968), pp. 179-211.

LECTURE IV

FLUIDS OBEYING AN EQUATION OF STATE

16. Equations of State

In this lecture we shall derive a simpler variational principle from which we may derive the Einstein field equations for a self-gravitating fluid that satisfies an equation of state of the form

$$p = p(w) \qquad (16.1)$$

where p is the pressure and w is the energy density of the fluid. That is, in terms of the quantities used previously we have

$$w = \rho(c^2 + \varepsilon) \qquad (16.2)$$

where ρ is the rest mass density and ε is the rest specific internal energy.

We have previously made use of the fact that function

$$\varepsilon = \varepsilon(p, \rho),$$

the caloric equation of state, describes the nature of the material with which we deal and serves to determine the temperature Θ and the entropy S by means of the equations

$$\Theta dS = d\varepsilon + p d(\frac{1}{\rho}). \qquad (16.3)$$

Hence for every fluid we may express the pressure as a function of two thermodynamic variables, say w and the entropy S.

Thus for every fluid we may write

$$p = p(w,S). \tag{16.4}$$

The assumption made above is that all thermodynamic variables are functions of one of them, say w. This assumption is satisfied in case the fluid motion is isentropic, that is

$$S = S_o$$

where S_o is a constant. It is also satisfied in other circumstances.

In the general situation when equation (16.4) holds the velocity of sound $c\alpha$ is determined by the equation

$$\alpha^2 = \left(\frac{\partial p}{\partial w}\right)_S \tag{16.5}$$

where c is the special relativistic velocity light and the entropy is kept constant in the differentiation occurring on the right-hand side of this equation. If there is a family of flows such that the thermodynamic variables are functions of a parameter e as well as the coordinates, then it follows from equations (16.4) and (16.5) that

$$p' = \alpha^2 w' + \left(\frac{\partial p}{\partial S}\right)_w S' \tag{16.6}$$

When we restrict this family by the condition that

$$S' = 0,$$

we shall say that the family of motions is an adiabatic family (or that the perturbations which distinguish one member of the family from another member are adiabatic perturbations).

For such families we have

$$\alpha^2 w' = p'. \tag{16.7}$$

17. Integration of the Equations of Motion

In case equation (16.1) holds we may derive from the equations of motion of the fluid,

$$T^{\mu\nu}_{F;\nu} = 0, \tag{17.1}$$

expressions for various components of the metric tensor in the comoving coordinate system in terms of the thermodynamic variables. In this coordinate system we have

$$U^\mu = \frac{1}{\sqrt{g_{44}}}\delta^\mu_4 \tag{17.2}$$

and

$$T^{\mu\nu}_F = (w + p)U^\mu U^\nu - pg^{\mu\nu} = \frac{(w + p)}{g_{44}}\delta^\mu_4\delta^\nu_4 - pg^{\mu\nu} \tag{17.3}$$

Equations (17.1) are in general equivalent to the equations

$$(w + p)U^\nu_{;\nu} + w_{,\nu}U^\nu = 0 \tag{17.4}$$

and

$$(w + p)U_{\mu;\nu}U^\nu = p_{,\nu}(\delta^\nu_\mu - U^\nu U_\mu) \tag{17.5}$$

When equation (17.1) holds there exists a thermodynamic function $\sigma(w)$ defined up to a constant by the equation

$$\frac{d\sigma}{\sigma} = \frac{dw}{w + p} \tag{17.6}$$

Hence equation (17.4) becomes

$$(\sigma U^{\mu})_{;\mu} = 0 \tag{17.7}$$

In case equation (16.1) is equivalent to the statement that the entropy is constant we have

$$\sigma = \rho$$

and equation (17.7) is the conservation of mass.

In the comoving coordinate system equation (17.7) becomes

$$\left(\frac{\sigma\sqrt{-g}}{\sqrt{g_{44}}}\right)_{,4} = 0$$

This equation may be integrated to give

$$\sigma\sqrt{-g} = \sqrt{g_{44}}\, f(x^{i}) \tag{17.8}$$

where $f(x^i)$ is an arbitrary function of the variables x^1, x^2 and x^3 but independent of x^4.

If equation (16.1) holds

$$\frac{dp}{w + p} = -\frac{d\sigma}{\sigma} + \frac{d(w + p)}{w + p}$$

Thus, in the comoving coordinate system equations (17.5) become

$$\left(\frac{(w + p)\sqrt{g_{44}}}{\sigma}\right)_{,\mu} = \left(\frac{w + p}{\sigma}\,\frac{g_{\mu 4}}{\sqrt{g_{44}}}\right)_{,4} \tag{17.9}$$

These equations are identically satisfied when $\mu = 4$. Their integrability conditions may be integrated to give

$$\left(\frac{w + p}{\sigma}\,\frac{g_{i4}}{\sqrt{g_{44}}}\right)_{,j} - \left(\frac{w + b}{\sigma}\,\frac{g_{j4}}{\sqrt{g_{44}}}\right)_{,i} = F_{ij}(x^{i})$$

where the $F_{ij} = -F_{ji}$ are arbitrary functions of the three variables x^i, x^2 and x^3 but not of x^4 and are such that

$$F_{ij,k} + F_{jk,i} + F_{ki,j} = 0$$

Hence we must have

$$\frac{g_{i4}}{\sqrt{g_{44}}} = \frac{\sigma}{w + p} c_i(x^j) + \frac{\sigma}{w + p} \phi_{,i} \tag{17.10}$$

where the c_i are arbitrary functions of x^i and ϕ may be a function of x^i and x^4. The solutions of equations (17.9) then become

$$\frac{w + p}{\sigma} \sqrt{g_{44}} = \phi_{,4} + k(x^4) \tag{17.11}$$

It is no restriction to take

$$\phi = \text{constant}$$

$$k = 1$$

for if these conditions are not satisfied we may make the coordinate transformation

$$\bar{x}^4 = \bar{x}^4(x^i, x^4)$$

$$\bar{x}^i = x^i$$

where

$$d\bar{x}^4 = (\phi_{,4} + k(x^4))dx^4 + \phi_{,i}dx^i$$

The $\bar{x}^\mu$ coordinate system is also a comoving one and in it we have

$$\bar{U}_i = \frac{\bar{g}_{i4}}{\sqrt{\bar{g}_{44}}} = \frac{\sigma}{w + p} c_i(x^j) \tag{17.12}$$

$$\bar{U}_4 = \sqrt{\bar{g}_{44}} = \frac{\sigma}{w + p} \qquad (17.13)$$

18. The Vorticity Vector

The three functions $c_i(k^j)$, which enter into the components of the metric tensor in the comoving coordinate system and which may be determined from the initial conditions satisfied by the motion of a self-gravitating fluid, determine and are determined by the amount of rotation in the fluid. This may be seen by examining the vorticity vector

$$v^\mu = \frac{1}{\sqrt{-g}} \varepsilon^{\mu\nu\sigma\tau} U_\nu U_{\sigma,\tau} \qquad (18.1)$$

in the comoving coordinate system. We have

$$v^k = \left(\frac{\sigma}{w + k}\right)^2 \frac{1}{\sqrt{-g}} \varepsilon^{kij} c_{i,j}$$

$$v^4 = \left(\frac{\sigma}{w + p}\right)^2 \frac{1}{\sqrt{-g}} \varepsilon^{kij} c_k c_{i,j} \qquad (18.2)$$

It is evident from these equations that the necessary and sufficient condition for $v^\mu = 0$ is that

$$c_{i,j} - c_{j,i} = 0,$$

that is c_i be the gradient of a scalar. In that case there exists a comoving coordinate system in which

$$U_\mu = \frac{\sigma}{w + p} \delta^4_\mu .$$

The world lines of the fluid particles are then orthogonal to the hypersurfaces x^4 = constant in the comoving coordinate system. In case the flow is isentropic, that is S is a constant, we have

$$\frac{\sigma}{w + p} = \frac{1}{c^2 + i}$$

where

$$i = \varepsilon + \frac{p}{\rho}$$

is the specific enthalpy of the fluid. Further we have in the comoving coordinate system

$$g_{44} = \frac{1}{(c^2 + i)^2} .$$

These are the results obtained earlier [1].

19. A Variational Principle

In the comoving coordinate system used above, we may use equation (17.13) to determine a thermodynamic variable such as p as a function of g_{44}, that is as a function of the coefficients of the metric tensor. Variations of the metric tensor in the comoving coordinate system will then produce variations in the pressure. Thus with the notation we have used earlier we have

$$p' = -\frac{1}{2}(w + p)\frac{g'_{44}}{g_{44}}$$

as follows from differentiating equation (17.13) with respect to e in the comoving coordinate system. This equation may be written as

$$p' = -\frac{1}{2}(w + p)\, g'_{\sigma\tau}U^{\sigma}U^{\tau} \tag{19.1}$$

Thus we have p as a function of $g_{\mu\nu}$ and an expression for the variation of this function.

Now consider the variational principle based on the integral

$$I = \int (R + 2Kp)\sqrt{-g}\, d^4x \qquad (19.2)$$

It follows from the results given earlier and equation (19.1) that

$$I' = -\int (G^{\mu\nu} + KT_F^{\mu\nu})g'_{\mu\nu}\sqrt{-g}\, d^4x + \int (g^{\mu\nu}g^{\rho\sigma} - g^{\mu\rho}g^{\nu\sigma})g'_{\mu\nu;\rho\sigma}\sqrt{-g}\, d^4x \qquad (19.3)$$

where

$$T_F^{\mu\nu} = (w + p)U^\mu U^\nu - g^{\mu\nu}p. \qquad (19.4)$$

This variational principle then leads to the field equations

$$G^{\mu\nu} = -KT_F^{\mu\nu} \qquad (19.5)$$

with equations (19.4) holding. In the latter equations $p = p(w)$ and hence one of the consequences of the Bianchi identities

$$0 = G^{\mu\nu}{}_{;\nu} = -KT^{\mu\nu}_{F;\nu}$$

is that

$$(\sigma U^\mu)_{;\mu} = 0$$

where σ is a function of w (or p) defined by equation (17.7).

Taub

LECTURE V

A VARIATIONAL PRINCIPLE FOR CHARGED FLUIDS

20. Introduction

It is the purpose of this lecture to discuss a variational principle from which the equations governing the motion of a gravitating charged electromagnetic fluid with dielectric permitivity, magnetic permeability and conductivity are derived. In case the conductivity is zero or infinite, we will have a 'holonomic' variational principle. In other cases the principle will be 'non-holonomic' in that not all terms in it will be derivable from the variation of a function of the dependent variables which enter into the problem. This is to be expected, since for non-vanishing and finite conductivity one is dealing with a non-conservative problem since ohmic heat is involved and for such problems holonomic variational principles do not exist.

We begin with a discussion of the electromagnetic fields. We shall use the notation given by A. Lichnerowicz [1] for the Minkowski formulation of the Maxwell equations governing these fields. They are described by two antisymmetric tensors $H_{\mu\nu}$ and $G_{\mu\nu}$. If u^μ is the four-velocity vector of the fluid satisfying

$$u^\mu u_\mu = u^\mu g_{\mu\nu} u^\nu = 1 . \tag{20.1}$$

Then the electric field e_α, the electric induction d_α, the magnetic field h_α and the magnetic induction b_α as measured by an observer whose world line is given by a solution of the equation

$$\frac{dx^\mu}{d\tau} = u^\mu ,$$

that is, one who is at rest with respect to the fluid, are given by

$$\begin{aligned} e_\beta &= u^\alpha H_{\alpha\beta} \qquad & d_\beta &= u^\alpha G_{\alpha\beta} \\ ib_\beta &= u^\alpha \check{H}_{\alpha\beta} \qquad & ih_\beta &= u^\alpha \check{G}_{\alpha\beta} \end{aligned} \tag{20.2}$$

where $i = \sqrt{-1}$ and

$$\begin{aligned} \check{H}_{\alpha\beta} &= \tfrac{1}{2} E_{\alpha\beta\gamma\delta} H^{\gamma\delta} \\ H^{\alpha\beta} &= \tfrac{1}{2} E^{\alpha\beta\gamma\delta} \check{H}_{\gamma\delta} , \end{aligned} \tag{20.3}$$

$E_{\alpha\beta\gamma\delta}$ and $E^{\alpha\beta\gamma\delta}$ are pure imaginary tensors defined by the equations

$$\begin{aligned} E_{\alpha\beta\gamma\delta} &= \sqrt{g}\, \varepsilon_{\alpha\beta\gamma\delta} \\ E^{\alpha\beta\gamma\delta} &= \frac{1}{\sqrt{g}} \varepsilon^{\alpha\beta\gamma\delta} \end{aligned} \tag{20.4}$$

and the ε's are the Levi-Civita alternating tensor densities. We have introduced the pure imaginary quantities in order to preserve the commutativity of various methods for manipulating indices. It follows that

$$u^{\alpha}e_{\alpha} = u^{\alpha}d_{\alpha} = u^{\alpha}h_{\alpha} = u^{\alpha}b_{\alpha} = 0 .$$

It follows from equations (20.2) and (20.3) and the properties of the ε's that

$$ib_{\beta}E^{\beta\lambda\mu\nu} = -u^{\lambda}H^{\mu\nu} - u^{\mu}H^{\nu\lambda} - u^{\nu}H^{\lambda\mu}$$

Hence

$$H^{\mu\nu} = u^{\mu}e^{\nu} - u^{\nu}e^{\mu} - ib_{\beta}u_{\lambda}E^{\beta\lambda\mu\nu} \tag{20.5}$$

Similarly we have

$$G^{\mu\nu} = u^{\mu}d^{\nu} - u^{\nu}d^{\mu} - ih_{\beta}u_{\lambda}E^{\beta\lambda\mu\nu} \tag{20.6}$$

For the purpose of simplifying the ensuing discussion we shall assume simple constitutive equations, namely

$$d_{\alpha} = \lambda e_{\alpha} \qquad b_{\alpha} = \mu h_{\alpha} \tag{20.7}$$

where λ, the dialectric permitivity of the matter and μ, the magnetic permeability are scalar functions which may depend on the coordinates. The situation in which λ and μ are functions of the vectors e_{α} and h_{α} respectively and of other variables characterizing the matter may be dealt with in a manner analogous to the discussion given below.

It follows from equations (20.5), (20.6) and (20.7) that

$$G^{\mu\nu} = \lambda^{\mu\nu\sigma\tau}H_{\sigma\tau} \tag{20.8}$$

where

$$\lambda^{\mu\nu\sigma\tau} = \frac{1}{2\mu}(g^{\sigma\mu}g^{\tau\nu}-g^{\sigma\nu}g^{\tau\mu}) + \frac{1}{2}(\lambda - \frac{1}{\mu})(u^{\mu}u^{\sigma}g^{\tau\nu}-u^{\nu}u^{\sigma}g^{\tau\mu} - u^{\mu}u^{\tau}g^{\sigma\nu} + u^{\nu}u^{\tau}g^{\sigma\mu}) \tag{20.9}$$

Hence

$$\lambda^{\mu\nu\sigma\tau} = \lambda^{\sigma\tau\mu\nu} = -\lambda^{\tau\sigma\mu\nu} \; . \tag{20.10}$$

It is a consequence of equations (20.5) and (20.6) that

$$\begin{aligned}
H^{\mu\nu}H_{\mu\nu} &= 2(e^{\nu}e_{\nu}-b^{\nu}b_{\nu}) & H^{\mu\nu}\breve{H}_{\mu\nu} &= 4ie^{\nu}b_{\nu} \\
G^{\mu\nu}G_{\mu\nu} &= 2(d^{\nu}d_{\nu}-h^{\nu}h_{\nu}) & G^{\mu\nu}\breve{G}_{\mu\nu} &= 4id^{\nu}h_{\nu} \\
G^{\mu\nu}H_{\mu\nu} &= 2(e^{\nu}d_{\nu}-h^{\nu}b_{\nu}) & G^{\mu\nu}\breve{H}_{\mu\nu} &= 2i(b^{\nu}d_{\nu}+h^{\nu}e_{\nu})
\end{aligned} \tag{20.11}$$

21. The Maxwell Equations

These equations are

$$\check{H}^{\mu\nu}_{;\nu} = 0 \qquad (21.1)$$

and

$$G^{\mu\nu}_{;\nu} = J^{\mu} \qquad (21.2)$$

where the semi-colon denotes the covariant derivative. The vector J^{μ} defined by equations (21.2) is the electric current and as a consequence of equations (21.2) satisfies the conservation equation

$$J^{\mu}_{;\mu} = 0 \qquad (21.3)$$

We may write

$$J^{\mu} = \varepsilon u^{\mu} + I^{\mu} \qquad (21.4)$$

where ε is the proper electric charge density and I^{μ} is the conduction current. Ohms law then is given by the statement that

$$I^{\mu} = \sigma e^{\mu} \qquad (21.5)$$

where σ is the conductivity of the fluid. Equation (21.5) is the third constitutive equation for the matter. It may be replaced by a more general (non-linear) law without affecting many of the results given below.

Equations (21.1) are equivalent to the statement that

$$H_{\lambda\mu;\nu} + H_{\mu\nu;\lambda} + H_{\nu\lambda;\mu} = 0 \ . \qquad (21.6)$$

These equations imply of course that there exists a four-potential such that

$$H_{\mu\nu} = \varphi_{\mu,\nu} - \varphi_{\nu,\mu} \tag{21.7}$$

The Minkowski stress-energy tensor is defined as

$$\tau^{\mu\nu} = H^{\mu}{}_{\rho}G^{\rho\nu} + \frac{1}{4}g^{\mu\nu}H^{\rho\sigma}G_{\rho\sigma} \tag{21.8}$$

and satisfies

$$\tau = g_{\mu\nu}\tau^{\mu\nu} = 0 \tag{21.9}$$

It follows from equations (20.5) and (20.6) that

$$\tau^{\mu\nu} = (\tfrac{1}{2}g^{\mu\nu} - u^{\mu}u^{\nu})(e^{\rho}d_{\rho} + b^{\rho}h_{\rho}) - e^{\mu}d^{\nu} - h^{\mu}b^{\nu} - u^{\mu}v^{\nu} - u^{\mu}w^{\nu} \tag{21.10}$$

where

$$v_{\nu} = ie^{\rho}h^{\sigma}u^{\tau}E_{\rho\sigma\tau\nu} \tag{21.11}$$

$$w_{\nu} = id^{\rho}b^{\sigma}u^{\tau}E_{\rho\sigma\tau\nu} \tag{21.12}$$

and hence satisfy

$$u^{\mu}v_{\mu} = u^{\mu}w_{\mu} = 0 \tag{21.13}$$

When the constitutive equations (20.7) hold we have

$$w_{\nu} = \lambda\mu v_{\nu}$$

and

$$\tau^{\mu\nu} - \tau^{\nu\mu} = (\lambda\mu - 1)(u^{\mu}v^{\nu} - u^{\nu}v^{\mu}) \tag{21.14}$$

It follows from the Maxwell equations, when these are written as equations (21.2) and (21.6), that

$$\tau^{\mu\nu}_{;\nu} = H^{\mu}{}_{\rho}J^{\rho} + G^{\rho\sigma}g^{\mu\nu}(H_{\nu\rho;\sigma} + \tfrac{1}{2}H_{\rho\sigma;\nu}) + \tfrac{1}{4}g^{\mu\nu}(H_{\rho\sigma}G^{\rho\sigma}_{;\nu} - G_{\rho\sigma}H^{\rho\sigma}_{;\nu})$$

In view of equations (21.6) and (20.8) we may write this equation as

$$\tau^{\mu\nu}_{;\nu} = H^{\mu}{}_{\rho}J^{\rho} + \frac{1}{4}g^{\mu\nu}\lambda^{\rho\sigma\alpha\beta}{}_{;\nu}H_{\rho\sigma}H_{\alpha\beta}$$

On making use of equation (20.9) we may in turn write this equation as

$$\tau^{\mu\nu}_{;\nu} = H^{\mu}{}_{\rho}J^{\rho} + \frac{1}{2}[\lambda_{,\rho}e^{\nu}e_{\nu} + \mu_{,\rho}h^{\nu}h_{\nu}]g^{\rho\mu} + u^{\lambda}_{;\rho}g^{\rho\mu}(w_{\lambda} - v_{\lambda}) \tag{21.15}$$

If we define

$$W = -\frac{1}{2}(e^{\rho}d_{\rho} + b^{\rho}h_{\rho}) \tag{21.16}$$

and

$$\sigma^{\mu\nu} = +(g^{\mu\nu} - u^{\mu}u^{\nu})W + e^{\mu}d^{\nu} - h^{\mu}b^{\nu} \tag{21.17}$$

then equation (2.10) becomes

$$\tau^{\mu\nu} = Wu^{\mu}u^{\nu} - u^{\mu}v^{\nu} - w^{\mu}u^{\nu} - \sigma^{\mu\nu} \tag{21.18}$$

and

$$\tau^{\mu\nu}_{;\nu} = Wu^{\nu}u^{\mu}_{;\nu} + (Wu^{\nu})_{;\nu}u^{\mu} - u^{\mu}v^{\nu}_{;\nu} - u^{\mu}_{;\nu}v^{\nu} - u^{\nu}_{;\nu}w^{\mu} - u^{\nu}w^{\mu}_{;\nu} - \sigma^{\mu\nu}_{;\nu} \tag{21.19}$$

Substituting from equation (21.19) into the left hand side of equation (21.15) and multiplying the resulting equation by u_{μ} we obtain

$$(Wu^{\nu} - v^{\nu})_{;\nu} + \sigma^{\mu\nu}u_{\mu;\nu} = (\sigma + \frac{1}{2}\lambda_{;\rho}u^{\rho})e^{\sigma}e_{\sigma} + \frac{1}{2}u_{,\rho}u^{\rho}h^{\sigma}h_{\sigma} + u^{\rho}u^{\mu}v_{\rho;\nu} \tag{21.20}$$

This is in the form of a conservation equation and relates the change in W, the energy density of the electromagnetic fields as measured by an observer at rest with respect to it, the divergence of v^{μ} and the electromagnetic stresses $\sigma^{\mu\nu}$. The vector w^{μ} is related to the momentum of the electromagnetic field.

22. A Variational Principle for the Maxwell Equations

Let us assume that the space-time with the metric tensor $g_{\mu\nu}$ is given and is not influenced by the presence of the electromagnetic fields described by $H_{\mu\nu}$ and $G_{\mu\nu}$. This is the case in special relativity where the space-time is the Minkowski space-time and in a galilean coordinate system the $g_{\mu\nu}$ are constants. We shall not restrict ourselves to galilean coordinate systems nor require that the curvature tensor of space-time vanish.

Consider the integral over an arbitrary region of space-time defined by

$$I_E = \int \tfrac{1}{4}G^{\mu\nu}H_{\mu\nu}\sqrt{g}\ d^4x = \int \tfrac{1}{4}\lambda^{\mu\nu\sigma\tau}H_{\sigma\tau}H_{\mu\nu}\sqrt{g}\ d^4x \qquad (22.1)$$

where $H_{\mu\nu}$ is assumed to satisfy equation(21.6), that is, it determines a four potential φ_μ. We shall assume that φ_μ in addition to depending on the space-time coordinate, also depends on a parameter e. For the present, we assume that the $g_{\mu\nu}$ and u^μ are independent of e. Then

$$I_E'(0) = \int \tfrac{1}{2}\lambda^{\mu\nu\sigma\tau}H_{\sigma\tau}H'_{\mu\nu}\sqrt{-g}\ d^4x$$

$$= \int \lambda^{\mu\nu\sigma\tau}H_{\sigma\tau}(\varphi'_\mu)_{;\nu}\sqrt{-g}\ d^4x$$

where

$$\varphi'_\mu = \left(\frac{d\varphi_\mu}{de}\right)_{e=0}$$

If we require that φ'_μ vanish on the boundary of integration then we have by integrating by parts

$$I'_E(0) = -\int G^{\mu\nu}_{;\nu}\varphi'_\mu\sqrt{-g}\, d^4x$$

The variational principle which requires that

$$I'_E(0) + \int J^\mu \varphi'_\mu \sqrt{-g}\, d^4x = 0 \qquad (22.2)$$

where J^μ is given by ercuations(21.4) and(21.5) is then equivalent to equations (21.2). In case I^μ is absent from the latter equation, equation(22.2) may be derived from looking for the extrema of the integral

$$I_E + I_I = \int (\tfrac{1}{4}G^{\mu\nu}H_{\mu\nu} + \varepsilon\varphi_\mu u^\mu)\sqrt{-g}\, d^4x \qquad (22.3)$$

The second term on the right hand side of equation(22.3) represents the interaction Lagrangean.

It is well known that if the Lagrangenan of a variational principle depends on the coordinates only because of its dependence on the field variables being varied, then there exists a second order, non-symmetric tensor t^σ_ρ which satisfies a conservation theorem. For the Lagrangean

$$L_E = \tfrac{1}{4}G^{\mu\nu}H_{\mu\nu}$$

we have

$$p^{\mu\nu} = G^{\mu\nu} = \frac{\partial L_E}{\partial \varphi_{\mu;\nu}}$$

and

$$t^\sigma_\rho = G^{\mu\sigma}\varphi_{\mu;\rho} - \tfrac{1}{4}\delta^\sigma_\rho G^{\mu\nu}H_{\mu\nu} + (G^{\mu\sigma}\varphi_\rho)_{;\mu} - G^{\mu\sigma}_{;\mu}\varphi_\rho$$

or

$$t^\sigma_\rho = -\tau^\sigma_\rho + (G^{\mu\sigma}\varphi_\rho)_{;\mu} - G^{\mu\sigma}_{;\mu}\varphi_\rho \qquad (22.4)$$

where τ is the Minkowski tensor. In case $J^\mu = 0$, and the space-time is flat the divergences of the tensors t and τ are equal and equal to zero as a consequence of the field equations.

23. The General Variational Principle

We now turn to the discussion of a variational principle from which we propose to derive the equations governing the motion of a perfect fluid which has a dielectric permitivity, a magnetic permeability, is self-gravitating and is subject to and electromagnetic field. We shall have to assume that the conductivity vanishes. If it does not do so we will have to resort to a non-holonomic variational principle. In the subsequent discussion we shall introduce as field variables, the metric tensor $g_{\mu\nu}$ of space-time, the particle paths of the elements of the fluid, the density of the fluid, a variable related to the temperature and the four-vector potential.

The variational principle will be described in terms of

$$I = I_g - 2\kappa(I_F + I_E - I_I) \qquad (23.1)$$

where κ is the Einstein constant of gravitation, I_E and I_I are given by equations (22.1) and (22.3) respectively,

$$I_g = \int \sqrt{-g}\, R d^4x \qquad (23.2)$$

where R is the scalar curvature determined by the metric tensor $g_{\mu\nu}$, and

$$I_F = \int \rho(c^2 + \varepsilon - \Theta S)\sqrt{-g}\, d^4x \qquad (23.3)$$

with ρ the rest density, ε the rest specific internal energy, Θ the rest temperature and S the rest specific entropy. I_F is the integral which entered into our previous discussions.

We have seen (cf. Lecture III) that

$$I_g' - 2\kappa I_F' = -\int[(R^{\mu\nu} - \tfrac{1}{2}g^{\mu\nu}R) + \kappa\phi^{\mu\nu})g_{\mu\nu}' + 2(\rho S u^{\mu})_{;\mu}\alpha']\sqrt{-g}\, d^4x$$

and if this is to vanish for arbitrary $g_{\mu\nu}'$ and α' which vanish on the boundaries, we obtain as the Euler equations

$$R^{\mu\nu} - \tfrac{1}{2}g^{\mu\nu}R + \kappa\phi^{\mu\nu} = 0 \tag{23.4}$$

and

$$(\rho S u^{\mu})_{,\mu} = 0 \tag{23.5}$$

Equation (23.5) may be shown to be a consequence of the fact that the conservation of mass holds and

$$\theta^{\mu\nu}{}_{;\nu} = 0 \tag{23.6}$$

as a result of equation (23.4).

We now turn to the evaluation of

$$I_E' = \tfrac{1}{4}\int(\lambda^{\mu\nu\sigma\tau}H_{\sigma\tau}H_{\mu\nu}\sqrt{-g})'\, d^4x\ .$$

We have evaluated the contribution to I_E' from the variation of the ϕ_μ in the preceeding section. Thus after an integration by parts

$$I_E' = \int[-G^{\mu\nu}_{;\nu}\phi_\mu' + \tfrac{1}{8}(G^{\mu\nu}H_{\mu\nu})g^{\sigma\tau}g_{\sigma\tau}' + \tfrac{1}{4}\lambda^{\mu\nu\sigma\tau\prime}H_{\sigma\tau}H_{\mu\nu}]\sqrt{-g}\, d^4x$$

where $\lambda^{\mu\nu\sigma\tau\prime}$ is to be computed from equation (20.9), with μ and λ assumed to be independent of e.

It may be verified that

$$I_E' = \int[-G^{\mu\nu}_{;\nu}\phi_\mu' + \tfrac{1}{2}E^{\mu\nu}g_{\mu\nu}']\sqrt{-g}\, d^4x$$

where

$$E^{\mu\nu} = (\tfrac{1}{2}g^{\mu\nu} - u^{\mu}u^{\nu})(e^{\rho}d_{\rho} + b^{\rho}h_{\rho}) - e^{\mu}d^{\nu}$$

$$- h^{\mu}b^{\nu} - u^{\mu}v^{\nu} - u^{\nu}v^{\mu} \qquad (23.7)$$

$$= E^{\nu\mu} .$$

The symmetric tensor $E^{\mu\nu}$ is similar to the Minkowski tensor. It is known as the Abraham tensor [2]. In fact we have

$$E^{\mu\nu} - \tau^{\mu\nu} = u^{\nu}(w^{\mu} - v^{\mu}) \qquad (23.8)$$

Hence on evaluating

$$I' = I'_g - 2\kappa(I'_F + I'_E + I'_I),$$

we find that

$$I' = - \int\{[(R^{\mu\nu} - \tfrac{1}{2}g^{\mu\nu}R) + \kappa T^{\mu\nu}]g'_{\mu\nu}$$

$$+ 2\kappa(\rho S u^{\mu})_{;\mu}\alpha' + 2\kappa(G^{\mu\nu}_{;\nu} - J^{\mu})\phi'_{\mu}\}\sqrt{-g}\, d^4x$$

where

$$T^{\mu\nu} = \theta^{\mu\nu} + E^{\mu\nu} . \qquad (23.9)$$

Thus the requirement that $I' = 0$ for arbitrary $g'_{\mu\nu}$, α' , and ϕ'_{μ} leads to the Euler equations

$$R^{\mu\nu} - \tfrac{1}{2}g^{\mu\nu}R + \kappa T^{\mu\nu} = 0 \qquad (23.10)$$

$$(\rho S u^{\mu})_{;\mu} = 0 \qquad (23.11)$$

$$G^{\mu\nu}_{;\nu} = J^{\mu} = \varepsilon u^{\mu} \qquad (23.12)$$

The latter equations are of course the Maxwell equations. Equations (23.10) are the Einstein field equations with the source of the gravitational field given by both the matter present and the electromagnetic field. Equations (23.11) are a consequence of equations (23.10) and (23.11)when matter is conserved. That is, the equations

$$(\rho u^{\mu})_{;\mu} = 0$$

hold and there are no ohmic losses. For, it is a consequence of equations (23.13)and the Bianchi identity that

$$T^{\mu\nu}_{;\nu} = 0 \qquad (23.14)$$

These four equations which are the generalization of the Lorentz pondermotive force equations contain the definition of the pondermotive force acting on the matter and a statement concerning the conservation of energy. The latter statement is equivalent to the statement about the rate of change of rest specific entropy along a world-line of the fluid vanishes.

24. Summary

The general variational principle given above, from which one may derive the Einstein field equations, the Maxwell equations, and the equations of hydrodynamics may be applied to the problem of general relativistic magnetohydrodynamics by setting $e^{\alpha} = 0$. The restriction to $\sigma = 0$ may be removed by using a non-holonomic variational principle, that is by replacing I_I' by an appropriate expression involving α' and ϕ'.

It is worthy of notice that the pondermotive force acting on the charged fluid with electric permitivity and magnetic permeability is not given by the divergence of the Minkowski tensor. Rather, it is given by the divergence of the symmetric tensor $E^{\mu\nu}$ the Abraham tensor defined by equation (23.7) related to $\tau^{\mu\nu}$ by equation (23.8). A similar result holding for less general circumstances has been given by Penfield and Haus [3]. It is reasonable to expect that the pondermotive force is to be derived from the symmetric tensor $E^{\mu\nu}$ even in non-conservative cases where the variational principle does not apply in the form discussed above in detail. This would resolve the old controversy over the appropriate stress-energy tensor to be used in describing general electromagnetic fields in moving bodies.

From the discussion of the variational principles it is clear as to why $E^{\mu\nu}$ replaces $\tau^{\mu\nu}$ in the calculation of the pondermotive force. This tensor describes the energy of the electromagnetic field alone as follows from the discussion at the end of section 22 above. However, because $\lambda^{\mu\nu\sigma\tau}$ depends on u^{μ}, as well as on λ and μ, the Lagrangean L_E contains interaction terms between the fluid and the electromagnetic field, even in the case of an uncharged medium. The presence of these terms then is responsible for the additional energy which changes $\tau^{\mu\nu}$ into $E^{\mu\nu}$

We conclude with the remark that the methods given above will apply to more general situations then those treated. Thus we need not assume that the electric properties of the matter are isotropic or that that matter is a fluid. All that is required is that a Lagrangean exists that is a function of the field quantities describing the various properties of the medium. Thus if in general

$$G^{\mu\nu} = G^{\mu\nu}(H^{\sigma\tau})$$

and if

$$\frac{\partial G^{\mu\nu}}{\partial H_{\sigma\tau}} \equiv \lambda^{\mu\nu\sigma\tau} = \lambda^{\sigma\tau\mu\nu} ,$$

we may define a Lagrangean for the electromagnetic fields. Similarly if the matter has a more general stress-energy tensor than that given by a fluid, but involving no viscous or similar forces, it too may be described by a more general Lagrangean.

References

1. A. Lichnerowicz, Relativistic Hydrodynamics and Magneto-hydrodynamics, W.A. Benjamin, Inc., New York (1967).

2. W. Pauli, Theory of Relativity, Pergamon Press, New York (1958), p. 110.

3. Paul Penfield, Jr. and Hermann A. Haus, The Physics of Fluids 9 (1966) 1195-1202.

LECTURE VI

STABILITY OF GENERAL RELATIVISTIC GASEOUS MASSES AND VARIATIONAL PRINCIPLES

25. The Second Variation

In this lecture we shall derive and then apply the well-known result that if a set of equations are the Euler equations of a variational principle based on an integral $I(e)$ then the perturbations of solutions to the Euler equations satisfy equations which may be derived from another variational principle. The latter principle is given by an integral equal to $I''(0)$ where the prime denotes the derivative of $I(e)$ with respect to e. The variational principle based on $I''(0)$ is called the second variation.

We shall apply these results to the discussion of the stability against radial perturbation of spherically symmetric static solutions of the Einstein field equations for a self-gravitating fluid which obeys an equation of state. In this case we may define I to be that given by equation (19.2). We shall show that the second variation problem for this integral is the same as the variational principle given by Chandrasekhar [1] for this problem.

Let $\mathcal{L}(\phi^A;\phi^A_{,\mu})$ be a scalar density formed from some scalar or tensor fields ϕ^A and the derivatives of these fields with respect to the coordinates in space time,

$$\phi^A_{,\mu} = \frac{\partial\phi^A}{\partial x^\mu} \qquad \begin{aligned} \mu &= 1, 2, 3, 4 \\ A &= 1, 2, \ldots, N. \end{aligned}$$

Then

$$I = \int \mathcal{L}(\phi^A;\phi^A_{,\mu})d^4x \tag{25.1}$$

where the integral is carried out over an arbitrary four volume in space time determines a variational principle in the following sense. We assume that the ϕ^A are functions of the x^μ and a parameter e, thus

$$\phi^A = \phi^A(x;e).$$

Then I is also a function of e and we may require that

$$I'(0) = \left.\frac{dI}{de}\right|_{e=o} = 0$$

for arbitrary

$$\phi'^A(0) = \left(\frac{d\phi^A}{de}\right)_{e=o}$$

We have

$$I'(e) = \int\left[\frac{\partial\mathcal{L}}{\partial\phi^A}\phi'^A + \frac{\partial\mathcal{L}}{\partial\phi^A_{,\mu}}(\phi^A_{,\mu})'\right] d^4x$$

$$I'(e) = \int\left[\frac{\partial\mathcal{L}}{\partial\phi^A}\phi'^A + \frac{\partial\mathcal{L}}{\partial\phi^A_{,\mu}}(\phi'^A)_{,\mu}\right] d^4x \tag{25.2}$$

since

$$\phi'^A = \frac{\partial\phi^A}{\partial e}$$

$$(\phi^A_{,\mu})' = \frac{\partial^2\phi^A}{\partial x^\mu\partial e} = \frac{\partial^2\phi^A}{\partial e\partial x^\mu} .$$

On integrating the above expression for $I'(e)$ by parts we obtain

$$I'(e) = \int F_A(e)\phi'^A(e)\, d^4x + \int\left(\frac{\partial\mathcal{L}}{\partial\phi^A_{,\mu}}\phi'^A\right)_{,\mu} d^4x . \tag{25.3}$$

where

$$F_A(e) = \frac{\partial \mathcal{L}}{\partial \phi^A} - \left(\frac{\partial \mathcal{L}}{\partial \phi^A_{,\mu}}\right)_{,\mu} \tag{25.4}$$

The second integral in Eq. (25.3) may be written as an integral over the hypersurface bounding the four-volume of integration.

The requirement that $I'(0) = 0$ for arbitrary $\phi'^A(0)$, in particular, for those that vanish on the boundary of the region of integration then leads to the Euler equations

$$F_A(0) = \left[\frac{\partial \mathcal{L}}{\partial \phi^A} - \left(\frac{\partial \mathcal{L}}{\partial \phi^A_{,\mu}}\right)_{,\mu}\right]_{e=o} = 0 \; . \tag{25.5}$$

The equations satisfied by the difference between two "almost equal" solutions of these equations, or the equations satisfied by perturbations of solutions to these Euler equations are

$$F'_A(0) = \left(\frac{dF_A}{de}\right)_{e=o} = 0$$

where $F_A(e)$ is given by Eq. (25.4). Thus

$$\begin{aligned} F'_A(0) = \Bigg[& \frac{\partial^2 \mathcal{L}}{\partial^2 \phi^A \partial \phi^B} \phi'^B + \frac{\partial^2 \mathcal{L}}{\partial \phi^A \partial \phi^B_{,\mu}} \phi'^B_{,\mu} \\ & - \left(\frac{\partial^2 \mathcal{L}}{\partial \phi^A_{,\mu} \partial \phi^B} \phi'^B + \frac{\partial^2 \mathcal{L}}{\partial \phi^A_{,\mu} \partial \phi^B_{\nu}} \phi'^B_{,\nu}\right)_{,\mu}\Bigg]_{e=o} \\ = {} & 0 \; . \end{aligned} \tag{25.6}$$

These are a set of linear equations for the variables $\phi'^A(0)$ whose coefficients depend on the $\phi^A(x;0)$ and their derivatives. The $\phi'^A(0)$ are called the perturbations and the $\phi^A(x;0)$ the unperturbed solutions.

Now it follows from Eq. (25.2) that

$$I'(e) = \int F_A \phi''^A d^4x + \int F'_A \phi'^A d^4x + \int \left(\frac{\partial \mathcal{L}}{\partial \phi^A_{,\mu}} \phi''^A\right)_{,\mu} d^4x$$

$$+ \int \left[\frac{\partial^2 \mathcal{L}}{\partial \phi^A \partial \phi^B_{,\mu}} \phi'^A \phi'^B + \frac{\partial^2 \mathcal{L}}{\partial \phi^A_{,\mu} \partial \phi^B_{,\nu}} \phi'^A_{,\mu} \phi'^B\right]_{,\nu} d^4x \tag{25.7}$$

or

$$I''(e) = \int F_A \phi''^A d^4x + \int \left(\frac{\partial \mathcal{L}}{\partial \phi^A_{,\mu}} \phi''^A\right)_{,\mu} d^4x$$

$$+ \int \left[\frac{\partial^2 \mathcal{L}}{\partial \phi^A \partial \phi^B} \phi'^A \phi'^B + \frac{2\partial^2 \mathcal{L}}{\partial \phi^A \partial \phi^B_{,\mu}} \phi'^A \phi'^B_{,} + \frac{\partial^2 \mathcal{L}}{\partial \phi^A_{,\mu} \partial \phi^B_{,\nu}}\right] \phi'^A_{,\mu} \phi'^B_{,\nu} d^4x. \tag{25.8}$$

From Eq. (25.8) we have

$$I''(0) = \int \left[\frac{\partial^2 \mathcal{L}}{\partial \phi^A \partial \phi^B} \phi'^A \phi'^B + \frac{2\partial^2 \mathcal{L}}{\partial \phi^A \partial \phi^B_{,\mu}} \phi'^A \phi'^B_{,\mu} + \frac{\partial^2 \mathcal{L}}{\partial \phi^A_{,\mu} \partial \phi^B_{,\nu}} \phi'^A_{,\mu} \phi'^B_{,\nu}\right] d^4x \tag{25.9}$$

when the $\phi^A(0)$ are such that $F_A(0) = 0$, that is the $\phi^A(0)$ are unperturbed solutions of the Euler equations associated with I, and the $\phi''^A(0) = 0$, on the boundary of the region of integration. If we now consider the ϕ'^A (not the ϕ^a) functions of x and a parameter f we may define

$$J(f) = I''(0)$$

and examine the Euler equations resulting from the condition

$$\left(\frac{dJ}{df}\right)_{f=0} = 0 .$$

This is the "second variation problem." We find

$$\delta J = 2\int F'_A \delta\phi'^A d^4x + 2\int \left(\frac{\partial^2 \mathcal{L}}{\partial\phi^A \partial\phi^B_{,\mu}} \phi'^A \delta\phi'^B\right)_{,\mu} d^4x$$

$$+ 2\int \left(\frac{\partial^2 \mathcal{L}}{\partial\phi^A_{,\mu} \partial\phi^B_{,\nu}} \phi'^A_{,\mu} \delta\phi^B\right)_{,\nu} d^4x \quad .$$

Hence for variations of the ϕ'^A such that ϕ'^A vanish on the boundary of the region of integration, $I''(0)$ takes on extreme values when the ϕ'^A satisfy the equations

$$F'_A(0) = 0 \quad ,$$

the equations satisfied by the perturbations.

Thus the solutions ϕ^A of the Euler equations $F_A(\phi) = 0$, when considered as $\phi^A(x;0)$ are such that $I'(0) = 0$, for ϕ'^A which vanish on the boundary of the region of integration, and the solutions ϕ'^A of the equations $F'_A(\phi;\phi') = 0$, where the ϕ^A satisfy the Eurler equations and are coefficients in the linear differential equations, are such that $I''(0)$ takes on extreme values.

26. The Spherically Symmetric Case

In a spherically symmetric space-time we may write the line element as

$$ds^2 = e^{2\phi}\, dt^2 - e^{2\psi}\, dr^2 - e^{2\mu}\, d\Omega^2 \tag{26.1}$$

where

$$d\Omega^2 \quad d\theta^2 + \sin^2\theta d\chi^2 \quad , \tag{26.2}$$

ϕ, ψ and μ are functions of r, t and a parameter e and these are comoving coordinates for each value of the parameter e. It then follows [2] that the non-vanishing components of $G^{\mu}{}_{\nu}$ are

$$- (R_4^4 - \frac{R}{2}) = e^{2\phi}(\mu_t^2 + 2\mu_t\psi_t) - e^{-2\psi}(2\mu_{rr} + 3\mu_r^2 - 2\mu_r\psi_r) + e^{-2\mu}$$

$$- (R_1^1 - \frac{R}{2}) = e^{-2\phi}(2\mu_{tt} + 3\mu_t^2 - 2\mu_t\phi_t) - e^{-2\psi}(\mu_r^2 + 2\mu_r\phi_r) + e^{-2\mu}$$

$$- (R_2^2 - \frac{R}{2}) = - (R_3^3 - \frac{R}{2}) \tag{26.3}$$

$$= e^{-2\phi}[\psi_{tt} + \mu_{tt} + \mu_t^2 + \psi_t^2 - \psi_t\phi_t + \mu_t(\psi_t - \phi_t)]$$

$$- e^{-2\phi}[\phi_{rr} + \mu_{rr} + \mu_r^2 + \phi_r^2 - \phi_r\psi_r + \mu_r(\phi_r - \psi_r)]$$

$$R_1^4 = 2e^{-2\phi}[\mu_{rt} - \mu_t\phi_r - \mu_r\psi_t + \mu_t\mu_r]$$

$$R_4^1 = - 2^{-2\psi}[\mu_{rt} - \mu_t\phi_r - \mu_r\psi_t + \mu_t\mu_r]$$

where the subscripts r and t denote the derivatives with respect to these variables.

The Einstein equations become

$$-F_\phi \equiv (R_4^4 - \tfrac{1}{2}R) + kw = 0 ,$$

$$-F_\psi \equiv (R_1^1 - \tfrac{1}{2}R) - kp = 0 , \tag{26.4}$$

$$-F_\mu \equiv (R_2^2 - \tfrac{1}{2}R) - kp = 0 ,$$

and

$$R_4^1 = 0 . \tag{26.5}$$

The four equations (26.4) and (26.5), are not all independent in view of the Bianchi identities. It may be shown that the solution of these equations is determined by the solution of equation (26.5) and $F_\psi = 0$ for a range of values of t and of $F_\phi = 0$ for $t = 0$.

The unperturbed solution we shall consider will be assumed to be static, that is ϕ, ψ and μ will be assumed to be functions of r alone. In that case it is no restrictior to take

$$\mu = \log r.$$

Equation (26.5) is identically satisfied and equations (26.4) reduce to

$$\frac{1}{r^2} - e^{-2\psi}\left(\frac{1}{r^2} - \frac{2}{r}\psi_r\right) = kw ,$$

$$\frac{1}{r^2} - e^{-2\psi}\left(\frac{1}{r^2} + \frac{2}{r}\phi_r\right) = -kp , \tag{26.6}$$

$$e^{-2\psi}[\phi_{rr} + \phi_r^2 - \phi_r\psi_r + \frac{1}{r}(\phi_r - \psi_r)] = kp .$$

It is a consequence of these equations that

$$2e^{-2\psi}(\phi_r + \psi_r) = k(w + p)r . \tag{26.7}$$

It is a further consequence of Eqs. (26.6) that

$$(w + p)\phi_r = -p_r . \tag{26.8}$$

The last equation also follows from the equation of state assumption.

The equations satisfied by the perturbations, ϕ', ψ' and μ' are obtained by differentiating Eqs. (26.4) and (26.5) with respect to e and setting $e = 0$. We then obtain from Eq. (26.5) and the last of (26.4) the equation

$$\mu'_{rt} - \mu'_t\phi_r - \frac{1}{r}\psi'_t + \frac{1}{r}\mu'_t = 0 \tag{26.9}$$

where now ϕ_r is determined by Eqs. (26.6). The solution of Eq. (26.9) is given by

$$\psi' - \psi'_o = e^{\phi}(e^{-\phi}r\mu')_r \tag{26.10}$$

where

$$\psi'_o = \psi'(r,0)$$

and we have chosen our comoving coordinates so that

$$\mu'_o = \mu'(r,0) = 0 . \tag{26.11}$$

This can always be achieved by a coordinate from transformation involving r alone.

The function ϕ' may be evaluated by using the integral of the field equations given by Eq.(17.8) which holds for all values of e. That equation may be written as

$$\sigma e^{\psi+2\mu} = \sigma_o e^{\psi o+2\mu o} \tag{26.12}$$

where now the subscript zero on $f(r, t, e)$ is defined by

$$f_o = f(r, 0; e) .$$

On differentiating Eq. (26.12) with respect to e, setting $e = 0$ and using Eqs. (17.6), (16.5), and (26.11) we obtain

$$\alpha^{-2}\phi' - \alpha_o^{-2}\phi_o' = (3\mu' + r\mu_r' - \phi_r r\mu') \qquad (26.13)$$

or

$$\alpha^{-2}\phi' - \alpha_o^{-2}\phi'_o = r^{-2}e^{\phi}(r^3\mu' e^{-\phi})_r \qquad (26.14)$$

where α is the velocity of sound in the unperturbed fluid, ϕ is given as above and ϕ'_o is $\phi'(r,0)$ for $e = 0$.

Thus ϕ' and ψ' are determined in terms of μ' . This function may be determined by solving the equation

$$F_{\psi}' = 0 \ .$$

The quantities ϕ' and ψ' enter into this equation but may be eliminated by means of Eqs. (26.10) and (26.14). We shall discuss this equation in the next section.

When the field equations, Eq. (26.4) and (26.5) are applied to a problem in which there exists a hypersurface in space-time across which the stress-energy tensor is discontinuous, the equations must be supplemented by conditions satisfied by the metric tensor, its derivatives and the stress energy tensor on this hypersurface. Thus for the problem we wish to consider, namely that of a gas occupying a limited region of space-time and bounded by a vacuum there exists the hypersurface Σ defined by

$$r = r_b$$

where r_b is the constant comoving coordinate of the boundary element of the material.

It is well known that the conditions referred to above become in this case

$$p(r_b, t) = 0$$

and that ϕ, ψ, and μ are continuous across the hypersurface Σ. In addition all first derivatives of these quantities except ψ_r must be continuous across Σ. These conditions must hold for the perturbed as well as for the unperturbed equations. Hence we must have

$$p'(r_b,t) = -[(w + p)\phi']_{r=r_b} = 0 \tag{26.15}$$

In view of Eq. (26.14) this condition becomes a boundary condition on the function μ'.

Another condition is the requirement that for the perturbed and the unperturbed solutions the function

$$R = e^{\mu} = r = 0$$

at the origin. This function is the analogue of the Eulerian coordinate of an element of the fluid which has the Lagrange coordinate r. Hence we must have

$$R' = e^{\mu}\mu' = r\mu' = 0 \tag{26.16}$$

at the origin.

Eqs. (26.15) and (26.16) provide boundary conditions for the second order partial differential equation $F_{\psi'} = 0$.

We close this section with a discussion of the implication of the Bianchi identities.

If we define

$$-K^{\mu}{}_{\nu} = R^{\mu}{}_{\nu} - \tfrac{1}{2}\delta^{\mu}{}_{\nu} R + kT^{\mu}{}_{\nu} ,$$

these identities are

$$K^{\mu}_{\nu;\mu} = \frac{1}{\sqrt{-g}} (\sqrt{-g}K^{\mu}_{\nu})_{,\mu} - K^{\mu}_{\rho}\, \Gamma^{\rho}_{\mu\nu} = 0$$

They hold for all values of e. If the above equations are differentiated with respect to e and then evaluated for $e = 0$, and if it is assumed that $K^{\mu}{}_{\nu}(x;0) = 0$, it follows that

$$\frac{1}{\sqrt{-g}} (\sqrt{-g}\; K'^{\mu}{}_{\nu})_{,\mu} - K'^{\mu}{}_{\rho}\Gamma^{\rho}_{\mu\nu} = 0 \qquad (26.17)$$

where now $g_{\mu\nu} = g_{\mu\nu}(x;0)$ is the unperturbed metric, and $\Gamma^{\rho}_{\mu\nu}$ is determined from this metric and this metric satisfies the field equations.

We now evaluate Eqs. (26.17) for the case considered above, when the unperturbed metric is spherically symmetric and static and the perturbed metric depends on time but is still spherically symmetric. In that case Eqs. (26.17) reduce to two equations corresponding to $\nu = 4$ and $\nu = 1$. These are

$$r^2 e^{\psi}(F'_{\phi})_t - (e^{\psi} r^2 R'^{1}_{4})_r = 0 \qquad (26.18)$$

and

$$-r^2 (R'^{4}_{1})_t + (r^2 e^{\phi} F'_{\psi})_r e^{-\phi} - 2rF'_{\mu} = 0 \qquad (26.19)$$

respectively.

Hence when $R'^{1}_{4} = (R'^{1}_{4})_t = 0$ as is the case when Eq. (26.10) holds, Eq. (26.18) becomes

$$F'_\phi(r,t) = F'_\phi(r,0) \qquad (26.20)$$

and Eq. (26.19) becomes

$$2rF'_\mu = (r^2 e^\phi F'_\psi)_r e^{-\phi} . \qquad (26.21)$$

The first of these equations implies that the equation $F'_\phi = 0$ is only a restriction on the functions ϕ'_o and ψ'_o . It may be verified that on substituting Eqs. (26.10) into the expression for F'_ϕ one obtains

$$-F'_\phi = \frac{2}{r^2}(re^{-2\psi}\psi'_o)_r + \frac{k}{\alpha^2}(w + p)\phi'_o = 0 \qquad (26.22)$$

27. The Equation $F'_\psi = 0$

The equation $F'_\psi = 0$ is derived by differentiating the second of Eqs. (26.4) into which Eqs. (26.3) have been substituted setting $e = 0$, and making use of the values of the unperturbed solution. One then obtains

$$F' = 2[e^{-2}\mu'_{tt} - \mu'_r e^{-2\psi}(\frac{1}{r} + \phi_r) - \frac{\mu'}{r^2} - \frac{1}{r}e^{-2\psi}\phi'_r$$

$$+ e^{-2\psi}\psi'(\frac{1}{r^2} + \frac{2}{r}\phi_r)] - k(w + p)\phi' = 0 \qquad (27.1)$$

when Eqs. (26.10) and (26.14) are used to express ψ' and ϕ' in terms of μ', one finds that

$$\frac{re^{2\psi}(w+p)}{2} F'_{\psi} = e^{2\psi-2\phi}(w+p)\xi_{tt} + \frac{4}{r} p_r \xi - \frac{1}{w+p} p_r^2 \xi$$
$$- e^{-\psi-2\phi}[e^{3\phi+\psi}\frac{(w+p)\alpha^2}{r^2}(e^{-\phi}r^2\xi)_r]_r \qquad (27.2)$$
$$+ ke^{2\psi}(w+p)p\xi + \psi'_o(\frac{1}{r^2} + \frac{2}{r}\phi_r) - e^{-\psi-\phi}(e^{\phi+\psi}\phi'_o)_r$$

where

$$\xi = r\mu' . \qquad (27.3)$$

The equation $F'_{\psi} = 0$ where F'_{ψ} is given by equation (6.2), has a boundary condition indicated in Eqs.(26.15) and (26.16). It is the equation given in [2] for the case of the radial perturbations of a self gravitating fluid when the equation of state was such that the fluid was isentropic. In that case $\sigma = \rho$, the rest mass density of the fluid. When $\phi'_o = \psi'_o = 0$, the equation is the same as the equation given by Chandrasekhar [1] as may be seen by writing

$$(w + p)\alpha^2 = \gamma p,$$

and thus defining γ. This definition of γ is that given by Chandrasekhar as may be verified by writing

$$w = N(1 + u(p,N))$$

where u is the internal energy. If one then computes $(\partial p/\partial w)_{\beta}$ and remembers that

$$TdS = du + pd\,(\frac{1}{N})$$

one verifies that the definition of γ given above is that used by Chandrasekhar.

28. The Evaluation of I"(0)

In this section we shall use the results obtained above to express $I''(0)$ in terms of μ', ψ'_o, and ϕ'_o. We begin by observing that when I is defined by Eq. (19.2) and when the perturbed and unperturbed metrics are of the form given by Eq. (26.1), then it is sufficient for the purpose of calculating $I'(e)$ and $I''(e)$ to evaluate $I(e)$ in the coordinate system in which Eq. (26.1) holds.

Thus we have

$$\frac{1}{8\pi} I(e) = -\iint\{e^{\phi+\psi} + e^{\phi-\psi+2\mu}(\mu_r^2 + 2\mu_r\phi_r) - e^{-\phi+\psi+2\mu}(\mu_t^2 + 2\mu_t\psi_t) + kpe^{\phi+\psi+2\mu} - (e^{-\phi}(e^{2\mu+\psi})_t)_t + (e^{-\psi}(e^{2\mu+\phi})_r)_r\}dr\,dt \quad . \tag{28.1}$$

Hence

$$\frac{1}{8\pi}I'(e) = -\iint\{e^{\phi+\psi+2\mu}(F_\phi\phi' + F_\psi\psi' + 2F_\mu\mu')dr\,dt - S(e) \tag{28.2}$$

where F_ϕ, F_ψ, and F_μ are defined by Eqs. (26.4) and (26.3) and

$$S(e) = \iint(A_t - B_r)\,dr\,dt \tag{28.3}$$

with

$$A = e^{-\phi+\psi+2\mu}(-\phi'(2\mu_t + \psi_t) + \psi'\psi_t + 2\mu'\mu_t + 2\mu'_t + \psi'_t), \tag{28.4}$$

$$B = e^{\phi-\psi+2\mu}(-\psi'(2\mu_r + \phi_r) + \phi'\phi_r + 2\mu'\mu_r + 2\mu'_r + \phi'_r) \quad . \tag{28.5}$$

The integration in Eqs. (28.1) to (28.3) may be taken to be the region bounded by the inequalities

$$0 \leqq r \leqq \infty \qquad 0 \leqq t \leqq t_1 \quad . \tag{28.6}$$

Across the boundary

$$r = r_b \ . \tag{28.7}$$

There is a discontinuity in the stress energy tensor. The pressure p must be continuous at $p = r_b$ but the energy density w need not be. The requirement that

$$\frac{1}{8\pi} I'(0) = 0$$

for arbitrary ϕ', ψ', and μ' which vanish together with their derivatives on the boundary of the region given by the inequalities (28.6) and such that ψ', ϕ', μ' and ϕ'_r and μ'_r may take on arbitrary values on the interior boundary given by Eq. (28.6) leads to the field Eqs (26.4) and the boundary conditions discussed in Section 26 (cf.[4]).

We also have

$$\frac{1}{8\pi} I''(0) = - \iint (F'_\phi \phi' + F'_\psi \psi' + 2F'_\mu \mu') e^{\phi+\psi} r^2 \, dr \, dt - S'(0)$$

where ϕ, ψ and μ are evaluated for $e = 0$ and these functions satisfy the unperturbed equations. In view of Eqs. (26.10), (26.7), and (27.3) we have

$$e^{\phi+\psi} r^2 F'_\psi \psi' = e^{\phi+\psi} r^2 \psi'_o F'_\psi + (e^{\phi+\psi} r^2 \xi F'_\psi)_r$$
$$- e^{\phi+\psi} r^2 \left(\frac{\mu'}{r} e^{-\phi} (e^\phi r^2 F'_\psi)_r + \frac{\xi}{2} r e^{2\psi} k(w + p) F'_\psi\right).$$

On using Eq. (26.21) we obtain

$$e^{\phi+\psi} r^2 (F'_\psi \phi' + 2F'_\mu \mu') = e^{\phi+\psi} r^2 (\psi'_o F'_\psi - \frac{\xi}{2} r e^{2\psi} k(w + p) F'_\psi)$$
$$+ (e^{\phi+\psi} r^2 \xi F'_\psi)_r \ .$$

This equation holds for all values of r, however for $r \geqq r_b$ $w = p = 0$ and for $r \leqq r_b$ we may use Eq. (27.2). Hence we have

$$\frac{1}{8\pi} I''(0) = J + J_1 + \Sigma \tag{28.8}$$

where

$$J = k\int_0^{t_1}\int_0^{r_b} [(w+p)r^2 e^{3\psi-\phi}\xi\xi_{tt} + e^{3\phi+\psi}\frac{(w+p)\alpha^2}{r^2}(re^{-\phi}\xi)_r^2$$

$$+ r^2 e^{\phi+\psi}\xi^2(ke^{2\psi}(w+p)p + \frac{4}{r} p_r - \frac{1}{w+p} p_r{}^2)] \, dr \, dt \tag{28.9}$$

$$J_1 = -\int_0^{t_1}\int_0^{\infty} r^2 e^{\phi+\psi}[F'_\phi\phi' + F_\psi\psi'_o] \, dr \, dt$$

$$+k\int_0^{t_1}\int_0^{r_b} [r^2 e^{\phi+\psi}\xi\psi'_o(\frac{1}{r^2} + \frac{2}{r}\dot{\phi}_r) - r^2\xi(e^{\phi+\psi}\phi'_o)_r] \, dr \, dt \tag{28.10}$$

and

$$\Sigma = -\int_0^{t_1}\int_0^{\infty}(e^{\phi+\psi}r^2\xi F'_\psi)_r \, dr \, dt - S'(0)$$

$$-\int_0^{t_1}\int_0^{r_b}(k\xi e^{2\phi+\psi}(w + p)\alpha^2(e^{-\phi}r^2\xi)_r)_r dr \, dt \tag{28.11}$$

with F'_ψ given in terms of ϕ', ψ' and ξ' by Eq. (27.1).

If the functions ϕ', ψ' and ξ are to be such that $I'(0) = 0$, that is if they and their derivatives are to vanish on the ex erior boundaries, and if the boundary conditions on ξ are to hold ar $r = r_b$ and $r = 0$ we must have

$$\frac{1}{8\pi} I''(0) = J \tag{28.12}$$

where J is given by Eq. (28.9).

The Euler equations of the variational principle

$$\frac{1}{8\pi}\,\delta I''(0) = \delta J = 0 \tag{28.13}$$

is the equation

$$\frac{re^{2\psi}(w + p)}{2}\,F'_{\psi} = 0 \tag{28.14}$$

where the explicit form of this equation is given by equation (27.2). This equation is equivalent to $F'_{\psi} = 0$ and the variational principle defined by Eqs. (28.13) and (28.9) was of course to be expected in view of the general discussion given in the introduction.

29. The Stability Criterion

The variational principle defined by Eqs. (28.13) and (28.9) may be related to that given by Chandrasekhar in [2] by observing that if one writes

$$\xi = \sin(\sigma t + a)\zeta(r) \tag{29.1}$$

we have

$$\frac{1}{8\pi}\,I''(0) = J = \int_0^{t_1}\sin^2(\sigma t + a)\mathcal{J}\,dt \tag{29.2}$$

where

$$\mathcal{J} = \int_0^{r_b} - \sigma^2 e^{3\psi-\phi} r^2 (w + p)\,\xi^2\,dr + \mathcal{J}_1 \tag{29.3}$$

where

$$\mathcal{J}_1 = \int_0^{r_b} [r^2 e^{\phi+\psi}\xi^2 (ke^{2\psi}(w+p)p + \frac{4p_r}{r} - \frac{1}{(w+p)}\,p_r{}^2) + e^{+3\phi+\psi}(w+p)\,\frac{\alpha^2}{r^2}\,(re^{-\phi}\xi)_r^2]\,dr\ . \tag{29.4}$$

The variational problem

$$\delta \mathcal{J} = 0 \qquad (29.5)$$

has as its Euler equation, Eq. (28.14) with ξ given by Eq. (29.3). The functions $\zeta(r)$ satisfying this Euler equation, that is the extremal $\zeta(r) = \zeta_e(r)$ are such that

$$\mathcal{J}[\zeta_e] = 0$$

Chandrasekhar has pointed out (cf. [2]) that the variational problem given by Eq. (8.4) expresses a minimum principle for the determination of the lowest value of σ^2 and that a sufficient condition for the dynamical instability of a mass is that $\mathcal{J}_1 = 0$ for some "trial function" ξ which which satisfies the required boundary conditions.

However, if such a trial function exists we shall have $\mathcal{J} \leqq 0$ and in view of Eq. (29.2), for this trial function

$$\frac{1}{8\pi} I''(0) \leqq 0 \qquad (29.6)$$

Thus the sufficient condition for instability used by Chandrasekhar is equivalent to the condition that there exists a trial function such that the inequality (29.6) holds. The latter criterion may be applied to discussion of the stability of general solutions of the Einstein field equations. We need not restrict ourselves to a static unperturbed solution and consider perturbations of such solutions which depend on the then defined time coordinate in an exponential manner.

References

1. Chandrasekhar, S. The dynamical instability of gaseous masses approaching the Schwarzschild limit in general relativity. Astrophys. J. 140, 417-433 (1964).

2. Taub, A.H. Small motions of a spherically symmetric distribution of matter. Les Theories Relativistes de la Graviation, pp. 173-191. Centre National de la Recherches Scientific, Paris (1962).

3. ________. Singular hypersurfaces in general relativity, Illinois J. Math. 1, 370-388 (1957.

CENTRO INTERNAZIONALE MATEMATICO ESTIVO

(C.I.M.E.)

GENERAL-RELATIVISTIC KINETIC THEORY OF GASES[1]

J. EHLERS

Corso tenuto a Bressanone dal 7 al 16 Giugno 1970

GENERAL-RELATIVISTIC KINETIC THEORY OF GASES [1]

Introduction

The relativistic kinetic theory of gases, which will be presented in the following lectures, is of interest for a number of reasons: It offers a simple, microscopic model for matter in bulk which is sufficiently general to provide a basis for hydrodynamics and thermodynamics of simple and multicomponent systems. Definite conservation laws, balance equations, equations of state, transport and reactions can be derived from it, and if cross from a microscopic scattering theory are fed in, kinetic theory gives transport and reaction coefficients. As in the non-relativistic theory, the arbitrariness of the constitutive equations and the indefiniteness of the transport coefficients inherent in the phenomenological continuum approach are overcome by the kinetic theory.

Moreaver, kinetic theory provides a description of gases under conditions where fluid dynamics does not apply, e.g., when collisions are rare and the mean free path is large.

---

(1) This work was supported in part by NSF - grant GP 20033

Applied to macroscopic particles like stars or galaxies, kinetic theory offers a method of treating systems or the sustem of galaxies, the "gas" of cosmology.

Another asset of relativistic kinetic theory is its uniform treatment of gases consisting of particles with positive mass and those having zero mass particles; its application to photons gives the cosmologically and astrophysically important theory of the transport of radiation.

Specific applications of relativistic kinetic theory to astrophysical problems which illustrate the usefulness of this theory will be mentioned later.

Although the domains of applicability of fluid dynamics and kinetic theory overlap, neither of them contains the other one. Nevertheless, kinetic theory may be considered as the more fundamental of the two theories, since within it one can derive from simple microscopic laws and plausible statistical assumptions and approximation methods the general forms of all the laws which are postulated in fluid dynamics; only the numerical values of (e.g.) transport coefficients have to be changed on leaving the domain of validity of kinetic theory.

Ideally, one would like to derive both kinetic theory and fluid dynamics from statistical mechanics; at the relativistic level, this has not yet been achieved. Therefore, we have to introduce the basic concepts and laws of kinetic theory on the basis of plausibility considertations as did Boltzmann.

There are many unsolved problems in relativistic kinetic theory, questions concerning the foundations, the mathematical structure, and specific physical applications. We shall refer to some of them in the following lectures.

Several systematic expositions of relativistic kinetic theory exist which naturally have much in common with the following lectures, in particular those by N. A. Chernikov (1963, 1964), C. Marle (1969), J. Ehlers and R. K. Sachs (1968), and J. Ehlers (1969). The elementary aspects of the special-relativistic theory which precede the Boltzmann equation (or sidestep it) are contained in the well-known book by J. L. Synge (1957) whose geometrical spirit has strongly influended the present lectures. (More specific references will be given at appropriate places in the lectures.)

In order to free equations of inessential factors, we shall use the following convention regarding physival dimensions and units: We put

$$c = 8\pi G = \hbar = 1,$$

where c is the speed of light in vacuo, G Newton's constant of gravitation, $\hbar$ the quantum of angular momentum, and k Boltzmann's constant. All physical quantities are then measured by pure numbers.

1. Assumptions on spacetime. Notation

Let X denote spacetime which we assume to be a real, four-dimensional, connected, differentiable Hausdorff manifold. In addition, we assume X to be oriented, and take always oriented local coordinate systems (x^a), $a = 1, \ldots, 4$.

The tangent space to X at p is denoted as $T_p(x)$; its dual, $T^*_p(X)$. Natural, dual bases in T_p and T^*_p are $(\frac{\partial}{\partial x^a})$ and (dx^a), respectively. X carries a normal hyperbolic metric whose signature we take as + 2. The metric tensor or gravitational potential is written g_{ab}, the Riemannian connection is Γ^a_{bc}, and the Riemanni curvature tensor is R^a_{bcd}. The Ricci tensor is given by $R_{ab} := R^c{}_{acb}$, and the Einstein tensor by $G_{ab} := R_{ab} - \frac{1}{2} R g_{ab}$, where $R := R^a{}_a$. The sign of the curvature tensors is fixed by the Ricci identity.

$$2 v_{a;[bc]} = v_d R^d{}_{abc}.$$

We assume that X is time-oriented with respect to g_{ab}, so that it is meaningful to distinguish between future directed and past directed timelike and lightlike vectors, respectivbly [1].

An orthonorlaal basis (e_j) of T_p is always chosen to be oriented and such that e_4 is future-directed.

---

[1] An example of a pair (X, g_{ab}) which is not time-orientable is given in appendix I of Ehlers (1969).

A coordinate-system (x^a) is said to be <u>inertial</u> <u>at</u> p, $p \in X$, if $\Gamma^a_{bc}\big|_p = 0$ and $\left(\frac{\partial}{\partial x^a}\big|_p\right)$ is orthonorlaal.

The physical interpretation of general relativity theory is largely based on the correspondence principle that physical laws in the presence of gravitation retain their special relativistic form at p if expressed with respect to coordinates which are inertial at p. This guiding principle is not unambigious, however.

The assumption that spacetime is oriented is not necessary for kinetic theory; it is made here only for convenience. Without this assumption several quantities appearing in kinetic theory would have to be defined with respect to oriented domains of X, and it would have to be shown that a change of orientation preserves all relevant equations. This can be done.

The assumption that spacetime is time-oriented is also not necessary for those parts of kinetic theory which are independent of the Boltzmann equation. Without it, some quantities would have to be defined relative to time oriented domains of X, and the relevant equations would have to be shown to be insensitive to changes of the time orientation; that can easily be done. The Boltzmann equation, however, can only be formulated in a time-oriented spacetime, and its form is not preserved under a change of that orientation. The reason is that the occupation numbers of initial and final states enter the collision integral in a non-symmetrical manner, as will be seen later and as is known from ordinary kinetic theory. The <u>arrow of time</u> built into the Boltzmann equation shows up particularly clearly in the H-theorem, to be derived later.

2. Some facts about differential forms and integration [1]

Kinetic theory deals with various kinds of averages which are expressed as integrals. The domain of integration is sometimes a hypersurface in X, sometimes a specetime region, sometimes a hypersurface or a region in phase space (to be defined below). The appropriate tools for forming such integrals-volume elements, hypersurface elements etc. - are differential forms. We assume that the elements of the theory of differential forms on manifolds are known, and collect here a number of facts which we need later.

On an n-dimensional manifold N, the differential form fields can be expressed, with respect to local coordinates, as sums of homogeneous forms like $\psi = \frac{1}{r!} \psi_{a_1 \dots a_r} dx^{a_1 \dots a_r}$; where the components $\psi_{a_1 \dots a_r}$ are real functions and

$$dx^{a_1 \dots a_r} := dx^{a_1} \wedge dx^{a_2} \dots \wedge dx^{a_r}$$

are exterior products of the coordinate differentials. With respect to the operations of addition, multiplication with real numbers and exterior multiplication the form fields from an associative algebra. The exterior differentiation operator d maps this algebra into itself.

An r-form ψ can be contracted, like any covariant tensor, with a vector A; the result is a (r - 1)-form $\varphi = A \cdot \psi$ with components $\varphi_{a_2 \dots a_r} = A^{a_1} \psi_{a_1 a_2 \dots a_r}$. A coordinate-independent definition of this operation

---

(1) See the "references"about mathematical tools" in the bibliography at the end of these lecture notes.

is contained in the following assertion:
For any system of r-1 vectors $A_2, \dots, A_r$, we have

$$\varphi(A_2, \dots, A_r) = \psi(A_1, A_2, \dots, A_r).$$

For a fixed vector A the mapping $\psi \to A\cdot\psi$ of the algebra of forms into itself in an antiderivation, i. e., it is linear and satisfies the product rule

$$A\cdot(\varphi\wedge\psi) = (A\cdot\varphi)\wedge\psi + (-1)^{\deg\varphi}\varphi\wedge(A\cdot\psi).$$

Moreover,

$$A\cdot(A\cdot\psi) = 0.$$

A trivial, but useful consequence of the definitions is the

Lemma 1. If Ω is a nonzero n-form at some point p of N, then the map $L \to L\cdot\Omega =: \omega$ is a vector space isomorphism of $T_p(N)$ onto the space of (n-1) -forms at p.

This lemma immediately leads to

Lemma 2. If Ω is an n-form at p, $\Omega \neq 0$, and $L \in T_p(N)$, $L \neq 0$, then the most general (n-1)-form ω at p such that $\omega(A_1, \dots, A_{n-1}) \neq 0$ whenever $(L, A_1, \dots, A_{n-1})$ is linearly independent, is given by $\omega = a\, L\cdot\Omega$ where $a \neq 0$.

Corollary. If ω has the property stated in Lemma 2, then $L\cdot\omega = 0$, and $\omega(A_1, \ldots, A_{n-1}) = 0$ whenever $(L, A_1, \ldots, A_{n-1})$ is linearly dependent.

Lemma 2 and its corollary should be visualized by considering Ω and ω as volume-functions for n-dimensional and (n-1)-dimensional parallelotopes, respectively.

Another useful fact needed later the proof of which is left as an exercise is

Lemma 3. If Ω is an n-form field on N, L a vector field and f a function, then

$$df \wedge (L \cdot \Omega) = L(f)\,\Omega. \tag{1}$$

We here recall that a vector is (identified with) a linear differential operator acting on functions: $L(f) = L^a f_{,a}$.

Finally we recall the fundamental theorem (of Stokes):
If M is an oriented, compact, m-dimensional submanifold-with-boundary ∂M of an n-manifold N, and φ is a (smooth) (m-1)-form field of N defined on M, then

$$\int_M d\varphi = \int_{\partial M} \varphi \tag{2}$$

The assumption that M is compact can be omitted provided φ decreases to zero sufficiently strongly at infinity of M; also, ∂M may be allowed to have "corners".

3. Volume elements in spacetime

Under the assumptions about spacetime stated in section 1 the expression

$$\eta = \sqrt{-g}\,dx^{1234}, \tag{3}$$

where $g := \det(g_{ab})$ and $\sqrt{-g} > 0$, is a nonvanishing 4-form such that $\eta(e_1, e_2, e_3, e_4) = 1$ for any orthonormal basis (e_j); it is the volume element of spacetime.

Let A be a vector field on X and D a 4-dimensional, oriented, compact submanifold-with-boundary of X - henceforth called a region. Then, according to (2),

$$\int_D d(A\cdot\eta) = \int_{\partial D} A\cdot\eta \tag{4}$$

The integrand on the left can be rewritten as

$$d(A\cdot\eta) = (\sqrt{-g}\,A^a)_{,a}dx^{1234} = A^a{}_{;a}\,\eta$$

and that on the right as

$$A\cdot\eta = A^a\,\sigma_a, \quad \sigma_a := \frac{1}{6}\,\eta_{abcd}dx^{bcd}, \tag{5}$$

where

$$\eta_{abcd} = \eta_{[abcd]}, \quad \eta_{1234} = \sqrt{-g} > 0 \tag{6}$$

are the components of η.

σ_a are the components of the (vectorial) hypersurface element in X; the latter is a vector-valued 3-form.

With this notation, (4) goes over into

$$\int_D A^a{}_{;a}\, \eta = \int_{\partial D} A^a\, \sigma_a, \tag{7}$$

the familiar metric-dependent version of Gauss's theorem in Riemannian space.

We shall heneeforth use the term hypersurface for "oriented hypersurface".

Since each tangent space T_q of X is itself a (flat, oriented) pseudoriemannian space, it has its own volume element

$$\pi = \sqrt{-g}\, dp^{1234}. \tag{8}$$

g is to be evaluated at q with respect to coordinates (x^a), and the p^a from $p = p^a \frac{\partial}{\partial x^a}$ define an oriented coordinate-system on T_q.

Physically important hypersurfaces of T_q are the mass-shells for masses $m \geqslant 0$. The mass-shell $P_m(q)$ con sists of all future directed (4 momentum) vectors p at q which belong to (proper) mass m; $p^2 = - m^2$. An oriented coordinate-system on $P_m(q)$ is defined as follows. Take coordinates on X around q such that $\frac{\partial}{\partial x^\nu}$, $\nu = 1, 2, 3$, are spacelike and $\frac{\partial}{\partial x^4}$ is future-directed and timelike at q. Then the restrictions of the natural coordinates p^ν to $P_m(q)$ form an oriented coordinate-system on $P_m(q)$, and p^4 (> 0) is determined by

$$g_{ab}(x^c)p^a p^b = - m^2. \tag{9}$$

$P_0(q)$ is the <u>future light cone</u> of q.

In order to obtain a <u>scalar volume element</u> on $P_m(q)$, consider the T_q-analogue of (5),

$$\tau_a := \frac{1}{6}\eta_{abcd}dp^{bcd}. \tag{10}$$

Its restriction to $P_m(q)$ has values proportional to the normal of $P_m(q)$, hence there exists a 3-form π_m such that

$$\tau_a \Big|_{P_m} = p_a \pi_m, \tag{11}$$

since p_a is a normal of P_m. Setting a = 4 in (10) and (11) gives explicitely

$$\pi_m = \frac{\sqrt{-g}}{|p_4|} dp^{123}. \tag{12}$$

The same volume element is formally obtained from

$$\pi_m = 2 \quad H(p) \quad \delta(p^2 + m^2) \pi , \tag{13}$$

in which H is the Heaviside function of p^4 and δ is the Dirac distribution.

For $m > 0$, $m\,\pi_m$ is the induced Riemannian volume element of $P_m(q)$ as a hypersurface of T_q.

In inertial coordinates at q, we have the familiar expression

$$\pi_m = \frac{dp^{123}}{E} \tag{14}$$

where $E = p^4$ is the energy. Taking polar coordinates in $\vec{p}$-space we obtain

$$\pi_m = \frac{\vec{p}^{\,2}}{\sqrt{m^2 + p^2}}\, d|\vec{p}| \wedge (\sin\vartheta\, d\vartheta \wedge d\varphi) \tag{15}$$

or also

$$\pi_m = \sqrt{E^2 - m^2}\; dE \wedge (\sin\vartheta\, d\vartheta \wedge d\varphi) \tag{16}$$

The consideration which led to the volume element π_0 on the tangent null cone $P_o(q)$ can be generalized to the actual null cone of q in X. We leave it as an exercise to the reader to verify

Lemma 4. Let $N_q^{\downarrow}$ be the past null cone of q, and let u_q be a future-directed timelike unit vector at q. A normal k_a of $N_q^{\downarrow}$ is obtained by drawing null geodesics through q, choosing tangent vectors k to them such that, at q, $k \cdot u_q = 1$, and parallely propagating these k's along the null geodesics. Also, put $v = 0$ at q, and put $k^a = \frac{dx^a}{dv}$, obtaining a field of affine distances v on $N_q^{\downarrow}$. Denote as $d\Omega_q$ the solid angle obtained by projecting a small bundle of null rays through q into that 3-space through q which is orthogonal to u_q, and call D the distance from apparent size of an arbitrary point $r \in N_q^{\downarrow}$ from q, as measured by an observer at q with 4-velocity u_q. Then

$$\sigma_a \Big|_{N_q} = k_a D^2 d\Omega_q \wedge dv \ , \qquad (17)$$

so that $D^2 d\,\Omega_q \wedge dv$ is a natural scalar volume element on N_q.

4. <u>Basic assumptions about a relativistic gas. Geometry of phase space.</u> (1)

The history of a system of many (classical) particles of negligible size is represented in relativity theory as a complex of timelike or lightlike wordlines. The particles may be thought of as being macroscopic (stars, galaxies) or microscopic (molecules, atoms, ions, nuclei, photons, . . .), and they may be interacting through long-range and/or short range forces.

Without attempting to give a detailed description of the dynamics of such a general system, we lay down a special, simple model for some systems which we call <u>gases</u>. In these systems, the particles are assumed to move like test particles in a <u>mean</u> <u>gravitational field</u> g_{ab} and <u>elec-</u>

---

(1) The geometric treatment given in this section follows essentially that of Bichteler (1965). See also Chernikov (1963), Lindquist (1966) (Appendix), and Marle (1969).

tromagnetic field F_{ab}, except during encounters due to short range interactions which are idealised as point collisions. (I. e., the range of these interactions must be much smaller than the mean free path.) The mean fields may be external fields - we then speak of a test gas - or may be collectively generated by the gas particles themselves, in which case we have a selfgravitating gas (or a Vlasov plasma).

We proceed to formalize this qualitative picture of a gas.

A particle of mass m ($\geqslant 0$) and charge e has a worldline $x^a(v)$ which obeys the Lorentz-Einstein equations of motion

$$\frac{dx^a}{dv} = p^a, \qquad \frac{Dp^a}{dv} = e\, F^a{}_b p^b, \tag{18}$$

if radiation reaction is neglected. The parameter v is so chosen that the tangent vector p^a is the (future-directed) 4-momentum. If $m > 0$, mv is proper time. $\frac{D}{dv}$ denotes, here and in the sequel, the absolute derivative along the world line,

$$\frac{Dp^a}{ds} = \frac{dp^a}{ds} + \Gamma^a{}_{bc} p^b p^c. \tag{19}$$

If a particle participates in a collision at $x \in X$, its world line may have a corner at x, or the world line may end or begin at x, if the particle is annihilated or created in the collision.

In the case of many particles the spacetime figure of a gas is a complicated network of curves, since several trajectories with different directions can pass through the same event, and the trajectories through nearby events can have quite different directions.

A simplification of the geometrical representation is achieved, as in nonrelativistic kinetic theory, by introducing a phase space. Since in relativity no preferred space sections $t = \text{const.}$ exist, the relativistic phase space cannot be defined in strict analogy to the ordinary $(\vec{x}, \vec{p})$ phase space (of one particle), but will correspond to the $(\vec{x}, t, \vec{p}, E)$-space. We define the (relativistic) one particle phase space for particles of arbitrary mass m to be the manifold

$$M: = \left\{ (x, p): x \in X, \quad p \in T_x(X), \; p^2 \lessgtr 0, \; p \text{ future directed.} \right\} \tag{20}$$

This set is indeed a 8-dimensional manifold, if we agree to take as local coordinates (x^a, p^a), where (x^a) is a coordinate-system on X and p^a are the corresponding natural vector components.
M is, in fact, a manifold with boundary, the boundary ∂M being the set of states (x, p) having mass zero, $p^2 = 0$.

M is a fiber bundle with base X. The fiber at x is the set of non-spacelike, future-directed vectors at x, i.e., the 4-momentum space at x. (If all vectors had been admitted, M would be the tangent bundle $T(X)$ over spacetime.)

M is obviously oriented, the (x^a, p^a)-systems being oriented coordinate-systems.

The equations of motion (18), (19) define on M a vector field

$$L = p^a \frac{\partial}{\partial x^a} + (e\, F^a{}_b p^a - \Gamma^a_{bc} p^b p^c) \frac{\partial}{\partial p^a} \tag{21}$$

called the Liouville vector (or operator). The oriented integral curves $(x^a(v), p^a(v))$ form a congruence in M, the phase flow generated by L. Physically, the phase flow represents the set of all test particle motions which are possible in the combined gravitational and electromagnetic fields occuring in L.

The rest mass m as given by equation (9) is a scalar function on M. It is constant on each phase orbit,

$$L(m) = 0. \tag{22}$$

Hence the restriction L_m of L to the hypersurface M_m of M defined by m = const. is tangent to M_m. We note that

$$M_m = \bigcup_{x \in X} P_m(x). \tag{23}$$

M_m, with its Liouville vector L_m and its phase flow, is the phase space for particles of fixed mass m; it is seven-dimensional and corresponds to the Newtonian $(\vec{x}, t, \vec{p})$ -space.[1] It is also a fiber

(1) In classical mechanics, this space is sometimes called "augmented phase space". See e.g. Liboff (1969). p. 16.

bundle with base X, the fiber over x now being $P_m(x)$, the mass-shell at x.

M_m, being the boundary of the oriented submanifold of M given by $p^2 \leq - m^2$, is also orientable. We orient it by choosing a coordinate system (x^a, p^a) on M such that $p_4 p^4 < 0$ whenever $p \in P_m(x)$, and then take (x^a, p^ν) as an oriented coordinate-system on M_m. We then have

$$L_m = p^a \frac{\partial}{\partial x^a} + (e F^\nu{}_b p^b - \Gamma^\nu_{bc} p^b p^c) \frac{\partial}{\partial p^\nu} \tag{24}$$

We know from ordinary statistical mechanics the usefulness of a measure on phase space which is invariant under canonical transformations and, in particular, under the phase flow.

Let us consider, therefore, the coordinate-independent 8 form

$$\Omega := \eta \wedge \pi = - g \, dx^{1234} \wedge dp^{1234} \tag{25}$$

on M (formed by means of (3) and (8)) and the 7-form

$$\Omega_m := \eta \wedge \pi_m = \frac{-g}{|p_4|} dx^{1234} \wedge dp^{123} \tag{26}$$

on M_m. Obviously, at each point.

$$\Omega \neq 0, \; \Omega_m \neq 0. \tag{27}$$

Ω and Ω_m are related as follows (exercise):

$$\Omega = m\ dm \wedge \Omega_m . \tag{28}$$

To see whether Ω is invariant with respect to the phase flow we compute $\mathcal{L}_L \Omega$, the Lie derivative of Ω with respect to L. Because of the identity[1]

$$\mathcal{L}_L \Omega = d\,(L \cdot \Omega) + L \cdot d\Omega$$

and $d\Omega = 0$, we get $\mathcal{L}_L \Omega = d\omega$, if we put

$$\omega := L \cdot \Omega = p^a\ \sigma_a \wedge \pi + \frac{1}{6}\ \eta_{abcd}(eF^a{}_d p^d - \Gamma^a{}_{de} p^d p^e) dp^{bcd} \wedge \eta . \tag{29}$$

The differential of this 7-form vanishes. This is really verified by using inertial coordinates at some (arbitrary) event x.
Hence,

---

(1) See, e.g. Hicks (1965), p. 94.

$$\mathcal{L}_L \Omega = d\omega = 0; \tag{30}$$

i.e., Ω is invariant under the phase flow (Liouville's theorem).

The 7-form ω which arose here rather naturally as a tool will be seen in the next section to be important in itself; let us note some of its properties. From its definition (29) and from $\mathcal{L}_L \omega = d(L\cdot\omega) + L(d\omega)$ we infer:

$$L \cdot \omega = 0, \quad \mathcal{L}_L \omega = 0. \tag{31}$$

These properties express that ω induces a nonzero 7 form on the quotient manifold M/L; ω can be consedered as a measure on the 7-manifold of phase orbits. Indeed, if we introduce on M comoving local coordinates ξ^A with respect to L, i.e., such that $L = \frac{\partial}{\partial \xi^8}$, then (31) means that $\omega = \sum_{A=1,\dots,7} \omega_A(\xi^1, \dots \xi^7)\, d\xi^A$, which "is" a form on M/L. If $\mathfrak{T}$ is a tube of phase orbits and Σ a cross section of $\mathfrak{T}$, $\int_\Sigma \omega$ measures, loosely speaking, the "number" of orbits contained in $\mathfrak{T}$; it is independent of the cross section.

The preceding conseiderations can be carried over straightforwardly from M, L, to M_m, L_m (exercise); one obtains

$$\omega_m := L_m \cdot \Omega_m = p^a\, \sigma_a \wedge \pi_m + \frac{1}{2|p_4|}\, \eta_{\lambda\mu\nu 4} (F^\lambda{}_b p^b - \Gamma^\lambda{}_{bc} p^b p^c) dp^{\mu\nu} \wedge \eta \tag{32}$$

$$\mathcal{L}_{L_m}(\Omega_m) = d\,\omega_m = \mathcal{L}_{L_m}(\omega_m) = L_m \cdot \omega_m = 0 \qquad (33)$$

5. Distribution function, collision density, Liouville's equation

An individual gas-history - a particular complex of world-lines - is too complicated to be useful; we are interested only in the typical, average properties of gases. Therefore, we imagine a large collection of microscopically different, but macroscopically indistinguishable gas histories, a Gibbs-ensemble of gases. The average properties of such an ensemble are the subject of kinetic theory. (The averaging may have the additional merit that it disposes of certain all-too-classical features of our gas model like sharply defined worldlines and collision events; the average properties may well provide an approximate macroscopic description of a gas whose particles obey quantum laws. [1])

Consider, then, a gas consisting of particles of different species. Concentrate on one component the particles of which have mass m and charge e. A definite microstate, or history, of the gas can be represented, as far as the specified component is concerned, as a collection of

---

(1) Nonrelativistically the Boltzmann equation, e.g., can be "derived" from classical as well as from quantum mechanics; see Kadanoff and Baym (1962); Lowry (1970).

segments of phase orbits in M_m, the states occupied by particles between collisions. (We do not assign phase-orbits to particles during collisions; hence there are no particle orbits in M_m transverse to the phase flow.)

The distribution of occupied states in M_m can be fully characterized by the functional $\Sigma \to N_m[\Sigma]$ which assigns to any compact hypersurface Σ the number of occupied orbit segments intersecting it. By a hypersurface in M_m we mean here and henceforth an oriented, 6-dimensional submanifold with boundary of M_m. The intersection of an orbit k with Σ is counted positively (negatively) if, at the event of intersection, the vector basis $(L_m, A_1, \dots, A_6)$ has the same (opposite) orientation as the basis of an oriented coordinate system of M_m, L_m being the tangent to k and $(A_1, \dots, A_6)$ an oriented basis tangent to Σ.

If D is any region in M_m, then $N_m[\partial D]$ is the number of collisions in D, if creations are counted positively, and annihilations negatively.

For a macrostate, let $\overline{N}_m[\Sigma]$ be the ensemble average of N_m. Since $\overline{N}_m[\Sigma]$ is a kind of flux through Σ of a fictitious fluid streaming in M_m with velocity L_m, we expect it to be expressible as an integral. We thus need a volume element for hypersurfaces in M_m.

It is natural to ask whether there exists a 6-form on M_m which could serve as such a volume element. From the meaning of $\overline{N}_m$ it is clear that this form would have to assign a nonzero volume to any hypersurface-element not tangent to L_m, since there could be a flux through it. Using the fact that Ω_m is a non-vanishing 7-form on M_m (see eq. (27)) and remembering lemma 2, we infer that such a 6-form must coincide with ω_m as defined in (32), except for a non vanishing factor. Because of Liouville's theorem, eq. (33), it is advisable to choose this factor to be constant on phase orbits in order that the 6-form is L_m-invariant ($0 = d(f\omega_m) = df \wedge \omega_m = df \wedge (L_m \cdot \Omega_m) = L_m(f)\Omega_m \Longrightarrow L_m(f) = 0$; we have used (1)). Hence, ω_m reoommends itself as an almost unique candidate for the required measure.

A hypersurface Σ in M_m whose projection into X is a spacelike hypersurface corresponds to a region of an "inttantaneous" ordinary $(\vec{x},\vec{p})$-phase space. On such a Σ (and, more generally, on any Σ whose projection into X is a hypersurface), ω_m from eq. (32) reduces to its first part,

$$\omega_m = p^a \, \sigma_a \wedge \pi_m \, . \tag{34}$$

If we choose at some point x with $(x,p) \in \Sigma$ an inertial coordinate system such that $\frac{\partial}{\partial x^4}$ is, at x, normal to Σ, (34) gives, at x,

$$\omega_m = - dx^{123} \wedge dp^{123} \tag{35}$$

which is, except for the (conventional) sign, the ordinary phase volume element of an observer at x with 4 velocity $\frac{\partial}{\partial x^4}$. Therefore, ω_m

is the appropriate 6-form we have been looking for.

We return to our study of M_m. According to its physical meaning, we make the following smoothness assumptions about N_m, i.e. about a macrostate of a gas:

D_1) On any fixed hypersurface $\Sigma \subset M_m$ there exists a continuous, nonnegative density function f_Σ such that for all compact parts $\Sigma' \subset \Sigma$

$$\bar{N}_m[\Sigma'] = \int_{\Sigma'} f_\Sigma \, \omega_m . \tag{36}$$

D_2) Every point $(x, p) \in M_m$ has a neighbourhood U such that for every region $D \subset U \subset M_m$

$$\left| \bar{N}_m [\partial D] \right| \leq A \int_D \Omega_m \tag{37}$$

for same constant A dppending on U.

D_1 asserts that on any fixed hypersurface Σ the measure defined by the expectation value of the number of occupied states contained in parts Σ' of Σ has a continuous derivative, or density function f_Σ with respect to the geometrical measure ω_m.

Equation (35) shows that $f_\Sigma(x, p)$ equals, for any observer whose worldline intersects the projection of Σ into X orthogonally at x, the ordinary density of states in his infinitesimal, ordinary $(\vec{x}, \vec{p})$ phase space.

D_2 asserts that the expectation value of the number of collisions in

D is at most of the order of the Ω_m volume of D; this assumption excludes, e.g., the possibility of having all (or particularly many) collisions occuring on one hypersurface of X.

The two assumptions D_1 and D_2 imply the existence of an invariant, i.e., hypersurface or observer-independent (one particle) distribution function [1] f_m on M_m such that for any hypersurface $\Sigma \subset M_m$

$$\bar{N}_m[\Sigma] = \int_{\Sigma} f_m \omega_m; \qquad (38)$$

To prove (38), we have to show that if a point $\xi \in M_m$ is contained in two hypersurfaces Σ_1 and Σ_2, then $f_{\Sigma_1}(\xi) = f_{\Sigma_2}(\xi)$. For that, consider a tube $\mathfrak{J}$ of phase-orbits having ξ on its boundary. Then $\Sigma_1 \cap \mathfrak{J}$, $\Sigma_2 \cap \mathfrak{J}$ are two cross sections of $\mathfrak{J}$. Without loss of generality we assume that these two cross sections together with the part Λ of the cylindrical boundary $\partial\mathfrak{J}$ which lies between $\Sigma_1 \cap \partial\mathfrak{J}$ and $\Sigma_2 \cap \partial\mathfrak{J}$ form the boundary ∂D of an orientable, compact region D of M_m. Since no phase orbits can intersect Λ we have $\bar{N}_m[\partial D] = \bar{N}_m[\Sigma_1 \cap \mathfrak{J}] - \bar{N}_m[\Sigma_2 \cap \mathfrak{J}]$. Because of

---

[1] The preceding introduction of f_m is a generalised and "rigorised" version of that given in Synge (1957), p. 12-14.

D_1 and with the mean value theorem for integrals this can be re-written as $\overline{N}_m[\partial D] = f_{\Sigma_1}(\xi_1)\int_{\Sigma_1\cap\mathfrak{F}}\omega_m - f_{\Sigma_2}(\xi_2)\int_{\Sigma_2\cap\mathfrak{F}}\omega_m$, where $\xi_i \in \Sigma_i \cap \mathfrak{F}$. But we know from Liouville's theorem that the two integrals on the right-hand side are equal. Hence, using also D_2, $\left| f_{\Sigma_1}(\xi_1) - f_{\Sigma_2}(\xi_2)\right| \leqslant A\int_D \Omega_m(\int_{\Sigma_i\cap\mathfrak{F}}\omega_m)^{-1}$. If one now lets $\mathfrak{F}$ shrink towards the orbit passing through ξ, the right-hand side tends to zero since the numerator is "one order smaller" than the denominator. Also, $\xi_i \rightarrow \xi$. Consequently, $f_{\Sigma_1}(\xi) = f_{\Sigma_2}(\xi)$. We call the common value $f_m(\xi)$, in order to emphasize that f_m is defined on M_m.

It is easy to verify that our orientation and sign conventions imply

$$f_m \geqslant 0. \tag{39}$$

It is technically desirable and physically not harmful to require also

D_3) f_m is continuously differentiable on M_m.

Having obtained a phase space density f_m which measures the average density of occupied states, we obtain straightforwardly a collision density in M_m. The average number of collisions in the region $D \subset M_m$, i.e., the difference between creations and annihilations of particles of the specified kind in D, is given by $\overline{N}_m[\partial D] =$
$= \int_{\partial D} f_m\,\omega_m = \int_D d(f_m\,\omega_m) = \int_D df_m \wedge \omega_m = \int_D df_m \wedge (L_m \cdot \omega_m) =$

$$= \int_D L_m(f_m)\, \Omega_m .$$

We have used equations (38), (2), (33), (32), and (1). Hence

$$L_m(f_m) = p^a \frac{\partial f_m}{\partial x^a} + (e\, F^\lambda{}_b p^b - \Gamma^\lambda{}_{bc} p^b p^c) \frac{\partial f_m}{\partial p^\lambda} \qquad (40)$$

is the collision density in M_m with respect to Ω_m (in the sense defined above).

Note that if $(x^a(v), p^a(v))$ is the phase orbit passing through (x, p) for $v = 0$, then the expression (40), evaluated at $(x, p,)$, equals $\left(\frac{d}{dv} f_m(x(v), p(v))\right)_{v=0}$, a fact that is often useful.

The preceding considerations prove the following theorem. The distribution function f_m of a component of a (possibly heterogeneous) gas satisfies Liouville's equation

$$L_m(f_m) = 0 \qquad (41)$$

in a region $D \subset M_m$ if and only if there is detailed balancing everywhere in D, i.e., if the average number of creations of particles of that component equals everywhere in D the average number of annihilations [1].

---

(1) Note that, in our terminology, even an elastic collision involves two annihilations and two creations.

Corollary 1.[1] If the particles of a particular species do not participate in any collisions in D, then the corresponding distribution function satisfied, in D, equation (41).

Corollary 2. If the assumptions of the theorem hold, then f_m is, in D, an integral of the motion defined by (18).

As an application of the invariance (observer-independence) of the distribution function, let us consider a radiation field as a photon gas with distribution function f_γ. Relative to an observer with 4-velocity u^a, it is customary to define a specific intensity I_ν of the radiation field, as the limit of the radio "(energy of photons with frequency in $d\nu$ and direction in solid angle $d\Omega$ passing in time dt normally through an area dA" / ($d\nu\, d\Omega\, dt\, dA$). It is related (exercise) to f_γ by

$$I_\nu = 2\pi \left| u_a p^a \right|^3 f_\gamma (x, p). \tag{42}$$

Since $\nu = (2\pi)^{-1} \left| u_a p^a \right|$, the observer-independence of f_γ im-

(1) For geodesic motion (e = o), this assertion has first been stated by Walker (1936).

plies that of I_ν/ν^3, a fact that is important, e.g., in cosmology; its direct, kinematical proof is somewhat cumbersome.

If the photons are emitted by a source S (galaxy, e.g.) and do not interact with matter on their journey to the observer O, Liouville's equation (41) for f_γ and (42) give the important relation

$$I_{\nu_0} = \frac{I_{\nu_s}}{(1+z)^3} \tag{43}$$

between I_{ν_s}, "measured" near the source by a fictitious comoving observer, and I_{ν_0}, the intensity actually measured by 0. z is the usual redshift of S relative to 0. (43) is basic for the derivation of observable relations in cosmology. Notice that the derivation just sketched holds in any spacetime, not only in the standard Robertson-Walker universes.

If one assumes that the famous 3° K "fireball" radiation was emitted thermally from the recombination hypersurface ($T \approx 3500^\circ$) in the early universe, one obtains from (43) the predicted intensity distribution in each direction in an arbitrary model universe, provided one can compute z from the null geodesics.(1)
This idea was used by R. K. Sachs and A. M. Wolfe (1967) to estimate the influence of material "lumps" on the radiation, and similar applications have been made more recently. The same method has been employed by W. L. Ames and K. S. Thorne (1968) to determine the optical appearance

(1) It is also assumed that no scattering occurred between emission and absorption.

of a collapsing star to a distant observer. Several other applications of (41) have been made, particularly in cosmology and stellar dynamics.

6. Macroscopic fluid variables, balance equations, conservation laws.

Let us rewrite (38) for a hypersurface Σ whose projection into X is a hypersurface G. We obtain, using (34),

$$\bar{N}_m[\Sigma] = \int_G \sigma_a \left\{ \int_{K_x} f_m p^a \pi_m \right\}. \qquad (44)$$

K_x is that part of the mass shell $P_m(x)$ which is contained in Σ.

In particular, the integral $\int_G \sigma_a \left\{ \int f_m p^a \pi_m \right\}$ gives the average total number of particles of the species considered whose world lines intersect G. Here we have used the convention, to be maintained throughout the remainder, that $\int \pi_m \ldots$ denotes an integral over the whole mass-shell $P_m(x)$. Therefore, the spacetime vector field

$$N_m{}^a(x) := \int f_m p^a \pi_m \qquad (45)$$

is the particle 4-current density of the respective species. It is always timelike and future-directed under our assumptions. (If we would permit f_m to be a distribution, $N_m{}^a$ could be lightlike in one particular case: m = 0, and there is no 4-momentum dispersion at any event).

Similarly,

$$J^a := e\, N_m{}^a \tag{46}$$

is the electric 4-current density of the species considered.

In analogy with (45) we define

$$T_m{}^{ab}(x) := \int p^a p^b f_m\, \pi_m \tag{47}$$

as the kinetic stress energy momentum or matter tensor of the species. (If is possible to define a 4-momentum flux through a hypersurface $G \subset X$ and to show that (47) is the corresponding 4-momentum flux density, but this has no further use and is therefore not treated in detail here.)

We have assumed here, and will do so throughout these lectures, that f_m vanishes at infinity on $P_m(x)$ so that integrals like (45), (47) exist. (Sufficient for this is exponential boundedness on $P_m(x)$, as defined at the end of section 8.)

Excluding the trivial case where f_m vanishes on $P_m(x)$ (and the singular distribution mentioned below (45)) we infer from (47)

Lemma 5. If v^a is not spacelike and $v^a \neq 0$, then

$$T_{m\,ab} v^a v^b > 0. \tag{48}$$

This lemma and a theorem due to J. L. Synge [1] imply

Lemma 6. Any kinetic stress energy momentum tensor is normal [2], i.e., admits a decomposition

$$T_m{}^{ab} = \mu u^a u^b + p^{ab} \tag{49}$$

with

$$u_a u^a = -1, \quad p_{ab} u^b = 0. \tag{50}$$

u^a can and will be chosen future-directed, and then (49) is unique.

The physical meaning of $N_m{}^a$, $T_m{}^{ab}$ for a local observer in terms of "3-dimensional" quantities is obtained by evaluating (45) and

---

(1) See Synge (1956), p. 292.

(2) See Lichnerowicz (1955).

(47) in an inertial coordinate system at x. We obtain, for an arbitrary observer at x:

$\underset{m}{N}^{4}$ is the number density

$\underset{m}{\vec{N}} := \underset{m}{N}^{\lambda} \frac{\partial}{\partial x^{\lambda}} = \underset{m}{N}^{4} \langle \vec{v} \rangle_{x}^{(m)}$ is the particle flux density,

$\underset{m}{T}^{44} = \underset{m}{N}^{4} \langle E \rangle_{x}^{(m)}$ is the energy density, (51)

$\underset{m}{\vec{T}} := \underset{m}{T}^{\lambda 4} \frac{\partial}{\partial x^{\lambda}} = \underset{m}{N}^{4} \langle \vec{p} \rangle_{x}^{(m)}$ is the momentum density,

$\underset{m}{\overleftrightarrow{T}} := \underset{m}{T}^{\lambda\mu} \frac{\partial}{\partial x^{\lambda}} \otimes \frac{\partial}{\partial x^{\mu}} = \underset{m}{N}^{4} \langle \vec{p} \otimes \vec{v} \rangle_{x}^{(m)}$ is the kinetic pressure tensor.

Here $\vec{v}$, $E \; (= p^{4})$, and $\vec{p}$ are the 3-velocity, energy, and 3-momentum, and $\langle \;\rangle_{x}^{(m)}$ denotes the conditional expectation value at x, evaluated by means of the probability distribution defined by f_{m} with respect to the chosen inertial system.

We also define mean kinetic pressure p by (tr : = trace)

$$p := \frac{1}{3} \operatorname{tr} \underset{m}{\overleftrightarrow{T}} = \frac{1}{3} \underset{m}{N}^{4} \langle \vec{p} \cdot \vec{v} \rangle_{x}^{(m)} \qquad (52)$$

and recognize the classical Bernoulli formula.

The rest mass density is $\rho := m \underset{m}{N}^{4}$. Writing μ for the energy density in $(51)_3$, we formulate

Lemma 7. For any observer and any distribution f_{m}, the inequalities

$$0 \leqslant 3p \leqslant \frac{3}{2}\,p + \sqrt{\left(\frac{3}{2}\,p\right)^2 + \rho^2} \leqslant \mu \leqslant \mu + 3p \qquad (53)$$

hold.

Most of these inequalities are obvious from (51), (52), and $E = \frac{m}{(1 - v^2)^{1/2}}$, $\vec{p} = E\,\vec{v}$; the only nontrivial inequality is the third one, due to A. H. Taub (1948). It follows by considering $(1 - v^2)^{-\frac{1}{4}}$ and $(1 - v^2)^{\frac{1}{4}}$ as elements of the Hilbert space $\mathcal{L}^2(P_m, f_m dp^{123})$, and applying Schwartz's inequality to them.

Equations (51) and (52) imply the well-known relations:

If $p \ll \rho$, then $\mu \approx \rho + \frac{3}{2}\,p$ (nonrel. monatomic gas),

If $p \gg \rho$, then $\mu \approx 3p$ (ultrarel. gas), (54)

If $m = 0$, then $\mu = 3\,p$ (photon or neutrino gas).

In order to obtain balance equations for various macroscopic fluid variables we observe that these latter quantities are <u>moments</u> of the distribution function in 4-momentum space, given by $\int p^{a_1} p^{a_2} \dots p^{a_r} f_m \pi_m$. The 0th moment is, at least for $m > 0$, essentially the trace of the matter tensor. Indeed, (47), (51), and (52) give

$$0 \leqslant m^2 \int f_m \pi_m = -\,T_m{}^a{}_a = \mu - 3p. \qquad (55)$$

The 1st moment is the particle current density $N_m{}^a$, and the 2nd moment is the matter tensor. We would like to evaluate the divergence of the r-th order moment. We first establish

<u>Lemma 8.</u> If g is a C^1-function on M_m, then

$$\left(\int p^a g\, \pi_m\right)_{;a} = \int L_m(g)\, \pi_m. \tag{56}$$

To prove this, take an arbitrary region $D \subset X$, and let $\hat{D} := \{ (x, p) : x \in D,\ p \in P_m(x) \}$ be the cylindrical region of M_m lying over D. Then, as in the derivation of the collision density above eq. (40), $\int_{\partial \hat{D}} g\, \omega_m = \int_{\hat{D}} L_m(g)\, \Omega_m$. We transform both these integrals into integrals over D:

$$\int_{\partial \hat{D}} g\, \omega_m = \int_{\partial D} \sigma_a \left\{ \int p^a g\, \pi_m \right\} = \int_D \eta \left(\int p^a g\, \pi_m \right)_{;a},$$

(Use (34) and (7))

$$\int_{\hat{D}} L_m(g)\, \Omega_m = \int_D \eta \left\{ \int L_m(g)\, \pi_m \right\}, \quad \text{from (26).}$$

Since D is arbitrary, (56) follows.

Next, we generalize this to obtain an "equation of transfer".

Lemma 9. [1] If f_m is an arbitrary distribution function on M_m, then (for $r \geq 2$)

$$\left(\int p^{a_1} p^{a_2} \dots p^{a_r} f_m \pi_m \right)_{;a} = \int p^{a_2} \dots p^{a_r} L_m(f_m) \pi_m + \sum_{\lambda=2}^{r} e F^{a_\lambda}{}_b \int p^{a_2} \dots p^{b} \dots p^{a_r} f_m \pi_m . \qquad (57)$$

(The integrand in the sum is to be understood such that b replaces a_λ in the sequence $a_2 \dots a_r$.)

Proof: Take an arbitrary tensor field $v_{a_2 \dots a_r}$ which satisfies $v_{a_2 \dots a_r ; b} = 0$ at some arbitrary event x_0. Put $p^{a_2} \dots p^{a_r} v_{a_2 \dots a_r} f_m = g$, and apply lemma 8, to obtain at x_0:

$$v_{a_2 \dots a_r} \left(\int p^{a_1} \dots p^{a_r} f_m \dot{\pi}_m \right)_{; a_1} = \int L_m (v_{a_2 \dots a_r} p^{a_2} \dots p^{a_r} f_m) \pi_m . \qquad (58)$$

Evaluate $L_m(\dots)$ at x_0 by taking the phase-orbit through x_0 and

---

(1) Tauber - Weinberg (1961)

differentiate (...) with respect to the parameter at x_0, getting

$$L_m(\dots) = \frac{D}{dv}(\dots) = v_{a_2 \dots a_r} \frac{D}{dv}(p^{a_2} \dots p^{a_r} f_m) =$$

$$= v_{a_2 \dots a_r}(p^{a_2} \dots p^{a_r} L_m(f_m) + e\, F^{a_2}{}_b p^b \dots p^{a_r} f_m + \dots),$$

where we have used (18). Insertion into (58) and "dividing" by v... gives (57).

(Generalizations of Lemma 9 have been given by Ph. M. Quan (1966) and C. Marle (1969), but they do not seem to have found applications yet.)

Applying (56) (with $g = f_m$) and (57) (with $r = 2$) and using the definitions (45), (46), and (47) we obtain

$$N_m{}^a{}_{;a} = \int L_m(f_m)\, \pi_m \tag{59}$$

and

$$T_m{}^{ab}{}_{;b} = F^a{}_b J^b + \int p^a L_m(f_m)\, \pi_m. \tag{60}$$

Equation (59) is the <u>balance equation for particles</u> of the species considered; since $L_m(f_m)$ has been shown to be the collision density in phase space, the right-hand side of (59) is the (spacetime) production density of these particles.

Equation (60) is the 4-momentum balance equation for the given species; the vectors on the right-hand side represent the electromagnetic and the collisional 4-force densities acting on the component of the gas with distribution function f_m. An example for the latter is the force exerted on an electron gas by photons due to Compton scattering.

So far, we always concentrated on one component of a gas which may contain other kinds of particles as well; all our equations are valid for any component of a mixture.

Let us now first specialize to the case of a monocomponent, or simple gas consisting of particles all having (proper) mass m and charge e. Then, assuming conservation of particles in collisions (59) gives the conservation law

$$N_m{}^a{}_{;a} = \int L_m(f_m)\,\pi_m = 0 \tag{61}$$

which, of course, implies also charge conservation, $J^a{}_{;a} = 0$. Assuming also 4-momentum conservation during collisions, (60) results in

$$T_m{}^{ab}{}_{;b} = F^a{}_b J^b, \quad \int p^a L_m(f_m)\,\pi_m = 0. \tag{62}$$

These equations give on the one hand the macroscopic conservation laws basic to fluid mechanics, and they impose restrictions on the evolution of f_m, required by the microscopic conservation laws.

For a simple gas, there are two sensible ways to define the mean 4-velocity. One can either use the fact that the particle current vector $N_m{}^a$ is timelike and put

$$N_m{}^a = n\, u_k{}^a, \quad u_k{}^a u_{ka} = -1, \quad u_k{}^a \text{ future-directed,} \tag{63}$$

or one can use the normality of the matter tensor $T_m{}^{ab}$ and use the u^a of Lemma 6, i.e., require

$$u_D{}^{[a} T_m{}^{b]}{}_c u_D^c = 0, \quad u_D^a u_{Da} = -1, \quad u_D{}^a \text{ future-directed} \tag{64}$$

$u_k{}^a$ is called the kinematic mean velocity.[1] An observer travelling with $u_k{}^a$ is characterized by the property that in his local inertial frames there is no particle flux density (see $(51)_2$); n from (63) is the proper particle number density.

u_D^a is called the dynamic mean 4-velocity.[1] An observer travell-

---

[1] The distinction and terminology is due to J. L. Synge; see Synge (1960)

ing with it will measure <u>no momentum density</u>, and this characterizes $u_D^{\ a}$ (see $(51)_4$). The energy density μ in (49) is the <u>minimum</u> of the energy densities measured by all possible observers; this property also characterizes $u_D^{\ a}$ (exercise).

The two mean velocities $u_k^{\ a}$ and $a_D^{\ a}$ are in general distinct, their equality characterizes (by definition) <u>adiabatic processes</u>. They are physically characterized by the existence of an observer u^a who finds neither a particle flux nor an energy flux in his local inertial systems. The necessary and sufficient condition for that is that $N_m^{\ [a} T_m^{\ b]}{}_c N_m^{\ c} = 0$, a very complicated restriction on the distribution function.

If one chooses <u>any</u> mean 4-velocity u^a, one can decompose the matter tensor uniquely according to the scheme (Eckart 1940):

$$T_m^{\ ab} = \mu u^a u^b + 2 u^{(a} q^{b)} + p h^{ab} + \pi^{ab}, \tag{65}$$

where

$$h^a_{\ b} = \delta^a_{\ b} + u^a u_b \tag{66}$$

projects $T_x(X)$ onto the 3-space orthogonal to u^a, and where

$$u_a q^a = u_a \pi^{ab} = \pi^a_{\ a} = 0. \tag{67}$$

μ is the mean energy density, q^a the mean energy flux density, p the mean kinetic pressure and π^{ab} the shear pressure tensor relative to u^a. These quantities change with u^a. If $u^a = u_D^a$, then $q^a = 0$. Adiabatic processes are characterized by the property that $q^a = 0$ for $u^a = u_k^a$. If, in addition, $\pi^{ab} = 0$, the gas behaves, in the process considered, as an ideal gas. We shall extend these mechanical considerations later on to the regime of thermodynamics.

Consider next a multicomponent gas. We distinguish the particle species by indices $A, B, \dots$; particles of species A have mass m_A, charge e_A, and (if we have microscopic particles) further characteristics like baryon number b_A etc. Each species has its phase space which we denote by M_A (instead of M_{m_A}), and its Liouville operator L_A; its distribution function f_A, current densities N_A^a, J_A^a, and its matter tensor T_A^{ab}, and the quantities which we have defined in terms of these, like $u^a_{D,A}$.

Requiring again 4-momentum conservation, we have instead of (62)

$$T_k^{ab}{}_{;b} = F^a{}_b J^b, \tag{68}$$

where

$$T_k^{ab} := \sum_A T_A^{ab} \tag{69}$$

is the <u>total kinetic stress energy momentum tensor</u> of the mixture, and

$$J^a := \sum_A J_A{}^a \tag{70}$$

is the <u>total electric 4-current density</u>. Moreover,

$$\sum_A \int p^a L_A(f_A)\, \pi_A = 0. \tag{71}$$

The individual particles will in general not be conserved during collisions, but certain combinations of the $N_A{}^a$ will have vanishing divergence. For example, if we define the <u>baryon current density</u>

$$B^a := \sum_A b_A N_A{}^a \tag{72}$$

and assume <u>conservation of baryon number</u> during collisions, we obtain

$$B^a{}_{;a} = 0 \tag{73}$$

and

$$\sum_A b_a \int L_A(f_A) \, \pi_A = 0. \tag{74}$$

Similarly, we will have

$$J^a{}_{;a} = 0 \tag{75}$$

and

$$\sum_A e_A \int L_A(f_A) \, \pi_A = 0. \tag{76}$$

Thus, we obtain macroscopic conservation laws for a mixture and corresponding integral conditions for the distribution functions.

Resonable mean 4-velocities for a mixture are the dynamic mean 4-velocity u^a_D defined as in (64), with T^{ab}_m replaced by T_k^{ab},
the barycentric mean velocity, defined by

$$\sum_A m_A N_A{}^a = \rho u^a_{CM} \quad (\rho > 0), \tag{77}$$

and the baryonic mean 4-velocity, defined by

$$B^a = b \, u^a_B \tag{78}$$

provided B^a is timelike, as it is for "ordinary"matter.

With <u>any</u> choice of mean 4-velocity, one can decompose T_k^{ab} according to (65), obtaining μ, p etc. for a mixture. Which 4-velocity is the most useful one depends on the circumstances; a careful investigation is not known to the author.

7. The selfconsistent Einstein-Maxwell-Liouville equations (1)

Consider a collisionless mixture of particles, so that (41) holds for each component, and consequently (71), (74), (76) are trivially satisfied. Then, we have the macroscopic conservation laws (73), (75) and the (generalized) Poynting equation (68). It is, therefore, permissible to assume that g_{ab}, F_{ab} are the mean fields produced by the gas, i.e., to require that they satisfy the Einstein-Maxwell field equations:

$$G^{ab} + \Lambda g^{ab} = T_K^{ab} + T_M^{ab}, \tag{79}$$

$$F_{[ab,c]} = 0, \qquad F^{ab}{}_{;b} = J^a,$$

(1) Compare with Tauber-Weinberg (1961) who apparently first advocated these equations.

where

$$T_M{}^{ab} = F^{ac}F^b{}_c - \frac{1}{4} g^{ab} F_{cd}F^{cd} \qquad (80)$$

is the Maxwell stress energy momentum tensor. Indeed, (80) and $(79)_2$ imply $T_M{}^{ab}{}_{;b} = - F^a{}_b J^b$, and if this is combined with (68), there results $(T_k{}^{ab} + T_M{}^{ab})_{;b} = 0$, as required by (79).

Hence, the equations (79) together with the Liouville equations

$$L_A(f_A) = 0 \qquad (81)$$

seem to provide a closed, consistent system of dynamical equations for a gravitating plasma (in the Vlasov approximation).

For neutral particles, $(79)_1$ (with $T_M{}^{ab} = 0$) and (81) give a relativistic version of the equations of stellar dynamics (for collisionless systems).

It is natural to pose the Cauchy initial value problem for the system (79), (81). Formally, there seems to be no obstacle to solving it in the usual way by separating initial constraints from evolutions equations, the former being propagated off the initial hypersurface in consequence of the evolution equations, which in turn can be solved for the highest derivatives off the initial hypersurface, provided that is not characteristic. A careful elaboration for the present system (79), (81) does not seem to have been performed, however. [2]

Examples of sulutions to the equations $(79)_1$ (with $T_M{}^{ab} = 0$), (81) are known, see E. D. Fackerell (1966, 1968), J. Ehlers, P. Geren and R. K. Sachs (1868), R. Hakim (1968), R. Berezdivin and R. K. Sachs (1970); see also Misner (1968), Stewart (1969), [1] Matzner (1969). Solutions with electromagnetic fields do not seem to be known at present (Problem).

8. The Boltzmann equation

Consider again a multicomponent gas with particle species A, B, ... If collisions occur, then the phase space density of all collisions in which particles of type A participate, $L_A(f_A)$, will be a sum (or integral) of various contributions due to different kinds of collisions, e.g., elastic and inelastic binary collisions, absorptions and emissions.

---

(1) For a series of papers on the stability theory of static, spherically symmetric solutions of (79), (81) (for Fab = 0), see J. R. Ipser and K. S. Thorne, Ap. J. 154, 251 (1968), and subsequent papers by Ipser in the same Journal.

(2) (Note added in proof) Meanwhile, the problem has been solved by Y. Choquet-Bruhat; see Journ. Math. Phys., 1970, and another forthcoming paper.

Let the symbol

$$(x;pp_A, p_B, \dots \quad p_C, \dots) \tag{82}$$

stand for a collisions in which particles of types A, B, ... with respective 4-momenta $p_A, p_B, \dots$ collide at $x \in X$ and produce particles C, ... with $p_C, \dots$; the numbers of incoming and outgoing particles may be arbitrary. (If, e.g., A = B, one has to write p_A, p'_A instead of p_A, p_A; this is tacitly assumed here and in the sequel.)

The set of all collisions (82) of a particular type, with $x \in X$, $p_A \in P_A(x), \dots$ is again a bundle over X, the <u>collision bundle</u>. It carries a measure, viz., $\eta \wedge \pi_A \wedge \pi_B \wedge \dots \wedge \pi_C \wedge \dots$.

Augmenting our former smoothness assumption D_2 concerning the probability distribution of collisions we make the hypothesis:

C_1) In any macrostate of a gas, the average number of collisions (82) in a compact region U of the collision bundle is

$$\int_U V\,(x; p_A, p_B, \dots \rightarrow p_C, \dots ;)\ \delta(\Delta p)\, \eta \wedge \pi_A \wedge \pi_B \dots \wedge \pi_C \dots \tag{83}$$

where V is a nonnegative (ordinary, measurable) function.(1) (In order to avoid ambiguities in the definition of V, U must be such that 4-momentum ranges $K_A(x), K_B(x), \dots$ of indistinguishable incoming or outgoing particles

(1) Because of the δ-factor in (83) the physically important domain of definition of V is $(X; p_A, p_B, \dots \rightarrow p_C, \dots) : \Delta p = 0$.

(A = B =. . .) do not overlap.

In (83), δ is the Dirac distribution (on R^4), and

$$\Delta p := p_A + p_B + \ldots - p_C - \ldots \tag{84}$$

is the 4-momentum difference between "in" and "out" states. The δ-factor in (83) expresses that collisions (82) occur only if they conserve 4-momentum.

It then follows that the distribution functions $f_A, f_B, \ldots$ of a gas satisfy equations of the form

$$L_A(f_A)(x, p_A) = \sum \Gamma^V_A \int V(x; p_A, p_B, \ldots \longrightarrow p_C, \ldots)\, \delta(\Delta p)\, \pi_B \ldots \wedge \pi_C \ldots \tag{85}$$

In this "collision balance" the sum is to be taken over all kinds of collisions in which A-particles participate, either as incoming or as outgoing collision partners. The integral goes over the mass - shells of all colliding particles except the one whose state occurs on the left-hand side of (85). Γ^V_A is a numerical factor depending on the type of collision and on whether the state p_A on the left-hand side of (85) is an "in" or an "out" state; it is defined thus: $\Gamma^V_A > 0$ (< 0) if A is an "out" ("in") state, and

$\left|\Gamma^V_A\right| = n_A (n_A ! \, n_B ! \ldots n_C !)^{-1}$, where $n_A, n_B, \ldots$ are the numbers of (indistinguishable) particles of types A, B, ... entering or leaving the V-collision, and n_A refer to the number of particles to which the left-hand state in (85) belongs.
(If we have a collision $(p_A, p'_A \rightarrow p''_A, p_B, p'_B)$ with $A \neq B$ and p_A is the state occuring on the left-hand side of (85), then $\Gamma^V_A = -2 (2! \; 1! \; 2!)^{-1} = \frac{-1}{2}$.) This factor is necessary in order that the various collisions involving identical in (or out) particles are not counted several times in the balance (85).

The equations (85) are useless as long as the dependences of the functions V on the state of the gas are not specified.[1] It is clear from nonrelativistic statistical mechanics that in a rigorous many-particle theory V will depend, not only on the one-particle distribution functions $f_A, f_B, \ldots$, but also (at least) on pair

---

[1] See, e.g., Liboff (1969).

correlations $g_{AB}(x, p_A; x', p'_B)$. No attempt will be made here to cope with these difficulties which pose important and interesting problems. Rather, I shall write down a "reasonable Ansatz" (as people say); then I shall make some remarks about the "philosophy" which is used to motivate that Ansatz; then modify it so as to account for the non-classical symmetry character of Bosons and Fermions; and then simply proceed on the basis of the resulting (generalized) Boltzmann equation.

Consider the hypothesis

$$C_2)\ V(x; p_A, p_B, \ldots \rightarrow p_C, \ldots) = f_A(x, p_A)\, f_B(x, p_B) \ldots R(p_A, p_B, \ldots \rightarrow p_C, \ldots) \tag{85}$$

in which the factors $f_A, \ldots$ refer to the "in" states only.

(C_2) is suggested by the assumptions that

(a) particles which are about to collide have uncorrelated momenta,

(b) the ranges of the collisional interactions are small in comparison with the scale on which the f_A's change appreciably with x,

(c) collisions take place in spacetime regions so small that the mean differential gravitational field $R^a{}_{bcd}$ (geodesic deviation etc. !) and the mean electromagnetic field $\mathbf{F}_{ab}$ do not affect their frequency appreciably.

(d) the presence of particles not participating directly in a collision does not affect the probability of occurence of that collision.

These assumptions, which essentially express that the gas is dilute ((d) and, for a gravitating gas, (c)), not too inhomogeneous in spacetime (b)), and in a state of high randomness ((a)), indicate the range of validity of the "Boltzmann collision hypothesis" C_2; each of them poses a problem of justification and indicates desirable generalizations. If, e.g., (c) were not true, then R might be expected (1) to depend on the principal directions and eigenvalues of $R^a{}_{bcd}$ and $F^a{}_b$.

In order to support the assumption C_2 further and to relate it to scattering theory, let us consider a collision $(p_A, p_B \rightarrow p_C, \ldots)$ with two incident and q emerging particles, and let us consider

(1) Compare Marle (1969), p. 88.

those collisions for which the momenta are contained in small ranges $K_A \subset P_A(x)$ etc.

According to (83) and (86), the number of those collisions per unit spacetime volume is

$$f_A(x, p_A)\, \pi_A(K_A)\, f_B(x, p_B)\, \pi_B(K_B)\ \delta(\Delta p)\ R\,(p_A, p_B \rightarrow p_C, \ldots)\, \pi_C(K_C) \ldots \tag{87}$$

Regarding the K_A-particles as a beam which hits the K_B-particles forming the target, we recognize that the number densities of projectiles and target particles, relative to any inertial frame with 4-velocity u, are given by (see (44))

$$n_A = f_A(x, p_A) \left| u \cdot p_A \right| \pi_A(\, K_A\,),$$

$$n_A = f_B(x, p_B) \left| u \cdot p_B \right| \pi_B(\, K_B\,),$$

whereas the relative velocity of these particles is

$$\left| \vec{v}_B - \vec{v}_A \right| = \frac{\left| (u \cdot p_A)\, p_B - (u \cdot p_B)\, p_A \right|}{(u \cdot p_A)\ (u \cdot p_B)} \ . \tag{88}$$

(Take $u^a = \delta^a_4$.)

Hence, we can rewrite (87) as

$$n_A n_B \left| \vec{V}_B - \vec{V}_A \right| dQ^u, \tag{89}$$

where

$$dQ^u = \frac{R(p_A, p_B \rightarrow p_C, \ldots)}{\left| (u \cdot p_A)\, p_B - (u \cdot p_B)\, p_A \right|} \, \delta(\Delta p)\, \pi_c(K_c) \ldots . \tag{90}$$

Equation (89) is recognized as the standard definition of the <u>differential scattering cross section</u> dQ^u for scattering of p_A, p_B-particles into the ranges $K_c, \ldots$, relative to the u-frame, and equation (90) is indeed the correct expression for that cross section which can be derived

α) in the nonrelativistic limit either from classical or from quantum mechanics, and

β) in the relativistic domain from quantum scattering theory.[1]

---

(1) See, e.g., Brenig and Haag (1959)

(In this case R is simply related to the S operator [1].)

(In the relativistic case, a classical derivation is not available, since there is no well developed theory of interacting particles.)

In a certain sense, we have now justified (85), since under the assumtions stated above the A- and B-particles in the gas should behave as if they were members of beams in a collision experiment.

One correction, or generalization, of (85) shall now be made. If the particles are atomic or sub-atomic, then assumption (d) is definitely wrong. In the case of Fermions, the presence of particles in the final states decreases, because of the Pauli principle, the collision probability, whereas for Bosons it enhances that probability (stimulated emission and scattering). This is incorporated by writing, instead of (86),

$$\begin{aligned} C_2') \quad & V(x;p_A, p_B, \ldots \rightarrow p_C, \ldots) = f_A(x, p_A)\, f_B(x, p_B) \ldots \\ & \times \left(1 \pm s_C f_C(x, p_C) \right) \ldots R(p_A, p_B, \ldots \rightarrow p_C, \ldots) \, . \end{aligned} \tag{91}$$

---

[1] See, e.g., Brenig and Haag (1959)

Here and in the sequel the upper sign refers to Bosons, the lower one to Fermions.

$$s_C := \frac{(2\pi)^3}{r_C} = \frac{h^3}{r_C}$$

is the volume of a phase-cell which corresponds asymptotically to a non-degenerate p-eigenstate of a free (quantum) particle[1] of spin degeneracy r_C. Hence, $s_C f_C(x, p_C)$ equals approximately the average occupation number of simple one-particle p_C-eigenstates localised near x. (In the "classical limit" $f_C \ll s_C^{-1}$, (91) reduces, of course, to (86).

A "pseudoproof" of (91) can be given within the Fock-space formalism, but that will not be reproduced here[2].

One simplification is possible and useful in (90).

If $u = \lambda p_A + \mu p_B$ - and these frames include the center - of - mass frame of the collision as well as the rest frames of the incoming particles-

---

[1] Weyl (1911), Peierls (1936)

[2] See Bichteler (1965), Ehlers and Sachs (1968), Ehlers (1969).

then $\left| (u\cdot p_A)p_B - (u\cdot p_B)p_A \right| = \sqrt{(p_A\cdot p_B)^2 - m_A^2 m_B^2}$, independently of u, whence the corresponding expression

$$dQ = \frac{R\,(p_A, p_B \rightarrow p_C, \dots)}{\sqrt{(p_A\cdot p_B)^2 - m_A^2 m_B^2}}\;\delta(\Delta p)\;\pi_C \wedge \dots \qquad (92)$$

is often called "the" (relativistic) cross section.
We have now suppressed the arguments $K_C, \dots,$ considering dQ henceforth as a (3q - 4) - form, where q is the number of final states and we imagine that $\delta(\dots)$ has been "absorbed" into four of the differentials $\pi_C \wedge \dots$:(If q = 1, dQ is a δ-"function": $dQ = \sigma_a(-p_A\cdot p_B)\,\delta\left(\frac{1}{2}\left[m_A^2+m_B^2-m_C^2\right] - (p_A\cdot p_B)\right)$; σ_a is the absorption cross section.)

Inserting (91) into (85) we obtain the generalized Boltzmann equation

$$L_A(f_A) = \sum \Gamma_A^R \int f_A f_B \dots f^C \dots \delta(\Delta p)\, R^{C\dots}_{AB\dots}\;\pi_B \wedge \dots \wedge \pi_C \dots, \qquad (93)$$

where we have simplified the notation in an obvious way. In particular, in (93) and henceforth,

$$f^C := s_C^{-1} \pm f_C; \qquad (94)$$

the factors s_C have been absorbed into $R_{...}$ (Γ^R_A equals the former Γ^V_A.)

The equation (93) has first been formulated in special relativity for a classical (i.e., Boltzmannian) gas with elastic binary interactions by Lichnerowicz and Marrot (1940); for other treatments and generalisations see Tauber-Weinberg (1961), Israel (1963), Bichteler (1965) and the papers mentioned in the introduction.

Henceforth we shall require the Boltzmann equation (93) to hold for the distribution functions of any gas. (Other "reasonable" alternatives for V which lead to different kinetic equations are possible, but will not be discussed in these lectures.)

One important symmetry needs to be mentioned. If the microscopic collision law (S-matrix) is invariant with respect to the total reflection, PT, then the collision "matrix" R :: is invariant with respect to an interchange of incoming and outgoing states(*):

(*) See Brenig and Haag (1959).

$$R^{C\dots}_{AB\dots} = R^{AB\dots}_{C\dots} \,. \tag{95}$$

We also add that, for Fermions, it is necessary that

$$s_A f_A \leqslant 1, \tag{96}$$

due to the exclusion principle.

The conservation law (71) is satisfied by (93), because of the $\delta(\Delta p)$ - factor. Other conservation laws like (74) can and have to be incorporated by similar restrictions on the R-functions; this will be assumed in the sequel.

It is now clear that we can generalize the selfconsistent field equations of section 7 so as to take into account collisions; we just have to replace equation (81) by (93). The remarks about the Cauchy problem made in section 7 still hold; a rigorous analysis for the system (79), (93) has not been performed, however.

In a given spacetime X (with a metric of class C^2), i.e., for a test gas, Bichteler (1967) has solved the (local) Cauchy problem for (93). Besides existence and uniqueness of exponentially bounded(1),

(1) on the mass-shell, i.e., $|f_A(x,p)| \leqslant b(x)\, e^{\beta_a(x) p^a}$, with b and β_a depending continuously on x.

nonnegative, continuous and a. e. differentiable solutions for given initial distributions of the same type, Bichteler has established the continuous dependence of the solution on the initial distributions, the metric, and the cross section (i. e., R$\dot{:}\dot{:}$). He assumes throughout that the total cross sections $\int_{\text{(all final states)}} dQ^{\cdot}$ are bounded. (This last assumption, though perhaps valid for strong interactions.[1], does not seem to hold, e. g., in the case of weak interactions.[2]) Bichteler obtained his results by applying Banach's fixed point theorem to an operator given naturally by means of (93), defined on a suitably chosen complete metric space of exponentially bounded distribution functions. As Bichteler pointed out, his results lend some credibility to the (formal) Chapman-Enskog approximation which will briefly be discussed later.

---

(1) See Eden (1966)

(2) See Bahcall (1964)

9. The second law of thermodynamics (H-theorem).

We define the entropy current density of a gas to be the 4-vector field

$$S^a := -\sum_A \int \left[f_A \log (s_A f_A) \mp f^A \log (s_A f^A) \right] p^a \pi_A \tag{97}$$

with s_A defined as before (below (91)).

The expression (97) can, in a sense, be derived from an information-theoretic point of view as indicated in Ehlers (1969). In the classical limit $s_A f_A \longrightarrow 0$ it reduces to

$$S^a = -\sum_A \left\{ \int p^a f_A \log (s_A f_A) \pi_A - N^a_A \right\}, \tag{98}$$

and one recognizes in the first term of S^4 the Boltzmann entropy density. Generally, $-S^a u_a$ is to be interpreted as the entropy density relative to an observer with 4-velocity u^a.

Using Lemma 8 one obtains

$$S^a{}_{;a} = \sum_A \int L_A(f_A) \log\left((s_A f_A)^{-1} \pm 1\right) \pi_A. \tag{99}$$

inserting $L_A(f_A)$ from the Boltzmann equation (93) and assuming the

PT - symmetry (95) for all collisions involvee one gets a sum of terms; one from each kind of collision and its inverse, of the form

$$\log\left(\frac{f_A f_B \dots f^C}{f^A f^B \dots f_C \dots}\right) (f_A f_B \dots f^C \dots - f^A f^B \dots f_C \dots) \, \delta(\Delta p) \times$$

$$\times R^{C\dots}_{AB\dots} \, \pi_A \wedge \pi_B \dots \wedge \pi_C \dots, \tag{100}$$

where we have again used the notation (94). Each such integral is nonnegative, since its integrand has the form $(\log \frac{a}{b})(a - b)$. Hence,

$$S^a{}_{;a} \geqslant 0. \tag{101}$$

This is the relativistic form of Boltzmann's H-theorem (Tauber-Weinberg (1961), Ehlers (1961), Chernikov (1963), which expresses locally the content of the second law of thermodynamics in the framework of kinetic theory.

(101) implies that the flux of S^a through any closed hypersurface in X is nonnegative. Hence, for an adiabatically enclosed

or isolated gaseous body the total entropy

$$S[\Sigma] := \int_{\Sigma} S^a \sigma_a, \qquad (102)$$

evaluated on a spacelike cross section of the world tube of the body, never decreases towards the future. (Notice that in the classical limit (98), $S[\Sigma]$ consists of the total number of particles and the Boltzmann contribution. If the total particle number is not constant, the Boltzmann S-term alone does not necessarily increase.)

Notice that (101) does not follow from (93) if the collisions are due to PT-violating interactions.

If (95) does hold, and if collisions occur frequently in a gas, then the competition between collisions of a certain kind and their inverses suggests the tendency of the gas to evolve in such a way that the difference in the integrand of (100) tends towards zero, so that ultimately the entropy production density $S^a{}_{;a}$ vanishes and the Liouville equations (41) holds, provided there are no disturbing external influences. Unfortunately, precise theorems supporting this physical expectation are so far missing in relativistic kinetic theory; even at the nonrelativistic level little is known. (For a brief discussion see, e.g., Uhlenbeek and Ford (1963), p. 81.) Any result in this direction would be of interest. It would also be of some interest to know whether in situations of gravitational collapse S may increase towards infinity,

10) Stationary states, equilibrium, and thermostatics

A gas given by g_{ab}, F_{ab}, f_A is said to be in a stationary state in a region $D \subset X$ if there exists, in D, a one-dimensional local group G of fixed-point free local isometries with timelike orbits which leaves F_{ab} and the f_A invariant. In terms of the generating vector field ξ^a of G the last two conditions can be expressed as

$$\mathcal{L}_\xi F_{ab} = 0, \tag{103}$$

$$\left(\xi^a \frac{\partial}{\partial x^a} + \xi^\lambda{}_{,a} p^a \frac{\partial}{\partial p^\lambda} \right) f_A = 0; \tag{104}$$

moreover , we then have Killing's equation

$$\xi_{(a;b)} = 0. \tag{105}$$

The last two equations imply

$$\mathcal{L}_\xi S^a = 0 \tag{106}$$

and similar statements for N^a_A, T^{ab}_A etc. Because of (105) it follows further that

$$(\mathcal{L}_\xi S^a)_{;a} = \mathcal{L}_\xi (S^a{}_{;a}) = 0, \tag{107}$$

i.e., the entropy production is constant on the G-orbits.

Let us assume now that an adiabatically isolated gas is in a stationary state in D, and that the boundary of the world tube $\mathfrak{F}$ of the gas is G-invariant; $\mathfrak{F} \subset D$. Let Σ be a spacelike cross section of $\mathfrak{F}$ and $a \in G$. Then $a(\Sigma)$ is again such a cross section, and because of the assumed stationarity $S[a(\Sigma)] = S[\Sigma]$. Applying Gauss's theorem to the part of $\mathfrak{F}$ between Σ and $a(\Sigma)$, using the adiabatic condition along the wall $\partial\mathfrak{F}$, and taking account of (101) we obtain in $\mathfrak{F}$

$$S^a{}_{;a} = 0. \tag{108}$$

This conclusion, combined with the expectation described at the end of the previous section, leads us to define:

A gas is in local equilibrium at $x \in X$ if, at x, $S^a{}_{;a} = 0$.

The formula (100) for a summand of $S^a{}_{;a}$ shows the validity of the first part of the

theoreme. If the collision functions $R^{\cdot\cdot\cdot}_{\cdot\cdot\cdot}$ of a gas are all strictly positive almost everywhere (w. r. t. the measure $\delta(\Delta p)\, \pi_A \wedge \ldots$) and continuous, then the gas is in local equilibrium at x if and only if at x

$$f_A f_B \dots f^C \dots = f^A f^B \dots f_C \dots \tag{109}$$

whenever $\Delta p = 0$, for all types of collisions which occur; or, equivalently, if and only if for each particle species on $P_A(x)$ there holds

$$L_A(f_A) = 0. \tag{110}$$

The second part of this theorem follows from the first part by means of equations (93) and (99).

The restriction $R\because > 0$ is not unsatisfactory from the physical point of view, since the R-functions are usually analytic functions of the momentum variables on the "collision fiber" $\Delta p = 0$, and hence they vanish only on sets of measure zero.

The problem of finding the general continuous solutions $(f_A, \dots)$ of (109) has been solved for binary elastic collisions between Boltzmann particles, where (109) reduces to

$$f_A f_B = f'_A f'_B \qquad \text{whenever} \qquad p_A + p_B = p'_A + p'_B. \tag{111}$$

In this case, the general solution is given by

$$f_A(x,p) = a_A(x)\ e^{\beta_a(x)p^a} \tag{112}$$

and a similar formula for f_B and with β_a the same for both species. (Chernikov (1964), Marle (1969) and, in the case where the f_A's are assumed C^1, Bichteler (1965), Boyer (1965). The nicest proof is that of Marle, the shortest that of **B**ichteler.)

If we consider elastic binary collisions between Bosons or Fermions (or a mixture) and assume that all factors in

$$f_A f_B f'^A f'^B = f'_A f'_B f^A f^B \tag{113}$$

are positive on their mass-shells, we may divide by $f_A f_B f'_A f'_B$ and obtain for $\frac{1}{s_A f_A} \pm 1$ etc. the same relation as for Boltzmann particles, so that we obtain

$$f_A(x,p) = \frac{r_A}{(2\pi)^3}\left[e^{-\alpha_A(x) - \beta_a(x)p^a} \mp 1\right]^{-1} \tag{114}$$

Whereas it is easy to deduce from (111) that $f_A f_B \neq 0$ <u>everywhere</u> provided that holds for <u>some</u> pair $p_A \neq p_B$, this does not seem so obvious in the case (113). Nevertheless I shall accept (114) as the general form of an equilibrium distribution at an event x for particles participating in some kind of binary elastic collision.

If particles in a gas undergo not only binary elastic collisions, but in addition other kinds of reactions, then (114) and (109) show that the α_A must obey

$$\alpha_A + \alpha_B + \ldots = \alpha_C + \ldots \tag{115}$$

for all permissible collisions $A + B + \ldots \rightleftarrows C +$

With (114) and (115) we have obtained the general local equilibrium distributions $(f_A, f_B, \ldots)$.

Since the f_A's have to vanish at infinity on the mass-shells, $\beta^a(x)$ must be a future-directed timelike vector. We put

$$\beta^a = \beta u^a, \quad u_a u^a = -1, \quad \beta > 0. \tag{116}$$

It is a straightforward matter to obtain from (114) the quantities N^a_A, T^{ab}_A, S^a_A, n_A, μ_A, p_A, u^a_K, u^a_D defined in eqs. (45), (47), (97), (63), (64), (49), (52), respectively. Working in the rest frame of u^a one gets

$$u^a_K = u^a_D = u^a, \tag{115}$$

$$S_A^a = s_A u^a, \tag{116}$$

$$T_A^{ab} = (\mu_A + p_A)\, u^a u^b + p_A\, g^{ab}, \tag{117}$$

with the scalars (we omit temporarily the index A) n, μ, p, s given in terms of α, β and the constants m, r by

$$n = \frac{r}{2\pi^2} \int_m^\infty \frac{\sqrt{E^2 - m^2}\, E\, dE}{e^{-\alpha+\beta E} \mp 1} \tag{118}$$

$$\mu = \frac{r}{2\pi^2} \int_m^\infty \frac{\sqrt{E^2 - m^2}\, E^2 dE}{e^{-\alpha+\beta E} \mp 1} \tag{119}$$

$$p = \frac{r}{6\pi^2} \int_m^\infty \frac{(E^2 - m^2)^{\frac{3}{2}} dE}{e^{-\alpha+\beta E} \mp 1} \tag{120}$$

$$s = \frac{r}{2\pi^2} \int_m^\infty \left\{ \frac{-\alpha + \beta E}{e^{-\alpha+\beta E} \mp 1} \mp \log(1 \mp e^{\alpha - \beta E}) \right\} \sqrt{E^2 - m^2}\; E\, dE. \tag{121}$$

These functions and further thermostatic relations obtained from them have been studied extensively; see, e;g., Landsberg and Dunning-Davies (1965) and the references given there.

The thermostatic meaning of the two parameters α, β is recognized thus: observe that

$$s = -\alpha n + \beta \mu \mp \frac{r}{2\pi^2} \int_m^\infty \log(1 \mp e^{\alpha - \beta E}) \sqrt{E^2 - m^2}\; E\, dE.$$

Transform the last term by partial integration and get, with (120),

$$s = -\alpha n + \beta\mu + \beta p. \tag{122}$$

Use (120) and compute, again integrating by parts,

$$dp = \frac{n}{\beta}\, d\alpha - \frac{\mu + p}{\beta}\, d\beta. \tag{123}$$

(122) and (123) give

$$d\mu = \beta^{-1} ds + \alpha\beta^{-1} dn \tag{124}$$

Now, $\mu(s,n)$ is a thermostatic potential, and $d\mu = T\,ds + \tilde{\mu}\,dn$, where T is the temperature and $\tilde{\mu}$ is the chemical potential (per particle). Hence we conclude

$$\beta = T^{-1}, \quad \alpha = \frac{\tilde{\mu}}{T}. \tag{125}$$

(125) can now be rewritten in terms of the $\tilde{\mu}_A$'s and reveals itself as the law of mass action.

For the thermodynamics of mixtures see Ehlers (1969), and for applications of the preceding theory to cosmology see Ehlers and Sachs (1968).

Let us now investigate which restrictions are imposed on the parameters α, β and on the mean velocity u^a by the requirement that there is global equilibrium, i.e., that there is local equilibrium at each event of a region $D \subset X$. According to the theorem above, the functions (114) must then obey Lionville's equation; i.e. $L_A(\alpha_A + \beta_a p^a) = 0$ in D. This equation is easily evaluated (see, e.g., Ehlers (1969) and leads to the

theorem. Global equilibrium requires that

(a) β^a is a conformal Killing vector and, if at least one component of the gas consists of particles with positive rest masses, a Killing vector, and

(b) the electric field strength $E_a := F_{ab} u^b$ is related to T and α by

$$T \, d\alpha = e \, E. \tag{126}$$

For a gas containing (also) ordinary particles ($m > 0$), equilibrium requires a stationary spacetime. Defining in such a spacetime a scalar gravitational potential U in terms of the Killingvector $\xi^a = T_o \beta^a$ by $e^{2U} = -\xi^2$ we obtain Tolman's law

$$e^U = \frac{T_o}{T}, \tag{127}$$

and if $E = 0$, then α = const., so that $\tilde{\mu}$ depends on the potential like the temperature. (For the general evaluation of (126) see Ehlers (1959).)

It is possible to characterize the global equilibrium solutions in a given, stationary spacetime by means of a variational principle in which S is maximised under certain constraints, see Marle (1969) pp. 107. For examples of equilibrium solutions, see Chernikov (1964).

By means of (42) and (114) it can be verified that Planck's distribution law results for $r_\gamma = 2$, $\alpha_\gamma = 0$, as it should be; $\alpha_\gamma = 0$ results from the relations (115), since there are always some processes which change the photon number but not the numbers of the other particles involved (ex.: e-e collisions).

A gas is said to be nondegenerate if the ∓ 1-term can be neglected without serious error, so that (112) holds. Otherwise, it is called degenerate.

One consequence of the last theorem is that a gas with $m > 0$ cannot maintain an equilibrium distribution if it expands isotropically, in contrast to an $(m = 0)$-gas (photons, neutrinos). A physical reason for this deviation from the nonrelativistic behaviour of a (m 0) gas will be given in the last section.

Since the thermostatic functions of a relativistic gas are explicitely known (cf. eqs. (118)-(121)) one can compute, e.g., the velocity of sound in such a gas, and one can check the validity of Weyl's condition for shock waves. For a Boltzmann gas with m 0 this has been done in detail by Synge (1957), with the result that the sound velocity increases monotonically with the temperature and approaches the limit $\frac{c}{\sqrt{3}}$ as $T \longrightarrow \infty$ (the value for a photon gas); shock speeds are always less than c. Shock waves in a gas of Fermions or Bosons have been investigate by israel (1960).

11. Irreversible processes in small deviations from equilibrium; hydrodynamics.

Whereas the equilibrium solutions of the Boltzmann equation can be written down exactly, there is not much hope to find rigorous solutions describing irreversible ($S^{a}{}_{;a} > 0$) processes-in fact no such (relativistic) solution is known at present. In physics, however, one is mostly interest in non-equilibrium situations. Therefore, in order to proceed one has to resort to approximations. We shall briefly describe such approximation methods in this section, and refer to research papers for details. Our main goal here will be to indicate how one may obtain from kinetic theory a complete system of equations for thermo-hydrodynamics which is sufficiently general to include heterogeneous systems in which transport processes and reactions take place, by applying suitable approximations to the Boltzmann equation. Partly our exposition will be a program rather than the exposition of a completed theory. For simplicity I shall consider here only neutral fluids, thus in the sequel "$e_A = J^a = F^a{}_b = 0$". Also, we shall only consider processes close to equilibrium, which will (for most of the sequel) mean states which are infinitesimal perturbations (first order variations) away from local equilibrium.

Two distributions f_A, $\bar{f}_A$ will describe nearly the same macrostate of a gas if their moments in p-space are everywhere nearly equal. This will be the case if $\bar{f}_A = f_A (1 + \varepsilon \Phi_A)$ provided Φ_A is a. e. bounded on M_A and the numerical "perturbation" parameter ε is small. With this motivation, we shall now consider a one-parameter family $f_A(\varepsilon)$ of states which is, for $\varepsilon = 0$, in local equilibrium, i. e., is such that for $\varepsilon = 0$ the f_A's have the form (114), with unspecified spacetime fields α_A, β_a, and we shall denote by f'_A the variations $\left.\frac{df_A}{d\varepsilon}\right|_{\varepsilon=0.}$ Notice that the "local equilibrium functions" α_A, $\beta^a = \frac{U^a}{T}$ are independent of ε. For "small" ε, the moments computed by means of the "perturbed distribution functions" $f_A(0) + \varepsilon f'_A$ will be considered to be the macroscopic variables describing a "state close to equilibrium".

It is clear that the perturbed macroscip variables will satisfy the conservation laws

$$T^{ab}{}_{;b} = 0, \quad B^{a}{}_{;a} = 0 \tag{128}$$

and similar ones, if we impose additional "scalar" conservation laws like b-conservation. Also, we shall have the "Clausius inequality"

$$\xi := S^{a}{}_{;a} \geq 0. \tag{129}$$

Again we can write the decomposition (65) for the total, perturbed tensor T^{ab}, with $\mu = \mu(0) + \varepsilon\mu'$, $p = p(0) + \varepsilon p'$, $q^a = \varepsilon (q^a)'$, $\pi^{ab} = \varepsilon (\pi^{ab})'$, because of (117) for $\varepsilon = 0$.

Similarly,

$$N^{a}_{A} = n_{A} u^{a} + i_{A}{}^{a}, \quad u_{a} i_{A}{}^{a} = 0, \tag{130}$$

and

$$S^{a} = s u^{a} + s^{a}, \tag{131}$$

with $n_A = n_A(0) + \varepsilon n'_A$, $i_A{}^a = \varepsilon (i_A{}^a)'$, $s = s(0) + \varepsilon s'$, $s^a = \varepsilon (s^a)'$, from (115) and (116) for $\varepsilon = 0$.

It is a straightforward matter to derive from (128) the <u>energy balance equation</u>

$$\dot{\mu} + (\mu + p)\vartheta + \pi_{ab}\,\sigma^{ab} + q^a{}_{;a} + q^a\,\dot{u}_a = 0, \tag{132}$$

where the kinematical quantities ϑ, σ_{ab}, $\dot{u}_a$, ω_{ab} are defined by

$$\begin{aligned} u_{a;b} &= \omega_{ab} + \sigma_{ab} + \frac{1}{3}\vartheta h_{ab} - \dot{u}_a u_b, \\ \omega_{ab} u^b &= \sigma_{ab} u^b = \omega_{(ab)} = \sigma_{[ab]} = \sigma^a{}_a = 0 \end{aligned} \tag{133}$$

and are interpreted as the rate of rotation (ω_{ab}), rate of shear (σ_{ab}), rate of expansion (ϑ), and 4-acceleration ($\dot{u}_a$) of the flow given by u^a (see, e.g., Ehlers (1961), Synge (1960)). Here and henceforth we write

$$(\ )^{\cdot} := (\ )_{;a}\, u^a. \tag{134}$$

Also, one obtains the <u>momentum balance equation</u>

$$(\mu+p)\,\dot{u}_a + h_a^b\,(\dot{q}_b + p_{,b} + \pi^c_{b;c}) + (\omega_{ab} + \sigma_{ab})\,q^b + \frac{4}{3}\,\vartheta q_a = 0 \qquad (135)$$

(h_{ab} has been defined in (66).)

Let us now assume that there are Q conserved scalar quantities, like b, which we call c_{qA}, where $1 \leqslant q \leqslant Q \leqslant N$ and where N is the number of species A of particles; the c_{qA} are given, constant "gharges". Then the reactions in the system are restricted by

$$\Big(\sum_A c_{qA}\, N^a_A\Big)_{;a} = 0, \quad 1 \leqslant q \leqslant Q. \qquad (136)$$

We assume the Q "vectors" $(c_{q1}, \ldots, c_{qN})$ to be linearly independent, and denote by $(r_{p1}, \ldots, r_{pN})$, $1 \leqslant p \leqslant R := N - Q$ a basis in the orthogonal space.(1) The vectors $(r_1, \ldots r_N)$ of the latter can be interpreted as (chemical or nuclear, e.g.) reaction coefficients, as is seen from the equations

$$N^a_{A;a} = \sum_p v_p\, r_{pA} \qquad (137)$$

which express the general solution of (136) in terms of the constants r_{pA} and the reaction rates v_p, giving the spacetime densities of reactions of type p.

From (130), (137) we obtain the particle balance equations

(1) I;e., $\sum_A c_{qA}\, r_{pA} = 0$ for $1 \leqslant q \leqslant Q$, $1 \leqslant p \leqslant R$.

$$\dot{n}_A + n_A \vartheta + i_A{}^a{}_{;a} = \sum_p v_p \, r_{pA} \tag{138}$$

Similarly, we rewrite (129), using (131), as

$$\dot{s} + s\vartheta + s^a{}_{;a} = \xi \geqslant 0. \tag{139}$$

To proceed further we vary the expression (97) for the entropy current density S^a; because of

$$\left[f \log(s\, f) \mp (s^{-1} \pm f) \log (1 \pm sf) \right]' = f' \log\left(\frac{1}{sf} \pm 1\right)^{-1}$$

and (114) we obtain

$$T s' = (\mu' - \sum_A \tilde{\mu}_A n'_A) \tag{140}$$

and

$$T s^{a'} = (q^{a'} - \sum_A \tilde{\mu}_A i_A^{a'}), \tag{141}$$

where the $\tilde{\mu}_A$ are the chemical potentials defined in (125).

If we combine (140) with the thermostatic <u>Gibbs relation</u>

$$d\mu = T\, d s + \sum_A \tilde{\mu}_A \, dn_A \tag{142}$$

which results from (124) by summing over the species A, and which holds for the unperturbed equilibrium functions (on the manifold $\{(s, n_1, \dots, n_A)\}$ of equilibrium states), we get the rather remarkable

Lemma 10. The perturbed thermodynamic variables μ, s, n_A satisfy

$$\mu = F(s, n_1, \dots, n_N) + 0(\varepsilon^2), \tag{143}$$

where F is the thermostatic potential of the system (as determined from the exact equilibrium relations of section 10).

It is, therefore, "reasonable" to use, for near-equilibrium processes, the ordinary Gibbs equation of state for the perturbed variables, neglecting the error term in (143), as we shall henceforth.

Also, we rewrite (141) for the perturbed variables:

$$T s^a = q^a - \sum_A \tilde{\mu}_A i_A{}^a. \tag{144}$$

We also recall that, from (122), the thermostatic pressure p_o associated with the perturbed state is

$$p_o = T s - \mu + \sum_A \tilde{\mu}_A n_A; \tag{145}$$

there is no reason why p_o should equal the total kinetic pressure p in (65).

We are now ready to derive an explicit expression for the entropy production density, ξ, in terms of appropriate thermodynamic and hydrodynamic quantities. Compute $\dot{s}$ from (142) for the perturbed state, which is permissible because of Lemma 10; insert $\dot{\mu}$ from (132), $\dot{n}_A$ from (138), and rearrange terms, using (145), (144) and the definition

$$\pi := p - p_o \tag{146}$$

for the volume viscosity π, to obtain the entropy inequality,

$$\begin{aligned} - T\xi = \pi_{ab}\,\sigma^{ab} + \pi\vartheta + \Big[q^a - \sum_A \tilde{\mu}_A\, i_A^a\Big]\Big[(\log T)_{,a} + \dot{u}_a\Big] + \\ + \sum_A i_A^{\ a}\,(\tilde{\mu}_{A,a} + \tilde{\mu}_A \dot{u}_a) + \sum_p \Big\{ v_p \sum_A \tilde{\mu}_A r_{Ap} \Big\} \leq 0. \end{aligned} \tag{147}$$

This expression has the usual form known from ordinary irreversible thermodynamics (see, e.g., de Groot and Mazur (1962)); in relativity, it has also been worked out on the basis of phenomenological assumptions by several authors (see, e.g. Stückelberg and Wanders (1953), Kluitenberg, de Groot and Mazur (1953), Kluitenberg and de Groot (1954), (1955).

We wanted to show that (147) and the previous formulae follow, in the sense we have specified, from kinetic theory, just as in the non-relativistic case; this does not seem to have been pointed out before with the generality we have retained here. The crucial fact is that equations (143) and (144) follow from the kinetic expression (97) for the entropy current.

The expression $-T\xi$ as given by (147) is bilinear in "fluxes" $\pi_{ab}, \pi, \ldots$ and "forces" $\sigma^{ab}, \vartheta, \ldots$. We have shown earlier that the "fluxes" vanish at an event x if there is local equilibrium at x, and that the "forces" vanish in a region if there is (global) equilibrium in that region. Hence, one is driven to conjecture that, in a near-equilibrium process, the fluxes (which are "caused" by the forces) depend linearly and homogeneously on the forces, with coefficients depending on the thermostatic variables s, n_A. This assertion is indeed used as an assumption in phenomenological approaches, and leads to (more or less) well-known relativistic linear transport and reaction equations for the shear viscosity π_{ab}, the volume viscosity π, the heat flow $w^a := q^a - \sum_A \tilde{\mu}_A i_A^a$, the diffusion currents i_A^a, and the reaction rates v_p. The corresponding matrix which transforms $(\sigma_{ab}, \vartheta, \ldots)$ into $(\pi_{ab}, \pi, \ldots)$ must be positive-semi-definite because of (147) . If one requires, as is natural for a fluid, that this matrix is invariant with respect to rotations (in the 3-

tangent-space orthogonal to u^a), the matrix reduces; and one obtains a further simplification by assuming Onsager-Casimir symmetry. All this follows strictly the standard theory.

However, we should not make these assumptions, but derive them from kinetic theory. This has not yet been done in the generality maintained here, but it will undoubtedly be done soon ([1]). Such a derivation will supposedly) give not only the form of the transport and reaction equations, but will also provide formulae for the transport and reaction coefficients in terms of thermostatic variables and cross sections.

Two classical methods for doing this offer themselves; the Chapman Enskog method, and the Grad method of moments. Both these methods have, in fact, been adapted to relativity; the former by Israel (1963) and, in a mathematically more complete form, by Marle (1969). (Israel, however, gives more detailed results, particularly for a special type of "Maxwellian" gas.) The method of moments has been taken over into relativity by Chernikov (1964) and in a more geometrical (and also analytically more powerful) manner by Marle (see Marle (1966), (1969)) and, independently, by Anderson and Stewart (see Stewart (1969), Anderson (1970). Mathematically, Marle's treatment is the most complete one as regards the discussion of the "relativistic Hermite-Grad polynomials", whereas Anderson and Stewart have

([1]) (Note added in proof) See a forthcoming paper by J. M. Stewart, to appear in Lecture Notes in Physics, Springer-Verlag.

gone further towards physical applications (transport coefficients from cross sections.) In all of this work, the gas is a simple Boltzmannian one; in that case, both methods give the transport equations for π_{ab}, π and q_a expected on the basis of (147). In particular, Israel (1963) and Anderson and Stewart (1969, 1970) both emphasized that a relativistic gas has (in general) a positive bulk viscosity, in contrast to a non-relativistic gas. The bulk viscosity vanishes both in the nonrelativistic ($T \rightarrow 0$) and in the ultrarelativistic ($T \rightarrow \infty$) limit. This result "explains" the difference between m = 0)-gases and ($m > 0$)-gases with respect to property a) of the theorem in section 10: A gas of the latter type behaves irreversibly if expanding isotropically, because of the term $\pi\vartheta$ in (147); a photon gas, however, behaves reversibly, since $T^a{}_a = 0$ implies that $\pi = 0$ always.

For more details concerning the transition from kinetic theory to thermo-hydrodynamics within the framework of relativity we refer to Chernikov (1964), and to the papers cited above.

The roles of temperature T, entropy S^a (or s) and of the main theorems of thermodynamics are completely clear within the framework of relativistic kinetic theory; there is no room for assumptions. (Of course, this changes if one wants to leave the domain of applicability which we have delineated above.) In particular, integration of (129) over a section of a world tube of streamlines, bounded by two spacelike cross sections Σ_1 and Σ_2, gives with the help of (131) and $s^a = \frac{w^a}{T}$ ($\Longleftrightarrow$(144)):

$$S[\Sigma_2] - S[\Sigma_1] \geq \int_{\Lambda} T^{-1} w^a \sigma_a \qquad (148)$$

where Λ is that part of the boundary of the world tube which lies between Σ_1, and Σ_2, and where Σ_2 is assumed to be later than Σ_1. This inequality is a precise version of the somewhat vague assertion $\delta S \geq \frac{\delta Q}{T}$ which has first been postulated in general relativistic thermodynamics by Tolman (1934). In a similar fashion one can derive other "global" thermodynamical laws for moving, finite systems enclosed in containers (timelike cylinders in X) from the basic differential relations discussed here; again, there is no ambiguity. (For another example of such a derivation, see Staruszkiewicz (1966).)

Last - but not least - I would like to mention that the long-discussed paradox concerning the acausal nature of temperature propagation (mathematically: the parabolic character of the corresponding system of equations) has been resolved by the observation that the general, "anormal" solutions resulting from the method of moments obey hyperbolic equations with non-spacelike characteristics (Stewart 1969), and only the special, so-called "normal" solutions give rise to the paradox, which is, therefore, due to an inadequate approximation.

Ehlers

REFERENCES

A. References about mathematical tools

R. ABRAHAM and J. E. MARSDEN, Foundations of Mechanics, Benjamin 1965

Y. CHOQUET-BRUHAT, Géométrie differentielle et systeme exterieurs, Dunod 1968

H. FLANDERS, Differential forms with applications to the physical sciences, Acad. Press 1962

N. J. HICKS, Notes on differential geometry, van Norstrand 1965

A. LICHNEROWICZ, Théories relativistes de la gravitation et de l'electromagnetism, Masson 1955

A. LICHNEROWICZ, Linear algebra and analysis, Holden-Day 1967

M. SPIVAK, Calculus on manifolds, Benjamin 1965

B. References on or related to relativistic kinetic theory

(The following list is not meant to be complete, but includes only papers which are closely related to the preceding lectures.)

J. L. ANDERSON, in: Relativity; M. Carmeli, S. J. Fickler, L. Witten (Ed.), Plenum Pr. 1970, p. 109

W. L. AMES and K. S. THORNE, Ap. J. 151, 659 (1968)

J. N. BAHCALL, Phys. Rev. 136, B 1164

R. BEREZDIVIN and R. K. SACHS, in: Relativity; M. Carmeli, S. J. FICKLER and L. WITTEN (Ed.), Plenum Press 1970, p. 125

K. BICHTELER, Beiträge zur relativistischen kinetischen Gastheorie,. Dissertation, Hamburg 1965

K. BICHTELER, Z. Physik 182 , 521 (1965)

" , Commun. math. phys. 4 , 352 (1967)

R. H. BOYER, Amer. Journ. of Physics 33 , 910 (1965)

W. BRENIG and R. HAAG, General Quantum Theory of Collision Processes, in: Quantum Scattering Theory, M. Ross (Ed.), Bloomington 1963

N. A. CHERNIKOV, Acta phys. Polon. 23 , 629 (1963)

" " " " " 26 , 1069 (1964)

" " " " " 27 , 465 (1964)

" " Physics Letters 5, no. 2, 115 (1963)

C. ECKART, Phys. Rev. 58 919 (1940)

R. J. EDEN, Lectures on High Energy Physics, University of Maryland Technical Report, 1966

J. EHLERS, Akad. Mainz Abh., Math. Natur. Kl., Jahrg 1961, 791

" " , General Relativity and kinetic theory, in Rend. d. Scuola Internaz. di Fisica "Enrico Fermi", Corso 47 (1969)

J. EHLERS, P. GEREN and R. K. SACHS, J. Math. Phys. 9, 1344 (1968)

J. EHLERS and R. K. SACHS, Kinetic theory and Cosmology, Brandeis Summer Institute in Theoretical Physics, 1968

E. D. FACKERELL, Ph. D. thesis, Univ. of Sidney 1966,

E. D. FACKERELL, Ap. J. 153, 643 (1968)

" " , Proc. Astron. Soc. Australia 1 , 86 (1968)

S. R. de GROOT and P. MAZUR, Non-equilibrium thermodynamics, North-Holland, 1962

R. HAKIM, Phys. Rev., 173 , 1235 (1968)

J. IPSER and K. S. THORNE, Ap.J. 154 , 251 (1968)

W. ISRAEL, Proc. Roy. Soc. 259 , 129 (1960)

" " , Journ. Math. Phys. 4 , 1163 (1963)

L. P. KADANOFF and S. BAYM, Quantum Statistical Mechanics, Benjamin 1962

G. A. KLUITENBERG, S. R. de GROOT and P. MAZUR, Physica 19, 689 (1953)

" " " " , Physica 19, 1079 (1953)

G. A. KLUITENBERG and S. R. de GROOT, Physica 20 , 199 (1954)

" " " , " 21, 148 (1955)

" " " , " 21, 169 (1955)

P. T. LANDSBERG and J. DUNNING-DAVIES, in: Statistical Mechanics of Equilibrium and Non-Equilibrium,

J. MEIXNER (Ed.), North-Holland 1965, p. 36

R. L. LIBOFF, Introduction to the theory of Kinetic Equations, Wiley 1969

A. LICHNEROWICZ and R. MARROT, C. R. Acd. Sc. Paris 210 , 759 (1940)

R. W. LINDQUIST, Ann. of Physics 37, 487 (1966)

J. LOWRY, Ph. D. Thesis, Univ. of Texas at Austin, 1970

C. MARLE, Ann. Inst. Henri Poincare A 10, 67 (1969)

" " " " " A 10, 127 (1969)

" , C. R. Acad. Sc. Paris 263, A, 485 (1966)

C. W. MISNER, Ap. J. 151, 431 (1968)

R. PEIERLS, M. N. 96, 780 (1936)

PHAM MAU QUAN, C. R. Acad. Paris 263, A, 106 (1966)

R. K. SACHS and A. M. WOLFE, Ap. J. 147, 73 (1967)

A. STARUSZKIEWICZ, Nuovo Cimento 45, ser. 10, p. 684 (1966)

J. M. STEWART, Ph.-D. thesis, Cambridge (Engl.), 1969

E. C. G. STUECKELBERG and G. WANDERS, Helv. Phys. Acta 26, 307 (1953)

J. L. SYNGE, Relativity: the special theory, North.- Holland 1956

" " , The relativistic gas, " " 1957

" " , Relativity: the general theory, " " 1960

A. H. TAUB, Phys. Rev. 74, 328 (1948)

G. E. TAUBER and J. W. WEINBERG, Phys. Rev. 122, 1342 (1961)

G. E. UHLENBECK and G. V. FORD, Lectures in Statistical Mechanics, Ann. Math. Soc., Providence 1963

A. G. WALKER, Proc. Edinb. Math. Soc. 4, 238 (1936)

H. WEYL, Math. Ann. 71, 441 (1911)

R. A. MATZNER, Ap. J. 157, 1085 (1969)

CENTRO INTERNAZIONALE MATEMATICO ESTIVO

(C. I. M. E.)

ABSTRACT

MINKOWSKI SPACES AS FIBRE BUNDLES

K. B MARATHE

Corso tenuto a Bressanone dal 7 al 16 Giugno 1970

ABSTRACT

MINKOWSKI SPACES AS FIBRE BUNDLES

K. B. Marathe
University of Rochester

A new definition is proposed for representation spaces of Special Relativistic physical events. These spaces - called here Minkowski Spaces - are given a structure of a fibre bundle over a four dimensional manifold. The Lorentz group appears as the structure group of this fibre bundle. It is shown that there is a unique torsion-free connection in this bundle which induces the usual pseudo-Riemannian structure on the base manifold.

MINKOWSKI SPACES AS FIBRE BUNDLES

K. B. Marathe
University of Rochester

1. Introduction

The study of Minkowski spaces arose early in this century in connection with the special theory of Relativity. The concept of four-dimensional space as a setting for physical phenomena was introduced by Hermann Minkowski. However, we do not find many investigations related to the discussion of the structure of these spaces from the mathematicians viewpoint. The questions of topology and differential structure if regarded asaa four-dimen sional manifold have not received due consideration.

The Lorentz group plays fundamental role in all considerations related to the special Theory of Relativity. It is then natural to ex pect that this group should arise in a characteristic manner in the consideration of the appropriate mathematical structure taken as a setting for the physical phenomena in the domaine of the special Theory of Relativity. However, it is known that regarding a physi-cal event as a point in $\mathbb{R}^4$ or in a suitable four-dimensional manifold does not lead to the Lorentz group (Ref. 1).

To overcome this difficulty alternative topologies have been proposed for Minkowski Space by Zeeman (Ref. 1) and Celnik (Ref. 2). The topology M^F proposed by Zeeman for $\mathbb{R}^4$ has among other properties the property that the Lorentz group is the group of homeomorphisms of M^F. Though this space has some interesting properties it is topologically quite complicated. In particular it is not locally compact and therefore standard methods of calculus cannot be employed in it.

In this paper we propose for the space of physical events (Special Relativistic) the structure of a fibre bundle over a four dimensional manifold M. The new definition is equivalent to saying that a physical event is characterized not just by a point P in the manifold but rather by the point P and a fixed coordinate chart at P. This coordinate chart corresponds to a unique linear frame in the bundle of linear frames L over M. The point P together with this frame give the point in the bundle of linear frames over M, which we shall call a representation of the physical event under consideration. Which points of the bundle L are to be regarded as representing physical events? The answer is provided by the following theorem which is the main -

stay of the proposed definition.

Theorem 1: Let M be an arbitrary four-dimensional manifold, L the bundle of linear frames over M, G the four-dimensional general linear group, H the Lorentz group and σ a cross-section of the associated bundle $E(M, G/H, G, L)$ then

1. There exists a unique (depending on σ) reduced subbundle $P(M,H)$ of L with structure group H,
2. There exists a unique torsion-free linear connection in P which makes M into a pseudo-Riemannian space of signature $---+$,
3. The holonomy group of this connection is a subgroup of H, and there exists a connection whose holonomy group is H.

We call $P(M,H)$ a representation space for physical events or in the standard terminology a Minkowski Space. With this definition two representatives correspond to the same physical event if and only if they belong to the same fibre of P. These two points of P are then related by a Lorentz transformation.

2. Geometrical Preliminaries

By a differentiable four-dimensional manifold M of class

C^r we mean a Hausdorff, connected, locally Euclidean topological space with a fixed four-dimensional C^r atlas. r need not be infinite, but in what follows we assume it to be large enough to ensure the smoothness of the operations involved.

We denote the tangent space at $x \in M$ by $T_x(M)$. A linear frame l_x at $x \in M$ is an ordered basis (X_1, X_2, X_3, X_4) of tangent vectors in the tangent space $T_x(M)$. We define the bundle of linear frames $L(M, G, \pi)$, written as L, as follows. As a set it contains all the linear frames l at all points of M. The group G acts freely on L by the following action. If $g \in G$ and $l_x \in L$ then gl_x is the frame $m_x \in L$ given by

$$Y_i = g_i^m X_m \quad \text{where} \quad l_x = (X_1, X_2, X_3, X_4)$$

$m_x = (Y_1, Y_2, Y_3, Y_4)$, and $g = (g_i^m)$, the summation convention being used.

The action of G defines an equivalence relation on L in a natural manner. Denoting the relation by $\sim$ we have $l \sim m$ if there exists $g \in G$ much that $m = gl$. Clearly $l_x \sim m_y$ if and only if $x = y$. Under this equivalence relation the factor space L/G can be identified with M.

Let $\pi : L \longrightarrow M$ be the natural projection given by

$$\pi(1_x) = x$$

A differentiable structure on L is introduced by making $\pi^{-1}(U)$ diffeomorphic with $U \times G$ where $x \in U \subset M$ and (U, d) is a coordinate chart at x. Let x^k $k = 1, 2, 3, 4$ be coordinate functions. Then $\frac{\partial}{\partial x^k}$ $k = 1, 2, 3, 4$ form a basis for $T_x(M)$. If $1_x = (X_i)$ then with respect to this basis

$$X_i = X_i^k \frac{\partial}{\partial x^k}$$

In L then (x^k, x_i^k) can be taken as local coordinates.

We regard H, the Lorentz group, as a subgroup of G. The bundle E with fibre G/H (considered as a right coset space) associated with L is constructed as follows: Consider the action of G on $L \times G/H$ as a given by

$$a.(l, \xi) = (a.l, \xi.a^{-1}) \in L \times G/H \quad \text{for } a \in G \text{ and } (l, \xi) \in L \times G/H.$$

The quotient space ot $L \times G/H$ under this action of G is denoted by $E(M, G/H, G, L)$, written as E. The map

$$L \times G/H \longrightarrow L \longrightarrow M \text{ induces the map } \pi_E : E \longrightarrow M,$$

and a differential structure is introduced in E in a natural manner by using π_E. (Ref. 3)
The surjective map $(l, \xi) \rightarrow \xi \cdot l$ of $L \times G/H \longrightarrow L/H$ factors through E and allows us to identify E with L/H. Consequently L can be regarded as a fibre bundle over E with structure group H. Let $\tau : L \longrightarrow E = L/H$ be the natural projection.

3. Proof of Theorem 1

The proof is divided into two lemmas.

Lemma 1: There exists a unique reduced subbundle P of L with the Lorentz group H as its structure group.
Proof: Define $P \subset L$ by

$$P = \tau^{-1} (\sigma M) \qquad \text{——— (0)}$$

For $x \in M$, $\sigma(x) \in E = L/H$ and Since τ is onto by definition, $\tau^{-1} [\sigma(x)]$ is non-empty. Therefore, for every $x \in M$ there exists $l \in P \subset L$ such that $\tau(l) = \sigma(x)$. Now if $\tau(l) = \tau(m)$, then l, m are H-related i-e. There exists $h \in H$ such that $m = hl$ and conversely. Thus the set of all H-related frames in a fibre of L form a fibre of P. The verification that P is a principle fibre subbundle of L is now straight forward.

Lemma 2: There exists a unique torsion-free connection in the bundle P which makes M into a pseudo-Riemannian space with fundamental tensor of signature ---+..

Proof: We first define a pseudo-inner product in $\mathbb{R}^4$ equipped with the fixed basis e_1, e_2, e_3, e_4 where $e_1 = (1, 0, 0, 0)$, $e_2 = (0, 1, 0, 0)$, $e_3 = (0, 0, 1, 0)$, $e_4 = (0, 0, 0, 1)$ by

$$\langle x, y\rangle = -x_1y_1 - x_2y_2 - x_3y_3 + x_4y_4 \qquad \text{———} (1)$$

where $x = e_1x_1 + e_2x_2 + e_3x_3 + e_4x_4$

and $y = e_1y_1 + e_2y_2 + e_3y_3 + e_4y_4$.

The product defined by (1) is invariant under H.

i.e. $h \in H$ implies $\langle hx, hy\rangle = \langle x, y\rangle$ ——— (2)

Now regarding each $1_x \in P \subset L$ as a linear isomorphism of $\mathbb{R}^4$ onto $T_x(M)$ we define a bilinear form on $T_x(M)$ by

$$g(X, Y) = \langle 1_x^{-1} X, 1_x^{-1} y\rangle \qquad \text{———} (3)$$

where $X, Y \in T_x(M)$.

Definition of P and (2) show that definition (3) is independent of choice of $1_x \in P$. (1) shows that the form g is symmetric. It then can be identified with a symmetric covariant tensor of order 2. If g_{ij} are the components of g with respect to the coordinate system (x^k) then the unique torsion-free connection Γ in P is defined by the usual relations

$$\Gamma^i_{jk} = \frac{1}{2} g^{ih} \left(\frac{\partial g_{jh}}{\partial x^k} + \frac{\partial g_{hk}}{\partial x^j} - \frac{\partial g_{jk}}{\partial x^h} \right) \quad - (4)$$

where g^{ij} are the components of the dual tensor g^* to g. Γ is the unique torsion-free metric connection for the pseudo-Riemannian structure on M defined by the tensor g. The signature $-,-,-,+$ of g follows from its definition and (1).

The theorem of Hano and Ozeki (Ref. 4) states that the structure group G of L can be reduced to a subgroup H if and only if there exists a connection in L whose holonomy group is H.

This last result and lemma 1 and lemma 2 constitute the proof of theorem 1.

4. Conclusion

In view of the results of theorem 1, it seems reasonable to define P as a representation space for Special Relativistic events. Each event is assigned an element of P. The assignment is not unique as is to be expected.

Two points $(x, 1_x)$ and (y, m_y) in P correspond to the same event if and only if $x = y$ and there exists $h \in H$ the Lorentz group such that $M_x = h\, 1_x$. Thus the Lorentz invariance of Special Relativistic Laws is built into the definition of P.

The result proved here applies to manifolds which are not necessarily flat. It may thus serve as a spring-board for a corresponding definition in the case of General Relativistic events. However, no definite results in this direction have yet been obtained.

Taking $M = \mathbb{R}^4$ we can precisely characterize the distinct representation spaces; they are in 1-1 correspondence with the elements of G/H. This follows from the fact that for $a \in G/H$ the map σ_a defined by

$$\sigma_a : x \longrightarrow (x, a) \qquad \text{of } M \longrightarrow E(M, G/H, G, L) \text{ is}$$

a cross-section and $\sigma_a \neq \sigma_b$ if and only if $a \neq b$ $a, b \in G/H$. The bundle obtained by using σ_a in definition (0) is denoted by P_{σ_a}. This may be interpreted as saying that within the same P the representatives of the same physical event are connected by transformations belonging to the Lorentz group, but an arbitrary non-singular transformation $a \in G$ will take representatives in P_{σ_e}

into $P_{\mathfrak{G}_a}$, e being the identity of G. If we define inertial frame as a cross-section of L then we have a unique bundle P_i whose fibres are inertial frames. For example a frame $e = (e_1, e_2, e_3, e_4)$ satisfying

$$g(e_1, e_1) = -1, g(e_2, e_2) = -1, g(e_3, e_3) = -1$$

$$g(e_4, e_4) = 1 \qquad \text{and} \quad g(e_i, e_j) = 0 \qquad i \neq j$$

may be defined as an inertial frame at each point of $\mathbb{R}^4$

For this last case a different approach has also been proposed by Eberlein (Ref. 5)

ACKNOWLEDGEMENT

My thanks are due to Professors Eberlein (Rochester), Stein (Rochester), Ehlers (Texas) and Lichnerowicz (Collège de France) for useful discussions.

REFERENCES

1. E. C. Zeeman, Topology, Vol. 6, pp. 161-170.

2. F. A. Celnik, Soviet Math. Dokl. Vol. 9 (1968) No. 5, pp. 1151-52.

3. S. Kobayashi, K. Nomizu, Foundations of Differential Geometry, Vol. 1, 1963, Interscience.

4. J. Hans, H. Ozeki, Nogoya. Math. J. 10 (1956), pp. 97-100.

5. W. F. Eberlein, Bull. Am. Math. Soc. Vol. 71;No. 5, pp. 731-736.

CENTRO INTERNAZIONALE MATEMATICO ESTIVO

(C. I. M. E.)

SUR LA PROPAGATION DE LA CHALEUR EN RELATIVITÉ

G. BOILLAT

Corso tenuto a Bressanone dal 7 al 16 Giugno 1970

SUR LA PROPAGATION DE LA CHALEUR EN RELATIVITÉ

G. Boillat

1.- Préliminaires.

Supposons qu'un champ $\underline{u}(x^\alpha)$ vérifie un système d'équations aux dérivées partielles quasi linéaire du premier ordre (ou ramené à cet ordre),

(1) $$A^\alpha(\underline{u})\underline{u}_\alpha = \underline{f}(\underline{u}, x^\beta), \quad \alpha = 0,1,2,3,$$

où les A^α sont des matrices carrées $N \times N$, x^0 une variable de temps, x^i $(i = 1,2,3)$ des variables d'espace. A la traversée de la surface d'onde $\varphi(x^\alpha) = 0$ le gradient du champ est discontinu et on a,

(2) $$[\underline{u}] = 0, \ [\underline{u}_\alpha] = \varphi_\alpha \delta\underline{u}, \ \varphi_\alpha = \partial_\alpha \varphi, \quad A^\alpha \varphi_\alpha \delta\underline{u} = 0,$$

d'où, en introduisant la vitesse normale λ de propagation dans la direction $\vec{n}$,

(3) $$\lambda = -\varphi_0/|\nabla\varphi| \ , \ \vec{n} = \nabla\varphi/|\nabla\varphi| \ ,$$

(4) $$\text{dét}(A_n - \lambda A^0) = 0, \quad A_n = A^i n_i \ .$$

Si A^0 est régulière (dét $A^0 \neq 0$) on peut supposer que $A^0 = I$. On dira alors que le système (1) est hyperbolique (au sens large) si les valeurs propres de la matrice A_n sont réelles (sans être nécessairement distinctes) et s'il existe un système complet de vecteurs propres, c'est-à-dire N vecteurs propres linéairement indépendants (ce qui est toujours vrai pour N valeurs propres distinctes). Dans ces conditions une perturbation initiale arbitraire peut toujours se décomposer en une somme de perturbations ($\delta\underline{u}$) sur la base des vecteurs propres, perturbations qui se propagent ensuite de façons différentes.

Lorsque le système n'est pas hyperbolique, distinguons trois cas.

a) D'après (4) il est clair que la somme S_k des produits k à k des racines du polynome caractéristique est proportionnelle à l'inverse du déterminant de A^0. En particulier,

$$|S_1| = |\Sigma \lambda| = |\text{coef.}|\,|\text{dét } A^o|^{-1} ,$$

$$|S_N| = |\Pi \lambda| = |\text{dét } A_n|\,|\text{dét } A^o|^{-1} .$$

Il en résulte que si, dans certaines conditions, dét $A^o \longrightarrow 0$, l'une au moins des sommes S_k tendra vers l'infini et il en sera par conséquent de même de l'une au moins des vitesses de propagation, ce qui n'est pas acceptable du point de vue physique. Ainsi on parle d'action instantanée à distance quand on considère l'équation de Laplace $\Delta u = 0$ comme limite de celle de d'Alembert $c^{-2}u_{tt} - \Delta u = 0$ lorsque $c \to \infty$.

b) Si une valeur propre est complexe il apparaît un manque de stabilité par rapport aux données initiales ([1]).

c) Si le nombre de vecteurs propres est inférieur à N, il est possible de produire une perturbation $\delta \underline{u}$ telle qu'un choc (discontinuité du champ lui-même : $[\underline{u}] \neq 0$) se produise immédiatement ([2]).

Nous verrons diverses circonstances se manifester suivant le choix de l'équation de la chaleur. Bien entendu, nous chercherons à obtenir un système hyperbolique.

2.- Tenseur d'impulsion-énergie.

Pour un fluide parfait quand on néglige la conduction de la chaleur ce tenseur s'écrit ([3]),

$$\tau^{\alpha\beta} = (\rho + p)u^\alpha u^\beta - pg^{\alpha\beta} = rfu^\alpha u^\beta - pg^{\alpha\beta} , \tag{5}$$

où r est la densité propre de matière, f l'index du fluide,

$$f = 1 + i, \tag{6}$$

i l'enthalpie spécifique. La température Θ et l'entropie S sont introduites par l'équation différentielle,

$$df = Vdp + \Theta dS, \qquad V = 1/r. \tag{7}$$

Nous écrivons l'équation de conservation de la matière,

$$\nabla_\alpha (ru^\alpha) = 0 \tag{8}$$

et celle du tenseur d'impulsion-énergie,

$$\nabla_\alpha T^{\alpha\beta} = 0, \qquad T^{\alpha\beta} = \tau^{\alpha\beta} - (q^\alpha u^\beta + q^\beta u^\alpha), \tag{9}$$

où q^α est le vecteur (transverse) courant de chaleur ([4]),

(10) $q_\alpha u^\alpha = 0, \quad (u_\alpha u^\alpha = 1).$

Compte tenu de (7),(8), les équations (9) s'écrivent,

(11) $(rfu^\alpha - q^\alpha)\nabla_\alpha u^\beta - \gamma^{\alpha\beta}\{\partial_\alpha p + ru^\gamma\nabla_\gamma(q_\alpha/r)\} = 0,$

(12) $r\theta u^\alpha \partial_\alpha S = \nabla_\alpha q^\alpha + u_\beta u^\alpha \nabla_\alpha q^\beta = \nabla_\alpha q^\alpha - q_\beta u^\alpha \nabla_\alpha u^\beta ,$

$\gamma^{\alpha\beta} = g^{\alpha\beta} - u^\alpha u^\beta .$

On ajoutera éventuellement les équations d'Einstein,

$$S^{\alpha\beta} = \chi T^{\alpha\beta}$$

et on supposera que l'on se donne les équations d'état, par exemple,

$$p = p(\theta,r), \qquad S = S(\theta,r).$$

Il reste encore à définir le courant de chaleur et nous étudierons plusieurs modèles dans les paragraphes qui suivent mais dès à présent nous pouvons obtenir des équations aux perturbations en faisant dans (8),(11),(12), le remplacement,

$$\partial_\alpha , \nabla_\alpha \longrightarrow \varphi_\alpha \delta .$$

Pour abréger nous poserons,

$$U = u^\alpha \varphi_\alpha , \; Q = q^\alpha \varphi_\alpha , \; C = \gamma^{\alpha\beta}\varphi_\alpha\varphi_\beta .$$

Alors,

(13) $r\, \varphi_\alpha \delta u^\alpha + U\, \delta r = 0,$

(14) $(rfU - Q)\delta u^\beta - \gamma^{\alpha\beta}\varphi_\alpha \delta p + Uq^\beta\, \delta r/r - U\gamma^{\alpha\beta}\delta q_\alpha = 0,$

(15) $r\theta U \delta S = \varphi_\alpha \delta q^\alpha + U u_\beta \delta q^\beta .$

Nous introduirons encore en un point le repère propre cartésien du fluide,

(16) $g_{\alpha\beta} = \mathrm{diag}(1, -1, -1, -1), \quad u^0 = 1, \; u^i = 0.$

Dans ce repère,

(17) $\gamma^{0\alpha} = 0, \quad \gamma^{ij} = -\delta^{ij} , \quad q^0 = 0$

et d'après (3) nous aurons la correspondance,

(18) $U \longrightarrow -\lambda , \; Q \longrightarrow \vec{q}.\vec{n} = q_n , \; C \longrightarrow -1.$

3.- Equation de Fourier.

On peut transposer directement l'équation classique de Fourier en relativité,

(19) $$q^{\alpha} = - K\gamma^{\alpha\rho}\partial_{\rho}\theta.$$

Cependant on voit immédiatement (Eq. 17) que dans le repère propre les équations (19) ne contiennent aucune dérivée par rapport au temps. On se trouve donc dans le cas (a) du §1. Ecrivons alors avec Eckart ([5]),

(20) $$q^{\alpha} = - K\gamma^{\alpha\rho}(\partial_{\rho}\theta - \theta u^{\gamma}\nabla_{\gamma}u_{\rho}) = - K\gamma^{\alpha\rho}\nabla_{\lambda}(\theta\gamma^{\lambda}_{\rho}).$$

Nous évitons ainsi l'inconvénient précédent et nous avons,

(21) $$\theta U\delta u^{\alpha} = \gamma^{\alpha\rho}\varphi_{\rho}\delta\theta.$$

Contractant cette équation avec φ_{α} on obtient en utilisant (13),

(22) $$U^2(\delta r/r) + C(\delta\theta/\theta) = 0.$$

Il nous suffit ici de calculer les vitesses de propagation dans une direction orthogonale au vecteur $\vec{q}$. Nous ferons donc $Q = 0$ et en contractant (14) avec φ_{ρ}, en supposant $U \neq 0$, en se servant de (13),(15) et en remarquant que (10),(21) entraînent,

$$u_{\alpha}\delta q^{\alpha} = - q_{\alpha}\delta u^{\alpha} = 0,$$

nous aurons.

$$fU^2\delta r + C\delta p + r\theta U^2\delta S = 0.$$

Avec (22), on en déduira,

$$\theta^2 S_{\theta}U^4 + U^2C(\theta V p_{\theta} - r\theta S_r - f) - C^2 p_r = 0.$$

Sous les hypothèses usuelles ([6]),

$$S_{\theta} > 0, \quad p_r > 0,$$

le produit des racines de cette équation bicarrée est négatif; il existera des racines imaginaires (§1,b).

Cependant il est intéressant de calculer ces racines. Avec les équations des gaz polytropiques(avec $p_o \neq 0$ et $G = 0$; voir § 13) on trouve

$$\lambda^4 - (\gamma - 2)\lambda^2 - (\gamma - 1) = 0.$$

En dehors de la solution imaginaire $\lambda^2 = -1$, on a,

$$\lambda^2 = \gamma - 1,$$

une valeur que nous retrouverons plus loin (§11).

4.- Equation de la chaleur.

En 1948, C. Cattaneo ([7]) proposa un nouveau modèle d'équation de la chaleur qui fut redécouvert dix ans plus tard par P. Vernotte ([8]) et mis récemment sous forme covariante par M. Kranyš ([9]). C'est ce modèle que nous étudierons après y avoir fait quelques changements.

Nous chercherons une modification simple des équations (19),(20) sans choisir a priori l'une ou l'autre,

$$q^{\alpha} = -\gamma^{\alpha\rho}(\bar{\nu}\,\partial_{\rho}\theta + \bar{\mu}\,u^{\gamma}\nabla_{\gamma}u_{\rho}). \tag{23}$$

(C'est l'expression classique dans le repère propre s'il est repère d'inertie.)

Ajoutons un terme T^{α} au premier membre. Comme il n'y a aucune raison pour que ce terme soit orthogonal à u^{α} nous devons récrire (23),

$$Q^{\alpha} + T^{\alpha} = \text{second membre}, \tag{24}$$

avec

$$Q^{\alpha} = \bar{q}u^{\alpha} + q^{\alpha}\,. \tag{25}$$

Nous avons introduit une nouvelle variable de champ $\bar{q}$. Pour que le champ soit hyperbolique il est nécessaire que sa dérivée par rapport au temps figure dans les équations du champ, c'est-à-dire dans T^{α} puisqu'elle n'est pas ailleurs. On voit tout de suite que l'on n'obtient rien d'intéressant en prenant T^{α} proportionnel à $g^{\alpha\rho}\partial_{\rho}\bar{q}$ ou $u^{\gamma}\nabla_{\gamma}(\bar{q}u^{\alpha})$. On est ainsi conduit à faire apparaître la dérivée temporelle de Q^{α} c'est-à-dire à prendre,

$$T^{\alpha} = \chi u^{\gamma}\nabla_{\gamma}Q^{\alpha} = u^{\gamma}\nabla_{\gamma}(\chi Q^{\alpha})$$

(si χ = cte). Nous écrirons ([10])([11])

$$Q^{\alpha} + u^{\rho}\nabla_{\rho}(\bar{h}Q^{\alpha}) = -\gamma^{\alpha\rho}(\bar{\nu}\,\partial_{\rho}\theta + \bar{\mu}\,u^{\gamma}\nabla_{\gamma}u_{\rho}). \tag{26}$$

On trouvera cette équation dans ([9]) à cela près que nous ne supposons pas, dorénavant, $\bar{h}$ et $\bar{\nu}$ constants.

Nous définirons,

$$h = \log\bar{h},\quad \bar{\mu}/\bar{h} = \mu,\quad \bar{\nu}/\bar{h} = \nu, \tag{27}$$

$$\eta = rf + \mu + \bar{q} = \varrho + p + \mu + \bar{q},\quad A = \eta U - Q,$$

et nous adjoindrons à (26) l'équation d'état,

(28) $F(\Theta, r, \bar{q}, h) = 0.$

5.- Perturbations.

De (26) on déduit,

(29) $- U\delta q^{\alpha} = Uu^{\alpha}(\bar{q}\,\delta h + \delta\bar{q}) + Uq^{\alpha}\delta h + \nu\gamma^{\alpha\beta}\varphi_{\beta}\delta\Theta + U(\bar{q} + \mu)\delta u^{\alpha}$,

que l'on porte dans (14) pour obtenir,

(30) $A\,\delta u^{\beta} - \gamma^{\alpha\beta}\varphi_{\alpha}(\delta p - \nu\delta\Theta) + Uq^{\beta}(\delta h + \delta r/r) = 0.$

Nous y ajoutons (13),(15) ainsi que les équations qui dérivent des équations d'état,

(31) $r\varphi_{\alpha}\delta u^{\alpha} + U\delta r = 0,$

(32) $r\Theta U\,\delta S = (\varphi_{\alpha} + Uu_{\alpha})\delta q^{\alpha}$,

(33) $u_{\alpha}\delta u^{\alpha} = 0, \quad q_{\alpha}\delta u^{\alpha} + u_{\alpha}\delta q^{\alpha} = 0,$

(34) $\delta p = p_{\Theta}\,\delta\Theta + p_{r}\,\delta r, \quad \delta S = S_{\Theta}\,\delta\Theta + S_{r}\,\delta r,$

(35) $F_{\Theta}\,\delta\Theta + F_{r}\,\delta r + F_{\bar{q}}\,\delta\bar{q} + F_{h}\,\delta h = 0.$

6.- La surface d'onde U = 0.

Cette surface se déplace à la vitesse du fluide. Il vient immédiatement,

$$\delta\Theta = \delta r = 0$$

et (si $Q \neq 0$),

$$\delta u^{\beta} = 0, \quad \varphi_{\alpha}\delta q^{\alpha} = u_{\alpha}\delta q^{\alpha} = 0, \quad F_{\bar{q}}\,\delta\bar{q} + F_{h}\,\delta h = 0.$$

Il en résulte qu'il correspond trois vecteurs propres à cette vitesse; δq^{α} a deux degrés de liberté et $\delta\bar{q}$ (ou δh si F_h était nul) peut être choisi arbitrairement.

7.- Déterminant caractéristique.

Nous supposons $U \neq 0$. Contractons (30) avec φ_{β} et utilisons (31),

(36) $U(A - Q)\delta r/r + C(\delta p - \nu\delta\Theta) - UQ\,\delta h = 0.$

(32) et (29) donnent ensuite,

(37) $r\Theta U^2\delta S + 2U^2(\bar{q}\,\delta h + \delta\bar{q}) + UQ\,\delta h + \nu C\delta\Theta - U^2(\mu + \bar{q})(\delta r/r) = 0.$

Enfin, multipliant (30) par q_{β} et tenant compte — d'après (33) et (29) — de,

(38) $q_{\alpha}\delta u^{\alpha} = - u_{\alpha}\delta q^{\alpha} = \bar{q}\,\delta h + \delta\bar{q} \qquad (U \neq 0),$

on obtient,

$$(39)\qquad A(\bar{q}\,\delta h + \delta\bar{q}) - Q(\delta p - \nu\delta\theta) - Uq^2(\delta h + \delta r/r) = 0,$$

$$q^2 = - q_\alpha q^\alpha .$$

Nous avons maintenant avec (35) un système linéaire et homogène de quatre équations pour les quatre inconnues $\delta W = (\delta\theta, \delta r/r, \delta\bar{q}, \delta h)$. Nous écrivons que le déterminant est nul,

$$(40)\qquad \begin{vmatrix} F_\theta & rF_r & F_{\bar{q}} & F_h \\ (p_\theta - \nu)C & rp_rC + U(A - Q) & 0 & - UQ \\ (p_\theta - \nu)Q & rp_rQ + q^2U & - A & q^2U - \bar{q}A \\ \nu C + r\theta S_\theta U^2 & (r^2\theta S_r - \mu - \bar{q})U^2 & 2U^2 & (2\bar{q}U + Q)U \end{vmatrix} = 0.$$

8.- Hyperbolicité. Existence de la surface d'onde $A = 0$.

Il est nécessaire d'examiner d'un peu près cette question afin de corriger une erreur que nous avons faite ([10]) et qui est d'ailleurs sans conséquence pour la suite sinon que les conditions qui seront imposées à F (28) ne le seront plus par un motif d'hyperbolicité.

Le système des équations du champ comporte 9 variables indépendantes: r, θ, u^α(3 composantes seulement sont indépendantes car $u_\alpha u^\alpha = 1$), Q^α. Nous devons donc trouver $N = 9$ vecteurs propres pour que le système soit hyperbolique. On pressent tout de suite que (40) est un polynôme de degré 5 (au plus) en λ : les deuxième et quatrième lignes du déterminant sont de degré 2, la deuxième de degré 1. On vérifie que le degré est bien égal à 5 (nous ferons le calcul plus loin).

1°) Si $A = 0$ ne vérifie pas (40), à toute solution $\delta W \neq 0$ correspondra un vecteur δu^α donné par (30) et un vecteur δq^α donné par (29). Nous aurons ainsi (certainement si toutes les racines sont distinctes et non nulles — différentes de $U = 0$) 5 vecteurs propres.

D'autre part si λ n'est pas une racine de (40) $\delta W = 0$ et $\delta u^\alpha = \delta q^\alpha = 0$ à moins que $A = 0$. Dans ce cas (cf. 29,31,33,38),

$$(41)\qquad - \delta q^\alpha = (\bar{q} + \mu)\delta u^\alpha , \quad u_\alpha \delta u^\alpha = \varphi_\alpha \delta u^\alpha = q_\alpha \delta u^\alpha = 0.$$

Une des composantes de δu^α est arbitraire. A la solution A = 0 correspond un vecteur propre. Le nombre total des vecteurs propres est donc

$$3(U = 0) + 5(\text{sol. de } 40) + 1(A = 0) = 9.$$

Le système est hyperbolique. (En supposant, bien entendu, que les valeurs propres sont réelles, ce qui reste à établir.)

2°) Supposons maintenant que (40) soit le produit par A d'un polynome P_4 du quatrième degré. Aux racines de ce polynome correspondront 4 vecteurs propres de la façon qui vient d'être dite. Pour A = 0 on pourra choisir arbitrairement une des composantes de δW, les autres seront déterminées de façon à satisfaire les équations (30),(35),(37) (nous verrons que cela est possible). Le vecteur δu^α sera seulement soumis aux trois conditions (31),(33),(38) :

$$(42) \qquad u_\alpha \delta u^\alpha = 0, \quad \varphi_\alpha \delta u^\alpha = - U\delta r/r, \quad q_\alpha \delta u^\alpha = \bar{q}\,\delta h + \delta \bar{q} \,;$$

une de ses composantes pourra encore être prise arbitrairement. Ce qui revient à dire qu'à la solution A = 0 correspondront 2 vecteurs propres dont le nombre total sera :

$$3(U = 0) + 4(P_4 = 0) + 2(A = 0) = 9.$$

(On a supposé, cela va sans dire, que P_4 ne s'annule ni avec U ni avec A.) Le système sera encore hyperbolique.

9.- Conditions d'existence de la solution double A = 0.

Nous allons montrer que par un choix convenable de F il est possible d'obtenir une solution double A = 0 ce qui permettra comme on l'a vu ci-dessus de simplifier le déterminant caractéristique (40).

On tire de (30) avec A = 0,

$$(43) \qquad \delta p = \nu \delta \theta, \quad \delta h + \delta r/r = 0,$$

soit, avec (35),

$$(44) \qquad (p_\theta - \nu)\delta\theta + r p_r \delta r/r = 0,$$

$$(45) \qquad F_\theta \delta\theta + (rF_r - F_h)(\delta r/r) + F_{\bar{q}}\,\delta\bar{q} = 0.$$

Supposons $F_{\bar{q}} \neq 0$. On peut alors (comme $p_r \neq 0$) à l'aide des équations ci-dessus exprimer δh, δr, $\delta\bar{q}$ proportionnellement à $\delta\theta$. Le premier membre de (37) devra être identiquement nul si l'on y porte ces expressions, compte tenu de l'égalité $Q = \gamma U$, ce qui est impossible en raison de la présence du terme en νC. On doit donc avoir

(46) $\qquad F_{\bar{q}} = 0.$

Alors (37) donne $\delta\bar{q}$ et (44),(45) l'équation aux dérivées partielles que doit satisfaire $F(\theta,r,h) = 0$,

(47) $\qquad a(F) = 0, \qquad a(F) = rp_r F_\theta + (\nu - p_\theta)(rF_r - F_h).$

Il est commode de prendre F sous la forme suivante résolue par rapport à h,

(48) $\qquad F = - re^h + \phi(\theta,r), \qquad \bar{h} = e^h = \phi/r.$

(47) devient,

(49) $\qquad (\nu - p_\theta)\phi_r + p_r\phi_\theta = 0,$

c'est-à-dire, si l'on écarte la solution ϕ = cte (qui n'est d'ailleurs pas l'hypothèse qui conduit aux résultats les plus simples; voir §11),

(50) $\qquad \left(\frac{\partial p}{\partial \theta}\right)_\phi = \nu .$

10.- <u>L'onde exceptionnelle A = 0.</u>

A = 0 c'est $\gamma U - Q = 0$ soit,dans le repère propre (cf. 18),

(51) $\qquad \lambda = - q_n/\gamma .$

En écrivant que $\lambda^2 \leqslant 1$, ou ce qui revient au même ici que la vitesse radiale $\gamma u^\alpha - q^\alpha$ est du genre temps (ou isotrope) on a l'inégalité,

(52) $\qquad q^2 \leqslant \gamma^2 .$

Cherchons maintenant s'il est possible que l'onde soit <u>exceptionnelle</u> [12][13] c'est-à-dire ne produise pas de chocs sur le front d'onde. Pour qu'il en soit ainsi on devra avoir [14],

$$\varphi_\alpha \delta(\gamma u^\alpha - q^\alpha) = 0,$$

soit,

$$U\delta\gamma + \gamma\varphi_\alpha\delta u^\alpha - \varphi_\alpha\delta q^\alpha = 0.$$

De (32),(42) on tire,

$$\varphi_\alpha\delta q^\alpha = U(r\theta\,\delta S - u_\alpha\,\delta q^\alpha) = U(r\theta\,\delta S + \bar{q}\,\delta h + \delta\bar{q}),$$

de sorte que l'équation ci-dessus s'écrit avec (31),

$$\bar{\delta}(rf + \mu + \bar{q}) - \gamma \bar{\delta} r/r - (r\theta \delta S + \bar{q} \delta h + \delta \bar{q}) = 0.$$

Mais, d'après (7),

$$r \delta f - r\theta \delta S = \delta p,$$

si bien que,

$$\delta p + \delta\mu - \mu \delta r/r = 0,$$

ou, puisque $\mu = \bar{\mu} e^{-h}$ et compte tenu de (43),

$$\bar{h} \delta p + \delta \bar{\mu} = 0, \quad \bar{\nu} \delta \theta + \delta \bar{\mu} = 0,$$

c'est-à-dire, en vertu de,

(53) $\delta \phi = 0$

— qui dérive de (48) et (43) —

(54) $\left(\frac{\partial \bar{\mu}}{\partial p}\right)_\phi = - \bar{h}, \quad \left(\frac{\partial \bar{\mu}}{\partial \theta}\right)_\phi = - \bar{\nu}$.

Ces équations (qui se déduisent l'une de l'autre avec (50)) ne sont pas satisfaites par $\bar{\mu} = 0$. Il en résulte que pour un tel choix l'onde A = 0 produira des chocs. (Les discontinuités des dérivées premières de variables de champ, discontinuités qui se propagent à la vitesse radiale $\gamma u^\alpha - q^\alpha$, deviendront infinies au bout d'un temps fini si l'onde est initialement accélérée ([13]).)

En revanche, supposons

(55) $\bar{\nu} = K(\phi)$,

une certaine fonction de ϕ, alors (54) donne,

$$\bar{\mu} = -K(\phi)\theta + \text{fonct}(\phi).$$

Ainsi l'onde A = 0 ne produira pas de chocs (semblable en cela aux ondes d'Alfvén de la magnétohydrodynamique). On remarque que si l'on prend simplement,

(56) $\bar{\mu} = - K(\phi)\theta$,

le second membre de l'équation de la chaleur est celui d'Eckart. Nous récrirons donc (26),

(57) $Q^\alpha + u^\rho \nabla_\rho (\phi Q^\alpha / r) = - K(\phi) \gamma^{\alpha\rho} \nabla_\lambda (\theta \gamma^\lambda_\rho)$,

avec la condition qui découle de (50),(55),

$$\frac{\phi}{r} = K(\phi)\left(\frac{\partial \theta}{\partial p}\right)_\phi \quad . \tag{58}$$

11.- Expression explicite de l'équation caractéristique.

Revenons au déterminant (40) et cherchons l'expression de l'équation du quatrième degré après mise en facteur de la solution A = 0.

Avant de considérer le cas général notons une approximation particulièrement simple ([11]). Elle consiste à prendre $\bar{h}$ $(= \phi/r) = \chi =$ cte. On obtient alors facilement, pour un gaz polytropique $(pV = p/r = R\theta)$,

$$Q^\alpha + \chi u^\rho \nabla_\rho Q^\alpha = - \chi R r \gamma^{\alpha\rho} \nabla_\lambda (\theta \gamma^\lambda_\rho), \tag{59}$$

et dans le repère propre (outre les vitesses $\lambda = 0$, $\lambda = - q_n/\pi$),

$$\lambda = - q_n/\pi \pm \left\{ (q_n/\pi)^2 + p/\pi \right\}^{\frac{1}{2}} \quad , \pi = \rho + \bar{q} \ , \tag{60}$$

$$\lambda = \pm (\gamma - 1)^{\frac{1}{2}} \quad , \tag{61}$$

où γ est la constante des gaz polytropiques, égale au rapport des chaleurs spécifiques : $\gamma = C_p/C_V$. On remarque que (61) atteint la valeur limite pour $\gamma = 2$ (voir §14).

Une combinaison linéaire de colonnes (indiquée par les têtes de colonnes) puis l'addition de la deuxième ligne à la dernière mettent (40) sous la forme,

$$\begin{vmatrix} F_\theta & a(F) + \bar{q}(\nu - p_\theta)F_{\bar{q}} & F_{\bar{q}} & F_h - \bar{q}F_{\bar{q}} \\ (p_\theta - \nu)C & (\nu - p_\theta)UA & 0 & - UQ \\ (p_\theta - \nu)Q & 0 & -A & q^2 U \\ r\theta S_\theta U^2 + p_\theta C & (\nu - p_\theta)B & 2U^2 & 0 \end{vmatrix} = 0, \tag{62}$$

avec,

$$B = r\left(\frac{\partial \rho}{\partial r}\right)_\phi U^2 - 2UQ + r\left(\frac{\partial p}{\partial r}\right)_\phi C. \tag{63}$$

En réalité, on n'obtient pas directement cette expression mais plutôt (à la place du terme $(\nu - p_\theta)B$),

$$rU^2 \left\{ r\theta p_r S_\theta + (\nu - p_\theta)(r\theta S_r + f)\right\} - 2(\nu - p_\theta)UQ + r p_r \nu C.$$

On met $\nu - p_\theta$ (supposé maintenant non nul) en facteur et on tient compte de (49),

$$\nu - p_\theta = - p_r \phi_\theta / \phi_r \, .$$

Le coefficient de $r(\nu - p_\theta)U^2$ devient,

$$r\theta(S_r \phi_\theta - S_\theta \phi_r)/\phi_\theta + f = r\theta(\partial S/\partial r)_\phi + f = (\partial \varrho/\partial r)_\phi \, ,$$

en raison de la relation,

(64) $$d\varrho = f dr + r\theta dS,$$

qui découle immédiatement de (7) (et $rf = \varrho + p$).

Pour ce qui est du coefficient de $r(\nu - p_\theta)C$, il s'écrit,

$$- \nu \phi_r / \phi_\theta = \nu \left(\frac{\partial \theta}{\partial r}\right)_\phi = \left(\frac{\partial p}{\partial \theta}\right)_\phi \left(\frac{\partial \theta}{\partial r}\right)_\phi = \left(\frac{\partial p}{\partial r}\right)_\phi \, ,$$

en utilisant (50). On a ainsi (63).

En vertu de (46),(47), le déterminant se développe facilement et il reste après mise en facteur de A et un calcul de coefficients analogue à celui que l'on vient de faire,

(65) $$P(X,Y) = a_2 X^2 + 2b_2 XY + c_2 Y^2 + a_1 X + b_1 Y + a_0 = 0,$$

$$X = - U^2/C, \quad Y = QU/C,$$

$$a_2 = 2q^2/\phi + \mathcal{H}\left(\frac{\partial \varrho}{\partial \phi}\right)_r , \quad 2b_2 = \left(\frac{\partial \varrho}{\partial \phi}\right)_{r\phi} - 2\left(\frac{\partial p}{\partial \phi}\right)_\theta , \quad c_2 = - 2/\phi,$$

$$a_1 = r\left(\frac{\partial \varrho}{\partial r}\right)_\phi \left(\frac{\partial p}{\partial \phi}\right)_\theta - \mathcal{H}\left(\frac{\partial p}{\partial \phi}\right)_r , \quad b_1 = 2\left(\frac{\partial p}{\partial \phi}\right)_\theta - \left(\frac{\partial p}{\partial \phi}\right)_{r\phi} ,$$

$$a_0 = - r\left(\frac{\partial p}{\partial r}\right)_\phi \left(\frac{\partial p}{\partial \phi}\right)_\theta \, .$$

Dans le repère propre (cf. 18),

$$X = \lambda^2, \quad Y = \lambda q_n \, .$$

On doit avoir,

(66) $$0 < X \leqslant 1, \qquad 0 \leqslant Y^2/X \leqslant q^2 \, .$$

Non seulement la surface d'onde doit être du genre espace mais la vitesse radiale doit encore être du genre temps. Ceci implique,

$$(\partial P/\partial \varphi_\alpha)(\partial P/\partial \varphi^\alpha) \geqslant 0;$$

$$C(\partial P/\partial \varphi_\alpha) = (-2UP_X + QP_Y)u^\alpha + UP_Y q^\alpha - 2(XP_X + YP_Y)(\varphi^\alpha - Uu^\alpha),$$

soit,

(67) $$4(X-1)(XP_X + YP_Y)P_X + (q^2 X - Y^2/X)P_Y^2 \leqslant 0,$$

pour toute solution de (65).

12.- Fluide incompressible.

Lorsque la conduction de la chaleur est négligée on sait que la vitesse du son dans le repère propre est égale à la racine carrée de $(\partial p/\partial \rho)_S$. On est donc amené ([15]) à définir le fluide incompressible comme correspondant au cas limite,

(68) $$\left(\frac{\partial p}{\partial \rho}\right)_S = 1 \implies \rho - p = \psi(S).$$

Ici nous procéderons de manière analogue. Nous chercherons une équation d'état reliant la pression p à deux variables thermodynamiques. Toutefois la vitesse normale $\lambda(q_n)$ dépend maintenant de la direction de propagation $\vec{n}$ par l'intermédiaire de la quantité q_n. Nous supposerons donc que la vitesse limite est atteinte à la frontière de l'intervalle de variation de q_n : $\lambda^2(\pm q) = 1$. Nous écrirons, par conséquent,

$$P(1, \varepsilon q) = 0, \; \varepsilon^2 = 1, \quad q = (q^2)^{\frac{1}{2}},$$

soit,

$$a_2 + a_1 + a_o + \varepsilon q(2b_2 + b_1) + c_2 q^2 = 0,$$

c'est-à-dire, en posant,

(69) $$\rho - p = \psi,$$

(70) $$\gamma\left(\frac{\partial \psi}{\partial \phi}\right)_r + r\left(\frac{\partial p}{\partial \phi}\right)_\theta \left(\frac{\partial \psi}{\partial r}\right)_\phi + \varepsilon q\left(\frac{\partial \psi}{\partial \phi}\right)_{r\phi} = 0.$$

L'équation d'état ne doit pas dépendre, bien entendu, de q ni de $\bar{q}$ qui figure dans γ. Il faut donc,

(71) $$\left(\frac{\partial \psi}{\partial r}\right)_\phi = 0 \implies \psi = \psi(\phi), \qquad (\gamma + \varepsilon q)\psi' = 0.$$

Afin de satisfaire la dernière équation on peut faire deux hypothèses:

a) ou ψ = cte,

b) ou bien supposer que la vitesse de la lumière est atteinte quand l'onde A = 0 atteint aussi cette vitesse ce qui a lieu quand $q^2 = \gamma^2$ (cf. 52). Dans ces conditions $\gamma + \varepsilon q = 0$ est vérifiée pour l'une des deux valeurs de $\mathcal{E}$ et il suffit de supposer $\psi = \psi(\phi)$.

Venons-en à une autre possibilité : $\lambda^2(q_n) = 1$, $q_n \in \left]-q, +q\right[$. Puisque cette valeur de λ doit être extrémale : $d\lambda/dq_n = 0$. En éliminant Y entre $P(1,Y) = 0$ et $P_Y(1,Y) = 0$, on trouve,

$$4c_2(a_2 + a_1 + a_o) = (2b_2 + b_1)^2,$$

qui ne fournit aucune équation d'état et renforce notre première façon de voir.

13.- Fluides polytropiques.

La comparaison de (68) et (71) nous invite à étudier le cas où ϕ ne dépendrait que de l'entropie,

(72) $$\phi = \phi(S).$$

En utilisant l'identité thermodynamique,

(73) $$(\partial V/\partial S)_p = (\partial \theta/\partial p)_S$$

nous pouvons récrire (58),

$$(1/V)(\partial V/\partial S)_p = \phi(S)/K(S),$$

soit,

(74) $$V = P'(p)e^{F(S)} \quad , \ F(S) = \int \frac{\phi}{K}\, dS.$$

Avec (73) et (7), on obtient,

(75) $$\theta = \frac{\phi}{K} Pe^F + G'(S), \qquad f = 1 + i = Pe^F + G,$$

où P et G sont des fonctions de p et S respectivement et où la dérivation est notée par un accent. D'après (27),(48),(56),

(76) $$\gamma = r(f - \frac{K}{\phi}\theta) + \bar{q} = r(G - \frac{K}{\phi}G') + \bar{q}.$$

En raison de l'arbitraire sur P on en déduit que

(77) $$\gamma = \bar{q} \Longleftrightarrow G = 0. \quad q^2 \leqslant \gamma^2 \Longrightarrow Q_\alpha Q^\alpha \geqslant 0.$$

Alors,

(78) $$\rho + p = f/V = P/P' \Longrightarrow \rho = \rho(p); \quad f = \frac{K}{\phi}\theta .$$

Si nous suposons $G' = 0$ (G = cte) nous avons, pour les chaleurs spécifiques à pression et volume constants,

(79) $$\frac{1}{C_p} = \frac{1}{\Theta}\left(\frac{\partial\Theta}{\partial S}\right)_p = \frac{K}{\phi}\left(\frac{\phi}{K}\right)' + \frac{\phi}{K} ,$$

(80) $$\frac{1}{C_V} = \frac{1}{C_p} - \frac{\phi}{K}\frac{P'^2}{PP''} .$$

Par intégration de (79),

$$\int\frac{dS}{C_p} = \mathrm{Log}\ \frac{\phi}{K} + F(S)$$

et le coefficient de la dérivée au fil du courant du vecteur Q^α dans l'équation de la chaleur (57) s'écrit encore,

(81) $$\frac{\phi}{r} = K(S)P'(p)\exp\int\frac{dS}{C_p} .$$

Si nous supposons de plus que C_V n'est fonction que de S,

(82) $$-\frac{P'^2}{PP''} = \mathrm{cte} = \gamma - 1 \implies P = a(p + p_o)^{(\gamma-1)/\gamma} ,$$

(83) $$\frac{1}{C_V} - \frac{1}{C_p} = (\gamma - 1)\frac{\phi}{K} .$$

En particulier, on obtient les équations des fluides polytropiques en prenant,

(84) $$\frac{\phi(S)}{K(S)} = \mathrm{cte} = \frac{1}{C_p} , \qquad \frac{C_p}{C_V} = \gamma .$$

Pour les gaz, $1 < \gamma < 2$ et si l'on prend,

(85) $$p_o = 0, \quad G = 1,$$

on aura,

$$i = C_p\Theta, \quad C_p = \gamma R/(\gamma - 1), \quad pV = R\Theta.$$

14.- <u>Le fluide incompressible comme fluide polytropique</u>.

$$rf = \rho + p = f/V.$$

Dans l'hypothèse la moins restrictive (b,§12),

$$\rho - p = \frac{P}{P'} - 2p + \frac{G}{P'}e^{-F} = \psi(S),$$

ce qui entraîne,

(86) $$P = a(p + p_0)^{\frac{1}{2}}, \quad G = 0.$$

Ainsi le fluide incompressible correspond à $\gamma = 2$. Comme ([16]),

(87) $$\rho - p = 2p_0 = \text{cte},$$

$$\rho - 3p > 0 \Longrightarrow p < p_0.$$

On se retrouve donc dans le cas (a) du §12; la vitesse de la lumière est atteinte dans une direction parallèle au vecteur courant de chaleur.

15.- Vitesses de propagation.

Avec (84), (65) donne, dans une direction orthogonale au vecteur $\vec{q}$,

(88) $$(2q^2 c_p \phi'/r\theta\phi + \eta)\lambda^4 - \{rf + (\gamma - 1)\eta\}\lambda^2 + \gamma(p + p_0) = 0.$$

Plaçons-nous dans le cas des gaz ($p_0 = 0$, $G = 1$) et supposons que l'on puisse négliger le terme en ϕ' (en admettant en particulier que ϕ varie peu). Alors,

(89) $$\lambda^4 - (\gamma - 1 + \frac{rf}{\eta})\lambda^2 + \frac{\gamma p}{\eta} = 0,$$

$$\eta = r + \bar{q}.$$

Si le dernier terme est(généralement)petit par rapport au carré du coefficient de λ^2,

(90) $$\lambda_1^2 \cong \gamma - 1 + \frac{rf}{\eta}, \quad \lambda_2^2 \cong \frac{\gamma p}{\eta} / (\gamma - 1 + \frac{rf}{\eta}).$$

On voit que

(91) $$\lambda_1^2 \longrightarrow \gamma - 1, \quad \lambda_2^2 \longrightarrow 0,$$

lorsque $\bar{q} \longrightarrow \infty$ (cf. (61)). D'autre part quand $\bar{q} \longrightarrow 0$, $rf/\eta = f$ qui est équivalent à 1 dans l'approximation non relativiste; par conséquent,

(92) $$\lambda_1^2 \cong \gamma > 1, \quad \lambda_2^2 \cong \frac{p}{r}.$$

La deuxième valeur coïncide sensiblement avec (60) ($q_n = 0$, $\rho \cong r$). C'est, toujours à la même approximation, v_s^2/γ où v_s est la vitesse duson quand on néglige la conduction de la chaleur,

$$v_s^2 = (\partial p/\partial \rho)_S = \gamma p/rf \cong \gamma p/r.$$

Passons au fluide incompressible. En négligeant encore ϕ' on cherche les solutions de $P^*(X,Y) = 0$ obtenu en remplaçant les coefficients de (65) par,

(93) $$a_2^* = \lim_{\phi' \to 0} \phi' a_2 \text{ , etc...}$$

On s'aperçoit alors que la vitesse limite ($\lambda = \pm 1$) est atteinte dans toutes les directions : $P^*(1,Y) = 0$. En conséquence, on obtient aisément les deux autres vitesses (en exprimant par exemple la somme et le produit des racines en fonctions des coefficients),

$$a_2^* \lambda^2 + 2b_2^* q_n \lambda - a_o^* = 0,$$

soit,

(94) $$\bar{q} \lambda^2 + 3q_n \lambda - C_p r\Theta = 0, \qquad C_p r\Theta = 2(p + p_o).$$

On notera que la condition $\lambda^2 < 1$ dans toutes les directions entraîne,

(95) $$\bar{q} > 3q + C_p r\Theta.$$

On sait que l'eau peut être considérée comme un fluide polytropique avec $\gamma = 7$, $p_o > 0$. Si jamais (88) peut être utilisée on se rend compte en tout cas qu'on ne peut plus négliger le terme en ϕ' (pour que $0 < \lambda^2 < 1$).

Remerciements

Nous exprimons notre gratitude à MM. C. Cattaneo et A. Lichnerowicz. Nos remerciements vont aussi à Mme Y. Choquet-Bruhat et M. J. Ehlers.

Références

(1) G. BOILLAT, C. R. Acad. Sc. Paris, 270 A (1970), 217.
(2) G. BOILLAT, Ibid., 1134.
(3) A. LICHNEROWICZ, dans ce livre.
(4) PHAM MAU QUAN, Ibid.
(5) C. ECKART, Phys. Rev., 58 (1940), 919.
(6) D. ter HAAR & H. WERGELAND, Elements of Thermodynamics, Addison-Wesley (Reading, 1966).

(7) C. CATTANEO, Atti del Seminario matematico e fisico della Università di Modena, 8 (1948), 1; C. R. Acad. Sc. Paris, 247 (1958), 431.

(8) P. VERNOTTE, C. R. Acad. Sc. Paris, 246 (1958), 3154.

(9) M. KRANYŠ, Nuovo Cimento, 42 B (1966), 51; 50 B (1967), 48.

(10) G. BOILLAT, Lett. Nuovo Cimento, 3 (1970), 521.

(11) G. BOILLAT, Velocities of Heat propagation in Relativistic polytropic Fluids, Ibid. (à paraitre en 1970).

(12) P. D. LAX, Ann. Math. Studies (Princeton), 33 (1954), 211; Comm. Pure & Appl. Math., 10 (1957), 537.

(13) G. BOILLAT, La propagation des ondes, Gauthier-Villars (Paris, 1965).

(14) G. BOILLAT, Journ. Math. Phys., 10 (1969), 452. Cf. Y. CHOQUET-BRUHAT, Journ. Math. Pures & Appl., 48 (1969), 117.

(15) A. LICHNEROWICZ, Théories relativistes de la gravitation et de l'électromagnétisme, Masson (Paris, 1955).

(16) C. CATTANEO, Lincei-Rend. Sc. fis. mat. e nat., 46 (1969), 698, a proposé de remplacer ρ par $\rho + p$ de façon à obtenir l'équation d'état ρ = cte au lieu de (87).

Editoriale Grafica - Roma